Environmental Chemistry for a Sustainable World

Other Publications by the Editors

Books
Environmental Chemistry
http://www.springer.com/978-3-540-22860-8

Organic Contaminants in Riverine and Groundwater Systems
http://www.springer.com/978-3-540-31169-0

Sustainable Agriculture
Volume 1: http://www.springer.com/978-90-481-2665-1
Volume 2: http://www.springer.com/978-94-007-0393-3

Book series
Environmental Chemistry for a Sustainable World
http://www.springer.com/series/11480

Sustainable Agriculture Reviews
http://www.springer.com/series/8380

Journals
Environmental Chemistry Letters
http://www.springer.com/10311

More information about this series at http://www.springer.com/series/11480

Sophie Fourmentin • Grégorio Crini
Eric Lichtfouse
Editors

Cyclodextrin Applications in Medicine, Food, Environment and Liquid Crystals

Springer

Editors
Sophie Fourmentin
UCEIV
Dunkerque, France

Grégorio Crini
Université de Bourgogne Franche-Comté
Besançon, France

Eric Lichtfouse
Aix-Marseille Univ, CEREGE, CNRS, IRD,
INRA, Coll France, Europole
Mediterraneen de l'Arbois
Aix en Provence, France

ISSN 2213-7114 ISSN 2213-7122 (electronic)
Environmental Chemistry for a Sustainable World
ISBN 978-3-030-09419-5 ISBN 978-3-319-76162-6 (eBook)
https://doi.org/10.1007/978-3-319-76162-6

Printed on acid-free paper

This Springer imprint is published by the registered company Springer International Publishing AG part of Springer Nature.
The registered company address is: Gewerbestrasse 11, 6330 Cham, Switzerland

Preface

> *"Cyclodextrins have been a source of fascination for over a hundred years as the heart of these molecules is easy to penetrate although they are hard to crack"*
>
> (Benito Casu, 1993)

The 2016 Nobel Prize in Chemistry was awarded to Jean-Pierre Sauvage, Sir James Fraser Stoddart and Bernard Lucas Feringa for the design and synthesis of molecular machines. This prize revealed a major recognition for supramolecular chemistry involving host-guest relationships such as cyclodextrin complexes. This recognition will surely contribute to a renewed interest for cyclodextrins. Cyclodextrins have been discovered more than 100 years ago by the pioneering work of the pharmacist and chemist Antoine Villiers in France. Cyclodextrins are obtained by enzymatic degradation of starch and still fascinate researchers because they are remarkable macrocyclic molecules with major theoretical and practical impacts in chemistry, biology, biochemistry, health science and agriculture. Cyclodextrins have broken barriers between different disciplines and, nowadays, various scientists work together to find new concepts and applications.

The most characteristic feature of cyclodextrins is their ability to form inclusion complexes with various molecules through host-guest interactions, which are at the origin of many applications (Fig. 1). Actually, cyclodextrins and their chemically modified derivatives have a wide variety of practical applications in almost all sectors of industry including pharmacy, medicine, foods, cosmetics, toiletries, catalysis, chromatography, biotechnology, textile industry, supramolecular chemistry and nanotechnology. New contributions and industrial developments are further expected in biomedicine, e.g., for magnetic resonance imaging and chemotherapy, in fermentation processing and enzymology, sensor applications, cosmeto-textiles, and agrochemistry.

This book is the second volume of two volumes on cyclodextrins published in the series *Environmental Chemistry for a Sustainable World*. This volume focuses on cyclodextrin applications. The first chapter by Divya Arora and Sundeep Jaglan

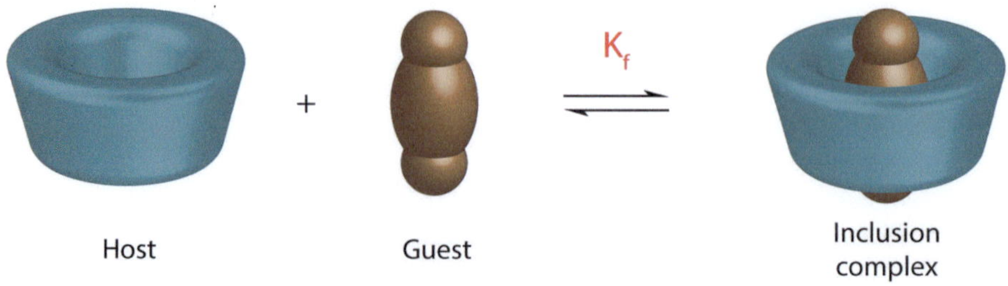

Fig. 1 Cyclodextrin hosts are able to trap a wide variety of guest chemicals. This complexing ability has led to many applications in various sectors

presents cyclodextrin-based carriers for delivery of dietary phytochemicals. The second chapter by Éva Fenyvesi et al. describes the interactions of steroids with cyclodextrins and their applications to pharmaceuticals, food, biotechnology, and environment. Nazli Erdoğar and Erem Bilensoy discuss cyclodextrin-based nanosystems in targeted cancer therapy. Miriana Kfoury et al. review the use of cyclodextrins for essential oils applications in chapter 4. Hiroshi Ikeda demonstrates in chapter 5 that chromophore-appended cyclodextrins are effective for chemosensors to detect organic molecules by fluorescence or absorbance changes. Then Grégorio Crini et al. describe silica materials-containing cyclodextrin for pollutant removal. The final chapter by Chang-Chun Ling et al. summarizes the synthesis and characterization of supramolecular liquid crystals based on cyclodextrins and their applications.

The editors extend their thanks to all the authors who contributed to this book. Its creation would not have been possible without the assistance of several friends and colleagues deserving acknowledgment. They have helped by choosing contributors, reviewing chapters, and in many other ways. Finally, we would like to thank the staff at Springer Nature for their highly professional editing of the publication.

Dunkerque, France — Sophie Fourmentin
Besançon, France — Grégorio Crini
Aix-en-Provence, France — Eric Lichtfouse

Contents

Contributors

Divya Arora Microbial Biotechnology Division, CSIR-Indian Institute of Integrative Medicine, Jammu, India

Academy of Scientific and Innovative Research (AcSIR), Jammu, India

Lizette Auezova Bioactive Molecules Research Laboratory, Faculty of Sciences, Lebanese University, Beirut, Lebanon

Erem Bilensoy Department of Pharmaceutical Technology, Hacettepe University Faculty of Pharmacy, Ankara, Turkey

Pier-Luc Champagne Alberta Glycomics Centre, Department of Chemistry, University of Calgary, Calgary, Alberta, Canada

Nazlı Erdoğar Department of Pharmaceutical Technology, Hacettepe University Faculty of Pharmacy, Ankara, Turkey

Éva Fenyvesi CycloLab Cyclodextrin R&D Laboratory Co., Budapest, Hungary

Marc Fourmentin Laboratoire de Physico-Chimie de l'Atmosphère (LPCA, EA 4493), Université du Littoral Côte d'Opale, Dunkerque, France

Hélène Greige-Gerges Bioactive Molecules Research Laboratory, Faculty of Sciences, Lebanese University, Beirut, Lebanon

Hiroshi Ikeda School of Life Science and Technology, Tokyo Institute of Technology, Yokohama, Japan

Sundeep Jaglan Microbial Biotechnology Division, CSIR-Indian Institute of Integrative Medicine, Jammu, India

Academy of Scientific and Innovative Research (AcSIR), Jammu, India

Miriana Kfoury Unité de Chimie Environnementale et Interactions sur le Vivant (UCEIV, EA 4492), Dunkerque, France

Bioactive Molecules Research Laboratory, Faculty of Sciences, Lebanese University, Beirut, Lebanon

Rajesh Kumar Alberta Glycomics Centre, Department of Chemistry, University of Calgary, Calgary, Alberta, Canada

Chang-Chun Ling Alberta Glycomics Centre, Department of Chemistry, University of Calgary, Calgary, Alberta, Canada

Nadia Morin-Crini UFR Sciences et Techniques, Laboratoire Chrono-environnement, Université de Bourgogne Franche-Comté, Besançon, France

István Puskás CycloLab Cyclodextrin R&D Laboratory Co., Budapest, Hungary

Ankit Saneja Product Development Cell-II, National Institute of Immunology, New Delhi, India

Lajos Szente CycloLab Cyclodextrin R&D Laboratory Co., Budapest, Hungary

Giangiacomo Torri Istituto di Chimica e Biochimica G. Ronzoni, Milano, Italy

About the Editors

Sophie Fourmentin was born in Bar/S/Aube (France) in 1968. She received her Ph.D. degree in Organic Chemistry from the University of Lille in 1994. In 1996, she became Assistant Professor at the University of Littoral-Côte d'Opale in Dunkerque. She developed a new subject area based on "Applications of cyclodextrins in the remediation of organic pollutants." In 2006, she completed her HDR (authorization to supervise research activities) entitled "Complexes Cyclodextrine/Polluants Organiques: Caractérisation, Application à la remédiation des Composés Organiques Volatils." In 2008, she was promoted Professor, and she managed the Laboratoire de Synthèse Organique et Environnement until 2010. Since January 2010, she supervises the supramolecular team in the UCEIV laboratory. Her research works are at the interface between host-guest chemistry and environmental chemistry. The aim is to take benefit of the properties of cyclodextrins to enhance conventional remediation processes (VOC absorption, Fenton oxidation). Her experience in the characterization of host-guest complexes in the case of complex VOC has been extended to the study of the encapsulation of flavors and essential oils, for applications in food, fragrance, or pharmaceutical industry. She is the author of 91 articles (h-index: 23), a patent, and coordinated two books on cyclodextrins.

Grégorio Crini 52, is researcher at University of Bourgogne Franche-Comté (UMR Chrono-environnement), Besançon. His current interests focus on the design of novel polymer networks and the environmental aspects of polysaccharide chemistry. He published over 180 papers in international journals and books, and he is a highly cited researcher. The total citation of his publications is over 7700 according to ISI Web of Science, h-index: 32.

Eric Lichtfouse born in 1960, is a biogeochemist at the University of Aix-Marseille, CEREGE, Aix-en-Provence, France. He got a Ph.D. in Organic Geochemistry at the University of Strasbourg in 1989 for the discovery of new fossil steroids in sediments and petroleum. He has invented the ^{13}C-dating method allowing to measure the dynamics of soil organic molecules, thus opening the field of molecular-level investigations of soil carbon sequestration. He is Chief Editor of the journal *Environmental Chemistry Letters*, of 3.6 impact factor. His former journal *Agronomy for Sustainable Development* has reached the second place in the category Agronomy, and has been awarded by the Essential Science Indicators. He is lecturing scientific writing and communication in universities worldwide. He has published the book *Scientific Writing for Impact Factor Journals*. This textbook describes in particular the micro-article, a new tool to identify the novelty of experimental results. Further details are available on ResearchGate, Slideshare, LinkedIn, ResearchGate, ResearcherID, Orcid, and Google Scholar citations.

Chapter 1
Cyclodextrin-Based Carriers for Delivery of Dietary Phytochemicals

Divya Arora, Ankit Saneja, and Sundeep Jaglan

Contents

Abstract The food and pharmaceutical industry is searching for innovative solutions to enhance the bioavailability and clinical efficacy of dietary phytochemicals. In this regard, cyclodextrins have gained widespread attention as functional excipients. Numerous studies have demonstrated that cyclodextrin inclusion complexes enhance apparent water solubility, physico-chemical stability and improve the bioavailability of the dietary phytochemicals. In addition, dual encapsulation, that is the complexation of dietary molecules with cyclodextrins followed by encapsulation into nanomaterials such as liposomes, nanoparticles, conjugates, has also been investigated. Here, we review the applications of natural and chemically modified cyclodextrins for the delivery of dietary phytochemicals. We focus mainly on outcomes of inclusion complexes for enhancing solubility, bioavailability and efficacy of the delivered phytochemicals. We also discuss recent trends in dual-encapsulation.

D. Arora · S. Jaglan (✉)
Microbial Biotechnology Division, CSIR-Indian Institute of Integrative Medicine, Jammu, India

Academy of Scientific and Innovative Research (AcSIR), Jammu Campus, Jammu, India
e-mail: sundeepjaglan@iiim.ac.in

A. Saneja
Product Development Cell-II, National Institute of Immunology, New Delhi, India

S. Fourmentin et al. (eds.), *Cyclodextrin Applications in Medicine, Food, Environment and Liquid Crystals*, Environmental Chemistry for a Sustainable World, https://doi.org/10.1007/978-3-319-76162-6_1

1.1 Introduction

A number of epidemiological and experimental investigations suggests that a regular consumption of fruits, whole grains, vegetables and other plant foods is related with reduced risks of developing chronic diseases such as cancer and cardiovascular diseases (Manach et al. 2009; Liu 2013; Arora and Jaglan 2016). This association has been partly ascribed to the presence of dietary phytochemicals which are bioactive compounds commonly found in plant-based foods. These dietary phytochemicals are generally classified into several categories as polyphenols, terpenoids, alkaloids and other nitrogen compounds, carbohydrates, lipids, phytosterols, and carotenoids (Fig. 1.1) (Manach et al. 2009; Liu 2013; Arora and Jaglan 2016). However, despite having good pharmacological activities of these phytochemicals, most of them are often associated with poor water solubility, poor stability due to gastric and colonic pH, metabolism by gut microflora, absorption across the intestinal wall, active efflux mechanism and first-pass metabolic effects which ultimately leads to poor bioavailability in humans (Aqil et al. 2013; McClements et al. 2015). Therefore, in order to tackle these challenges, food industry has shifted its considerable attention towards making inclusion complexes of these dietary phytochemicals with cyclodextrins (Astray et al. 2009).

Cyclodextrins are a series of natural cyclic oligosaccharides synthesized from the union of glucose monomers (glucopyranose) linked by α-1,4 glycosidic bonds (de Oliveira Makson et al. 2015). Depending on the number of glucopyranose units, the natural occurrence of cyclodextrins can be classified in to α, β and γ- cyclodextrins, which are composed of 6, 7 and 8 glucose units respectively (Zhang and Ma 2013). Cyclodextrins are shaped like a truncated cone instead of perfect cylinders (due to chair conformation of the glucopyranose units) with tapered cavity of 0.79 nm in depth, while both the top and bottom diameters are increased with the number of glucose units (Li and Purdy 1992; Jambhekar and Breen 2016). The hydroxyl groups are oriented to the outer space flanking the upper and lower rims, with the primary hydroxyl groups of the sugar residues towards the narrow edge of the cone and the secondary hydroxyl groups towards the wider edge (Mellet et al. 2011; Jambhekar and Breen 2016). The central cavity of the cone is lined with the skeletal carbons and ethereal oxygen of the glucose residues, which impart a hydrophobic character (Jambhekar and Breen 2016). The hydrophobic cavity of cyclodextrins exhibits the unique ability to trap a guest molecule inside its cavity and has been extensively exploited by the pharmaceutical industry to improve bioavailability of poorly aqueous soluble or biodegradable drugs, to prevent adverse effects or to enhance permeability of biological membranes (Mellet et al. 2011). It is noteworthy to emphasize that currently there are more than 30 marketed products based on cyclodextrin complexes (Loftsson et al. 2005; Loftsson and Duchêne 2007; Zhang and Ma 2013).

In recent years, cyclodextrins have been used in food industry due to various reasons: (i) to protect lipophilic dietary photochemicals that are sensitive towards degradation due to oxygen, light or heat; (ii) to solubilise food colourings and

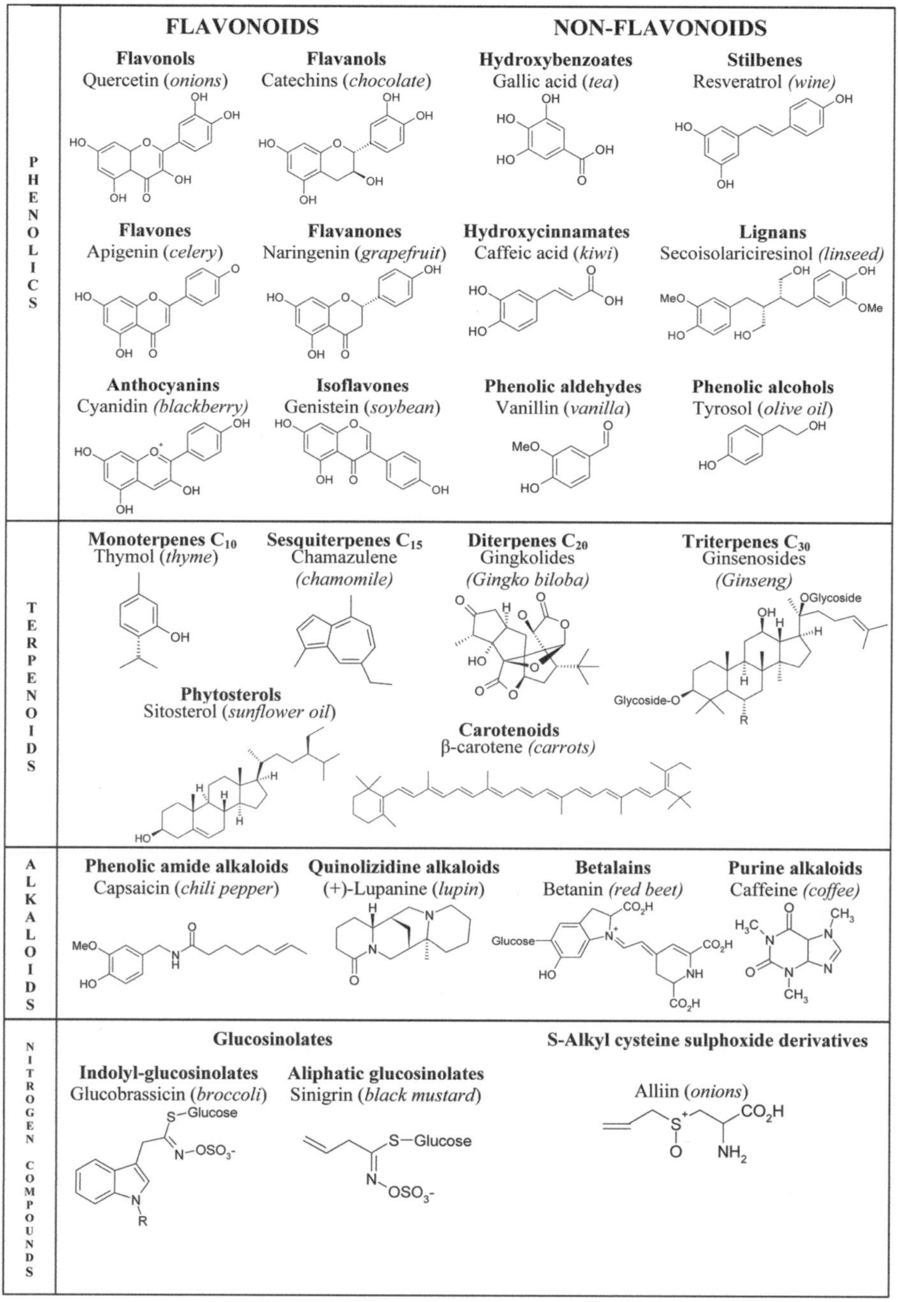

Fig. 1.1 Major classes of phytochemicals in food. Reproduced from Ref. (Manach et al. 2009) with permission from Wiley

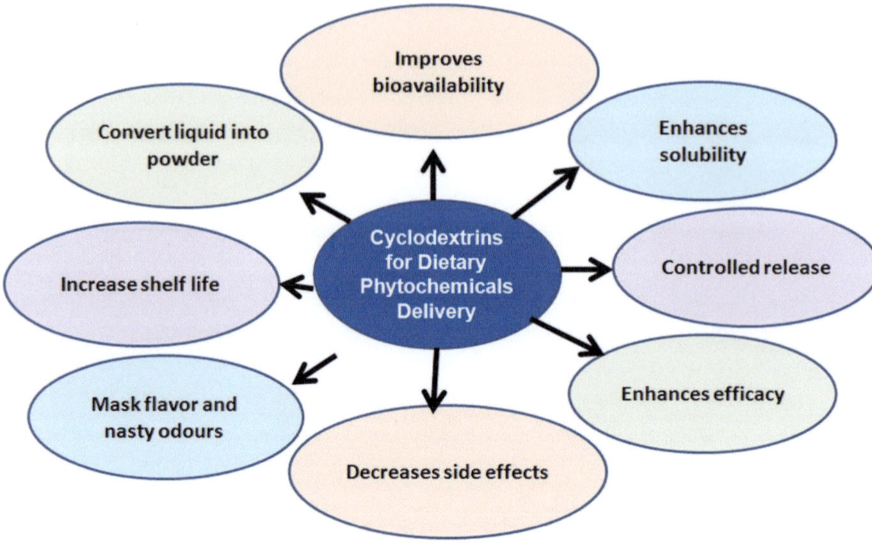

Fig. 1.2 Advantages of cyclodextrin based carriers for delivery of nutraceuticals. Cyclodextrins have been used for delivery of dietary photochemical with a variety of aims like enhancing solubility, bioavailability, shelf-life and efficacy

vitamins; (iii) to stabilize fragrances, flavours, vitamins, and essential oils against undesirable changes; (iv) to mask unpleasant odours or tastes; (v) to achieve a controlled release of dietary components and (vi) to enhance the bioavailability of dietary molecules (Fig. 1.2) (Loftsson et al. 2005; Astray et al. 2009; Pinho et al. 2014). However, the regulatory status of cyclodextrins in foods varies from country to country. For example, α-, β-, and γ- cyclodextrins have obtained the generally recognized as safe (GRAS) status as per United States Food and Drug Administration (USFDA) and can be commercialized as such. However, in Japan these cyclodextrins are recognized as natural products and their commercialization in the food sector is restricted subject to their purity. While in Australia and New Zealand γ- and α- cyclodextrins are classified as Novel Foods from 2003 and 2004, respectively (Cravotto et al. 2006; Martina et al. 2013). Several reviews have been published describing the mechanism of cyclodextrins complexation and methods to improve the complexation efficiency (Hirayama and Uekama 1999; Challa et al. 2005; Loftsson et al. 2005; Zhang and Ma 2013; Jambhekar and Breen 2016). In this chapter, we mainly focus on the recent advances in cyclodextrins based delivery of dietary molecules in order to enhance the solubility and efficacy of these molecules. In addition, the drug delivery systems consisting of dual systems of cyclodextrins and functional materials such as liposomes, nanoparticles have also been described.

Table. 1.1 Physicochemical properties of cyclodextrins and their derivatives for delivery of nutraceuticals. Reproduced from Ref. (Brewster and Loftsson 2007) with permission from Elsevier

Cyclodextrin	Substitution[a]	MW[b] (Da)	Solubility in water (mg/mL)[c]
α-Cyclodextrin	0	972	145
β-Cyclodextrin	0	1135	18.5
2-Hydroxypropyl-β-cyclodextrin	0.65	1400	>600
Sulfobutylether β-cyclodextrin sodium salt	0.9	2163	>500
Randomly methylated β-cyclodextrin	1.8	1312	>500
6-O-Maltosyl-β-cyclodextrin	0	1459	>1500
γ-Cyclodextrin	0	1297	232
2-Hydroxypropyl-γ-cyclodextrin	0.6	1576	>500

[a]Average number of substituents per glucose repeat unit
[b]MW: Molecular weight
[c]Solubility in pure water at approx. 25 °C

1.2 Chemically Modified Cyclodextrins

The natural cyclodextrins, in particular β-cyclodextrin, are of limited aqueous solubility which results in precipitation of the complexes in water and aqueous systems after their inclusion with hydrophobic molecules. Further, their use is limited to oral and topical formulations due to nephrotoxicity which results from the accumulation of cyclodextrins crystals or cyclodextrin-cholesterol complexes due to their low aqueous solubility (Zhang and Ma 2013). Therefore to circumvent these problems native cyclodextrins can be modified by hydroxyalkylation, alkylation or sulfoalkylation (Table 1.1). The ultimate goal of these modifications is to enhance the aqueous solubility of parent cyclodextrins. For example, the addition of the hydroxyl propyl group to β-cyclodextrin (HP-β-CD) enhanced its solubility by more than 27-folds (Gould and Scott 2005; Al-Rawashdeh et al. 2010). Similarly, sulfobutylether-β-cyclodextrin (SBE-β-CD) enhanced the aqueous solubility of the parent β-cyclodextrin and considered safe at relatively high doses for the parenteral route (Irie and Uekama 1997; Stella and He 2008). The detailed reviews on chemically modified cyclodextrins have already been presented in literature (Hirayama and Uekama 1999; Loftsson et al. 2005; Brewster and Loftsson 2007; Loftsson and Duchêne 2007; Loftsson and Brewster 2012; Jambhekar and Breen 2016).

1.3 Cyclodextrin in Dietary Phytochemicals Delivery

Cyclodextrins can be considered as empty capsules which act as "host" in which bioactive molecules "guest" molecules can be totally or partially incorporated. There are several methods for the formation of inclusion complexes between

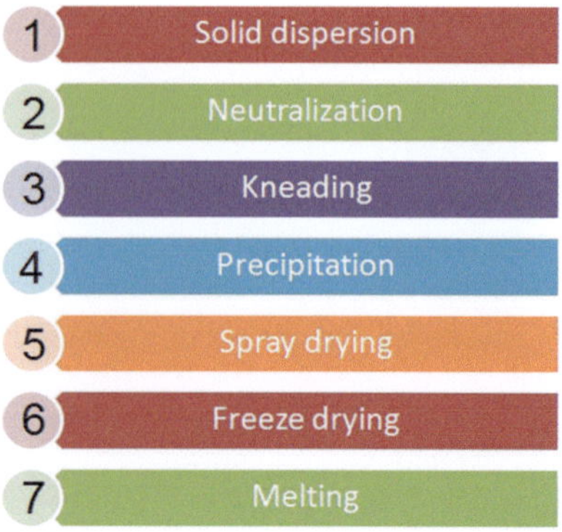

Fig. 1.3 Various methods used for cyclodextrin complexation with dietary phytochemicals. The most common used methods for these are neutralization, kneading, precipitation, spray drying, freeze drying and melting

cyclodextrins and dietary molecules, and the selection of the process is clearly based on physicochemical properties of the guest molecule, the facilities available and the cost involved. The most common methods are neutralization, co-precipitation, kneading, spray drying, freeze drying, melting and solid dispersion (Fig. 1.3) (Hedges 1998; Marques 2010; Pinho et al. 2014). The detailed reviews on approaches for cyclodextrin complexation, mechanisms for formation of inclusion complexes and methods to enhance complexation efficiency are already been present in literature (Loftsson and Brewster 2012; di Cagno 2016). The applications of cyclodextrin for phytochemicals delivery are described in following sections.

1.3.1 Cyclodextrin Complexation with Dietary Phytochemicals

The complexation of dietary molecules with cyclodextrins has emerged as promising delivery systems for overcoming the solubility and pharmacokinetic limitations of dietary phytochemicals. In recent years, a wide array of cyclodextrins inclusion complexes with dietary molecules has been prepared in order to improve their solubility and efficacy (Table 1.2). As discussed earlier, most of the dietary phytochemicals are poor aqueous soluble which leads to their incomplete absorption. For example, Apigenin (Api, 5,7,4′-trihydroxyflavone) which is a consumed in the human diet from the main sources German chamomile (*Matricaria chamomilla L.*) (Avallone et al. 2000), celery (*Apium graveolens L.*) (Popović et al. 2006) and parsley (*Petroselinum crispum L.*) (Meyer et al. 2006) is having poor aqueous solubility (1.35 μg/mL). Therefore, in order to enhance the solubility of apigenin different complexes with cyclodextrin and its derivates were developed (Pápay et al. 2016). The solubility studies demonstrated that RM-β-cyclodextrin (random methyl β-Cyclodextrin) enhanced much solubility as compare to other cyclodextrin derivates and solubility was found to be in order of RM-β-CD > SBE-β-CD (sulfobutyl ether-β-cyclodextrin) > γ-CD > HP-β-CD (Hydroxypropyl-β-cyclodextrin) > β-CD > α-CD.

Table 1.2 Various cyclodextrin-based inclusions developed for nutraceutical delivery along with their major outcomes

Nutraceutical	Outcome	Ref.
Apigenin	Apigenin inclusion complexes demonstrated higher solubility and increased antioxidant activity.	(Pápay et al. 2016)
Betulinic acid	Betulinic acid inclusion into the octakis-[6-deoxy-6-(2-sulfanyl ethanesulfonic acid)]-γ-cyclodextrin enhanced solubility as well as efficacy compared to native betulinic acid.	(Soica et al. 2014)
Carvacrol	Encapsulation of carvacrol in β- cyclodextrin enhanced antinociceptive effect as compared to carvacrol without producing motor deficit to orofacial pain rodent models.	(Silva et al. 2016)
Chrysin	Inclusion of chrysin in β-CD increased the solubility of chrysin as well as its antioxidant potential, antimicrobial activity and anti-tumor activity.	(Zhu et al. 2016)
Curcumin	Curcumin-β-cyclodextrin-loaded sponge demonstrated comparable wound healing rate as compared to that of marketed formulation with no sign of adverse consequence.	(Kaur et al. 2016)
	Curcumin- β cyclodextrin complex enhance the solubility of free curcumin and improve its antitumor activity.	(Zhang et al. 2016)
	Curcumin–β- cyclodextrin complex demonstrated improved color stability against sunlight, pH, storage and isothermal heating as that of the pure colorant.	(Mangolim et al. 2014)
Dentatin	Complexation of dentatin in hydroxypropyl-β-cyclodextrin enhanced aqueous solubility by >300-fold as compared to dentatin alone.	(Ashwaq et al. 2017)
Fisetin	Complexation of fisetin with sulfobutylether-β-cyclodextrin enhanced the aqueous solubility of fisetin without altering its *in vitro* antioxidant activity.	(Mohtar et al. 2017)
Gallic acid	Inclusion complex of gallic acid with hydroxypropyl-β-cyclodextrin had better antibacterial efficiency and also exhibited higher stability.	(Pinho et al. 2015)
Hinokitiol	Hinokitiol cyclodextrin inclusion complexes enhanced 4 times more antimicrobial activity than hinokitiol crystals.	(Suzuki et al. 2015)
	Pharmacokinetic study demonstrated honokiol-in-HP-β-CD-in-liposome significantly retarded the elimination of honokiol and prolonged the residence time in circulating system.	(Wang et al. 2011)
Naringenin	β-cyclodextrin inclusion complexes demonstrated enhanced water solubility and thermal stability of naringenin.	(Yang et al. 2013)
	Hydroxypropyl-β-cyclodextrin inclusion complexes enhanced the solubility of naringenin by over 400-fold, and its transport across a Caco-2 model of the gut epithelium by 11-fold. Further, the pharmacokinetic studies in rats demonstrated the developed inclusion complex enhanced the AUC by 7.4-fold and C_{max} by 14.6-fold.	(Shulman et al. 2011)
	Naringenin β-cyclodextrin complex enhanced water solubility and improved biological activity leading to more potent inhibitory effect on CNV formation in rats.	(Xu et al. 2014)

(continued)

Table 1.2 (continued)

Nutraceutical	Outcome	Ref.
Resveratrol	Resveratrol hydroxypropyl-β-cyclodextrin complex demonstrated a higher antioxidant efficacy both in terms of capacity and rate of scavenging DPPH radical as compare to resveratrol β- cyclodextrin complex.	(Lu et al. 2009)
	Pharmacokinetic studies after oral administration in BALB-c mice demonstrated hydroxypropyl-β-cyclodextrin complexation enhanced AUC_{0-120} and C_{max} by two and four-fold respectively as compared with resveratrol nanosuspension.	(Yang et al. 2016)
	Resveratrol complexation with methylated-β-cyclodextrin enhanced its aqueous solubility while retaining resveratrol's biological properties.	(Duarte et al. 2015)
Silibinin	Silibinin hydroxypropyl-β-cyclodextrin demonstrated enhanced solubility and cytotoxicity as compared to native silibinin.	(Kellici et al. 2015)
Simvastatin	Simvastatin hydroxypropyl-β-cyclodextrin inclusion complex demonstrated superior efficacy than simvastatin in reducing total cholesterol and TG levels due to improved solubility and dissolution.	(Jun et al. 2007)
Sulforaphane	Inclusion complex of sulforaphane with hydroxypropyl-β-cyclodextrin enhanced the thermal stability and the chemical stability of sulforaphane.	(Wu et al. 2010)
Thymoquinone	*In vitro* cytotoxicity on MCF7 cells demonstrated higher cytotoxicity of thymoquinone - cyclodextrin nanoparticles as compare to native thymoquinone.	(Abu-Dahab et al. 2013)
Ursolic acid	*In vitro* anti-proliferative activity of ursolic acid hydroxypropyl-γ-cyclodextrin complex demonstrated higher activity as compared to the native ursolic acid.	(Oprean et al. 2016)
Quercetin	β-cyclodextrin quercetin complex enhanced aqueous solubility of quercetin.	(Borghetti et al. 2009)

Table Abbreviations: *AUC* Area under curve; C_{max} Concentration maximum; *DPPH* 2,2-diphenyl-1-picrylhydrazyl

In another study, betulinic acid, a pentacyclic triterpene found to be an antimelanoma agent was complexed with octakis-[6-deoxy-6-(2-sulfanyl ethanesulfonic acid)]-γ-cyclodextrin (GCDG) in order to enhance its solubility and efficacy (Soica et al. 2014). The complex formed caused a reduction in tumor volume and weight *in vivo* in C57BL/6 J mice. Silva *et al.* developed β-cyclodextrin complex with carvacrol (5-isopropyl-2-methylphenol), an isoprenoid present in the essential oils of genera *Origanum* and *Thymus* belongs to Lamiaceae family (Silva et al. 2016). The developed complex was evaluated in formalin, capsaicin, and glutamate induced orofacial nociception in mice. The study demonstrated the developed complex showed superior efficacy than native carvacrol. The carvacrol-β-cyclodextrin complex (20 mg/kg, p.o.) produced 49.3% behavior pain while native carvacrol produced 28.7% analgesic inhibition at same dose in the second phase of formalin test. Zhu *et al.* prepared chrysin (5,7-dihydroxyflavone)-β-cyclodextrin inclusion complex in order to improve its solubility as well as efficacy (Zhu et al. 2016). The study demonstrated the solubility of chrysin and inclusion complex were 1.93 mmol/L and 7.32 mmol/L, respectively at pH 7.5. The antioxidant activity determined using scavenging oxygen free radicals, scavenging 2,2-diphenyl-1-picrylhydrazyl (DPPH) and scavenging hydroxyl free radical showed much stronger antioxidant activity of inclusion complex than in native chrysin. *In vitro* cytotoxicity in H22 mouse liver tumor cells after 72 h demonstrated the inclusion complex demonstrated almost equal cytotoxicity as compare to native chrysin.

Zhang *et al.* developed curcumin-β-cyclodextrin complex in order to enhance the solubility of curcumin and improve its antitumor activity (Zhang et al. 2016). *In vitro* cytotoxicity in lung cancer cells (A549, NCI-H446 and NCI-H520) demonstrated higher cytotoxicity of the complex as compare with native curcumin. Further, *in vivo* efficacy in H22 tumor cells induced Kunming mice demonstrated higher tumor inhibition rate of the complex 34.64% as compare with native curcumin 9.52%. In another study, curcumin was complexed with β-cyclodextrin using co-precipitation, freeze-drying and solvent evaporation methods (Mangolim et al. 2014). The inclusion complex increased the solubility of curcumin by 31-folds and exhibited a sunlight stability 18% higher than native curcumin. Kaur *et al.* developed curcumin-β-cyclodextrin complex and then loaded into gelatin sponge (Kaur et al. 2016). The curcumin-β-cyclodextrin-loaded sponge treated wound was found to heal in rate as compare to marketed formulation silver sulfadiazine with no sign of adverse consequence.

Ashwaq *et al.* developed dentatin-hydroxypropyl-β-cyclodextrin (HPβCD) complex for enhancing the solubility of dentatin (Ashwaq et al. 2017). The solubility of dentatin was enhanced more than 300-fold after complexation as compare to dentatin alone. Moreover, the complexation of dentatin did not reduce cytotoxicity on prostate cancer (LNCaP), human adenocarcinoma breast cancer (MDA-MB-231) and human gastric adenocarcinoma cell line (HDT). Mohtar *et al.* prepared complexation of fisetin with sulfobutylether-β-cyclodextrin with the aim of increasing the solubility of fisetin (Mohtar et al. 2017). Further, the developed complex was spray dried into dry powder inhaler (DPI) formulation with optimized aerodynamic properties and tested against the human lung adenocarcinoma cell line (A549). The study revealed fisetin-SBE-β-cyclodextrin complex improved the solubility of fisetin and was capable of delivering high amount of fisetin to the deep lung region for therapeutic applications.

Pinho *et al.* developed inclusion complex of gallic acid with β-cyclodextrin, (2-hydroxy) propyl-β-cyclodextrin and methyl-β-cyclodextrin (Pinho et al. 2015). The inclusion complex of gallic acid with HP-β-cyclodextrin had better antibacterial efficiency and also exhibited higher stability than other complexes. Suzuki *et al.* developed solid dispersion on mixtures of hinokitiol (HT) and γ-cyclodextrin (c-CD) and of hinokitiol and (2-hydroxypropyl)-γ-cyclodextrin (HP-γ-CD) (Suzuki et al. 2015). The developed ground mixtures enhanced the antimicrobial activity 4 times than the HT crystals which is due to increase in the solubility of HT as a result of the formation of HT/cyclodextrin inclusion complexes. In another study Wang *et al.*prepared honokiol with hydroxypropyl-β-cyclodextrin and encapsulated it into liposome (Wang et al. 2011). The pharmacokinetic study demonstrated that honokiol-in-HP-β-cyclodextrin-in-liposome significantly retarded the elimination and prolonged the residence time in circulating system as compare with native honokiol which was quickly removed from the circulating system after intravenous injection. Further, the *in vitro* cytotoxicity in A549 and HepG2 cells demonstrated the bioactivity of honokiol-in-HP-β-cyclodextrin-in-liposome was relatively a little weak as compare with free honokiol which maybe, due to the reason that honokiol was not completely released from honokiol-in-HP-β-cyclodextrin-liposome in 48 h.

Yang *et al.* developed β-cyclodextrin and its derivatives, heptakis-(2,6-di-O-methyl)-β-cyclodextrin (DMβCD) and heptakis (2,3,6-tri-O-methyl)-β-cyclodextrin (TMβCD) complexes of naringenin which is one of the most abundant flavonoids in grapefruits and citrus fruits in order to enhance its solubility (Yang et al. 2013). The solubility studies demonstrated that the water solubility of naringenin was increased after their inclusion with cyclodextrins. The native naringenin demonstrated solubility 4.38 μg/mL which was remarkably enhanced to approximately 1.34, 1.60 and 1.52 mg/mL by the solubilizing effects of β-cyclodextrin, DM-β-cyclodextrin and TM-β-cyclodextrin, respectively. Shulman *et al.* developed hydroxypropoyl-β-cyclodextrin (HPβCD) complex of naringenin and demonstrated its enhanced solubility by 400-fold, and transport across a Caco-2 model of the gut epithelium by 11-fold (Shulman et al. 2011). Further, the pharmacokinetic studies in rats demonstrated enhanced AUC values by 7.4-fold and C_{max} by 14.6-fold as compare to native naringenin. Xu *et al.* prepared β-cyclodextrin (β-CD) complex of naringenin and demonstrated its solubility was increased by more than ten-fold (Xu et al. 2014). Further, the prepared complex also significantly reduced choroidal neovascularization (CNV) area than native naringen in laser-induced CNV model in rats.

Borghetti *et al.* prepared β-cyclodextrin solid complex of quercetin (3,3′,4′,5,7-pentahydroxy flavone), which is a frequent component of major dietary constituents, such as onions, apples, red wine, and green tea (Borghetti et al. 2009). The enhancement of aqueous solubility of quercetin to 4.6-fold in the presence of 15 mM of β-cyclodextrin. Oommen *et al.* prepared inclusion complex of plumbagin with β-cyclodextrin and then encapsulated it into niosomes using a lipid layer hydration method (Oommen et al. 1999). The niosome entrapped drug complex had improved anticancer activity as compare to native plumbagin when administered subcutaneously to C57BL/6 J mice bearing melanoma B16F1 at a dose of 5 mg kg^{-1}, the as evidenced by the enhanced volume doubling time and growth delay.

Resveratrol (3,4′,5-trihydroxystilbene) which is found in a number of which are dietary components, such as mulberries, peanuts, and grapes is sparingly soluble in water, which may be responsible for its limited absorption upon oral administration (Frémont 2000; Alarcon De La Lastra and Villegas 2005; Arora and Jaglan 2017). Therefore, in order to enhance its efficacy its inclusion with β-cyclodextrin (β-CD) and hydroxypropyl-β-cyclodextrin (HP-β-CD) were prepared (Lu et al. 2009). The inclusion ability of HP-β-cyclodextrin is larger than that of β-cyclodextrin. Further, the antioxidant activity also demonstrated a higher scavenging capacity of HP-β-cyclodextrin as compare to β-cyclodextrin. Yang *et al.* compared the pharmacokinetics of resveratrol 2-hydroxypropyl-β-cyclodextrin (HP-β-CD) complex and resveratrol nanosupension (obtained by diluting a resveratrol ethanol solution with phosphate buffer saline, added of 0.05% hydroxyethylcellulose) (Yang et al. 2016).

The pharmacokinetic studies after oral administration in BALB-c mice demonstrated HP-β-cyclodextrin complexation enhanced two-fold augment of RVT AUC_{0-120} and ~ four-fold increment in C_{max} as compare with resveratrol nanosuspension. Duarte *et al.* developed inclusion complexes of resveratrol with methylated-β-cyclodextrin in order to enhance its aqueous solubility (Duarte et al. 2015). The resveratrol complexation enhanced 400-fold improvements in its aqueous dissolution. Further, the developed inclusion complex also preserved the potential of resveratrol in decreasing the cell viability of Caco-2 cells, as well as the very strong antioxidant activity of resveratrol.

Yee *et al.* prepared phenoxodiol- β-cyclodextrin complex *via* a modified co-evaporation method and demonstrated enhancement in aqueous solubility by ~nine-folds (Yee et al. 2017). Further the developed, complex demonstrated enhanced *in vitro* anti-proliferative activity against three different cancer cell lines, namely the neuroblastoma cells SKN-BE(2)C, the triple negative breast cancercells MDA-MB-231 and the glioblastoma cells U87.

Jun *et al.* prepared inclusion complex of simvastatin (SV) with hydroxypropyl β-cyclodextrin (HP-β-CD) using supercritical antisolvent (SAS) process in order to improve its efficacy (Jun et al. 2007). The study demonstrated that SV/HP-β-CD inclusion complex showed superior efficacy than SV in reducing total cholesterol and TG levels which was attributed to improved solubility and dissolution associated with inclusion complex between simvastatin and HP-β-CD.

Lee *et al.* prepared inclusion complexes of soy isoflavone extract (IFE) with β-cyclodextrin in order to improve its solubility and bioavailability (Lee et al. 2007). The study demonstrated the complexes of isoflavone extract with β-CD enhance the aqueous solubility by 26-folds than that of native isoflavone extract. Further, the pharmacokinetic studies in Sprague-Dawley rats demonstrated the bioavailablity of major components of isoflavone extract *i.e.* daidzein, genistein and glycitin increased to 126%, 180% and 170% respectively as compare to that of native IFE. Wu *et al.* prepared the inclusion complex of sulforaphane (SF) with hydroxypropyl-β cyclodextrin (HP-β-CD) using co-precipitation method in order to enhance its stability (Wu et al. 2010). The study demonstrated that inclusion complex of sulforaphane with HP-β- cyclodextrin enhanced the thermal stability and the chemical stability of SF. Oprean *et al.* developed ursolic acid complexes of 2-hydroxypropyl-β-cyclodextrin and 2-hydroxypropil-γ-cyclodextrin and demonstrated higher *in vitro* anti-proliferative activity of ursolic acid HP-γ- cyclodextrin complex as compared to the native ursolic acid (Oprean et al. 2016).

1.3.2 Dual Encapsulation of Cyclodextrin with Nanocarriers

In recent years, there have been considerable trends towards dual nano-encapsulation approach *i.e.* initially forming the inclusion complexes with cyclodextrins and then encapsulating it into nanocarrier. As in inclusion complexes there is no covalent association between host and guest molecules and the dissociation of nutraceuticals

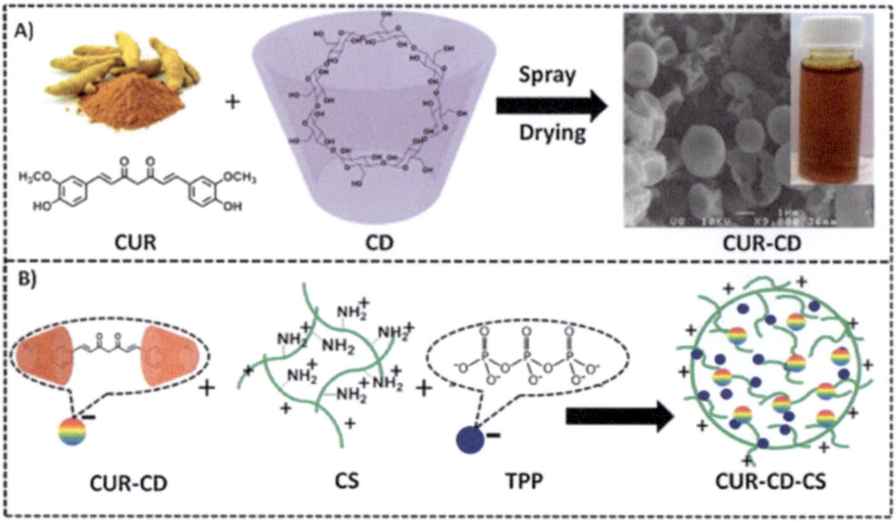

Fig. 1.4 (**A**) The preparation of highly soluble curcumin (CUR)-cyclodextrin (CD) complex by a novel spray drying method. The scanning electron microscopy image represents hollow microspheres after spray drying and inset shows a water solution of curcumin-cyclodextrin; (**B**) the preparation method of curcumin–cyclodextrin-chitosan (CS) nanoparticles. *TPP* triphosphate pentaanion. Reproduced from Ref. (Popat et al. 2014) with permission from Elsevier

occurs rapidly due to displacement by blood components or dilution by blood plasma/extracellular fluid (Chen et al. 2014). Further, these inclusion complexes do not provide any tumor-targeting benefit, thus limits their use less favorable for cancer treatment. Thus considering these aspects, Soo *et al.* developed 2-hydroxypropyl-β-cyclodextrin complex of resveratrol and then encapsulated it into liposomes. *In vitro* cytotoxicity studies in HT-29 colon cancer cell lines demonstrated that the developed liposomes showed dose dependent and enhanced cytotoxicity as compare to free resveratrol.

In another study, Popat *et al.* developed curcumin (CUR)-γ-hydroxypropyl cyclodextrin (CUR-CD) hollow spheres using spray drying method and then encapsulating it into positively charged biodegradable chitosan (CUR-CD-CS) nanoparticles (Fig. 1.4) (Popat et al. 2014). The developed CUR-cyclodextrin-CS nanoparticles are more effective than the native CUR and CUR-CS in human skin cancer SCC25 cell lines and induce cell cycle arrest of S phase and G2/M phase followed by complete apoptosis (~99.9%). Kellici *et al.* prepared inclusion complexes of silibinin (SLB) with 2-hydroxypropyl-β-cyclodextrin (HP-β-CD) in different molar ratios (Kellici et al. 2015). The silibinin solubility increased upto 10–100-folds upon interaction with HP-β-cyclodextrin as compare to native silibinin. Further, *in vitro* cytotoxicity studies in MCF7 cell line demonstrated enhanced cytotoxicity of SLB HP-β-cyclodextrin as compare to native SLB.

Serri *et al.* developed curcumin complex with (2-hydroxypropyl)-β-cyclodextrin and then encapsulated the complex in the poly(D,L-lactic-co-glycolic acid) nanoparticles using nanoprecipitation method (Serri et al. 2017). The purpose of formation of inclusion complex was to improve the loading efficiency due to more encapsulation of the inclusion complex in the internal aqueous phase of the emulsion used to produce the nanoparticles. More interestingly, the increase in the encapsulation did not cause significant changes in nanoparticle dimension, polydispersity index, zeta potential and yield.

Ji *et al.* developed a biodegradable nanocomplex from β-cyclodextrin grafted hyaluronic acid (HA) and phenylalanine based poly(ester amide) for gambogic acid (GA) delivery in order to treat multidrug resistant tumor (Ji et al. 2017). *In vitro* cytotoxicity results demonstrated the nanocomplex enhanced the therapeutic potency of GA in MDA-MB-435/MDR multidrug resistant melanoma cells, and induced enhanced level of apoptosis and mitochondrial depolarization.

In another study, Baek *et al.* developed combinatorial lipid nanoparticle for co-delivery of curcumin and paclitaxel for multidrug resistant breast cancer cells (Baek and Cho 2017). Initially, curcumin was encapsulated in 2-hydroxypropyl-β-cyclodextrin (HPCD) in order to improve its stability, aqueous-solubility and for providing faster release relative to the release of paclitaxel. The faster release of curcumin will lead to the sufficient p-gp inhibition for enhanced intracellular accumulation of paclitaxel against MCF-7/ADR cells. The developed nanoparticles were attached to folic acid in order to achieve targeted delivery. The results demonstrated that folate-conjugated curcumin and paclitaxel loaded lipid nanoparticles exhibited enhanced uptake of paclitaxel and curcumin into MCF-7/ADR cells *via* folate receptor-mediated internalization.

Aadinath *et al.* co-encapsulated curcumin-β-cyclodextrin inclusion complex (IC) and iron oxide nanoparticles (IONPs) within liposomes in order to achieve the synergistic antioxidant potential of curcumin and IONPs. The developed curcumin-in-β-cyclodextrin-innanomagnet liposomes demonstrated highest DPPH radical scavenging activity (IC_{50} value, 64.7791 μg/mL) as compared to IONPs and curcumin liposome and thus demonstrating synergistically enhanced radical scavenging property.

1.4 Conclusion

The use of dietary phytochemicals has gained a substantial interest during the last decade. However, their pharmaceutical applications are limited due to their poor aqueous solubility and bioavailability. The uses of cyclodextrin have demonstrated the capability of improving the aqueous solubility as well as bioavailability. Further, they have also demonstrated their potential in protecting these dietary phytochemicals from elevated temperatures, pH values or moisture induced degradation. Recent advances in dual drug delivery *i.e.* combination of cyclodextrins and other drug delivery systems assemblies such as nanoparticles, liposomes have also demonstrated promising results. Nevertheless, most of these studies are in the preclinical stage, and much effort is desirable for successful translation of these laboratory innovations to clinical reality.

Acknowledgments The authors thanks to Council of Scientific and Industrial Research (CSIR), New Delhi (PMSI-BSC-0117) for financial assistance. DA gratefully acknowledges Council of Scientific and Industrial Research (CSIR) for providing Senior Research Fellowship.

References

Abu-Dahab R, Odeh F, Ismail S, Azzam H, Al Bawab A (2013) Preparation, characterization and antiproliferative activity of thymoquinone-β-cyclodextrin self assembling nanoparticles. Die Pharmazie-an international. J Pharm Sci 68(12):939–944

Alarcon De La Lastra C, Villegas I (2005) Resveratrol as an anti-inflammatory and anti-aging agent: mechanisms and clinical implications. Mol Nutr Food Res 49(5):405–430

Al-Rawashdeh NA, Al-Sadeh KS, Al-Bitar M-B (2010) Physicochemical study on microencapsulation of hydroxypropyl-β-cyclodextrin in dermal preparations. Drug Dev Ind Pharm 36(6):688–697

Aqil F, Munagala R, Jeyabalan J, Vadhanam MV (2013) Bioavailability of phytochemicals and its enhancement by drug delivery systems. Cancer Lett 334(1):133–141

Arora D, Jaglan S (2016) Nanocarriers based delivery of nutraceuticals for cancer prevention and treatment: a review of recent research developments. Trends Food Sci Technol 54:114–126

Arora D, Jaglan S (2017) Therapeutic applications of resveratrol nanoformulations. Environ Chem Lett:1–7

Ashwaq A-AS, Rasedee A, Abdul AB, Taufiq-Yap YH, Al-Qubaisi MS, Eid EE (2017) Characterization, drug release profile and cytotoxicity of dentatin-hydroxypropyl-β-cyclodextrin complex. J Incl Phenom Macrocycl Chem:1–12

Astray G, Gonzalez-Barreiro C, Mejuto J, Rial-Otero R, Simal-Gándara J (2009) A review on the use of cyclodextrins in foods. Food Hydrocoll 23(7):1631–1640

Avallone R, Zanoli P, Puia G, Kleinschnitz M, Schreier P, Baraldi M (2000) Pharmacological profile of apigenin, a flavonoid isolated from *Matricaria chamomilla*. Biochem Pharmacol 59(11):1387–1394

Baek J-S, Cho C-W (2017) A multifunctional lipid nanoparticle for co-delivery of paclitaxel and curcumin for targeted delivery and enhanced cytotoxicity in multidrug resistant breast cancer cells. Oncotarget 8(18):30369

Borghetti GS, Lula IS, Sinisterra RD, Bassani VL (2009) Quercetin/β-cyclodextrin solid complexes prepared in aqueous solution followed by spray-drying or by physical mixture. AAPS PharmSciTech 10(1):235–242

Brewster ME, Loftsson T (2007) Cyclodextrins as pharmaceutical solubilizers. Adv Drug Deliv Rev 59(7):645–666

Challa R, Ahuja A, Ali J, Khar R (2005) Cyclodextrins in drug delivery: an updated review. AAPS PharmSciTech 6(2):E329–E357

Chen J, Lu W-L, Gu W, Lu S-S, Chen Z-P, Cai B-C, Yang X-X (2014) Drug-in-cyclodextrin-in-liposomes: a promising delivery system for hydrophobic drugs. Expert Opin Drug Deliv 11(4):565–577

Cravotto G, Binello A, Baranelli E, Carraro P, Trotta F (2006) Cyclodextrins as food additives and in food processing. Curr Nutr Food Sci 2(4):343–350

de Oliveira Makson G, Guimarães AG, Araújo Adriano A, Quintans Jullyana S, Santos MR, Quintans-Júnior LJ (2015) Cyclodextrins: improving the therapeutic response of analgesic drugs: a patent review. Expert Opin Ther Pat 25(8):897–907 di Cagno MP (2016) The potential of Cyclodextrins as novel active pharmaceutical ingredients: a short overview. Molecules 22(1):1

Duarte A, Martinho A, Luís Â, Figueiras A, Oleastro M, Domingues FC, Silva F (2015) Resveratrol encapsulation with methyl-β-cyclodextrin for antibacterial and antioxidant delivery applications. LWT-Food Sci Technol 63(2):1254–1260

Frémont L (2000) Biological effects of resveratrol. Life Sci 66(8):663–673

Gould S, Scott RC (2005) 2-Hydroxypropyl-β-cyclodextrin (HP-β-CD): a toxicology review. Food Chem Toxicol 43(10):1451–1459

Hedges AR (1998) Industrial applications of cyclodextrins. Chem Rev 98(5):2035–2044

Hirayama F, Uekama K (1999) Cyclodextrin-based controlled drug release system. Adv Drug Deliv Rev 36(1):125–141

Irie T, Uekama K (1997) Pharmaceutical applications of cyclodextrins. III. Toxicological issues and safety evaluation. J Pharm Sci 86(2):147–162

Jambhekar SS, Breen P (2016) Cyclodextrins in pharmaceutical formulations I: structure and physicochemical properties, formation of complexes, and types of complex. Drug Discov Today 21(2):356–362

Ji Y, Shan S, He M, Chu C-C (2017) Inclusion complex from cyclodextrin-grafted hyaluronic acid and pseudo protein as biodegradable nano-delivery vehicle for gambogic acid. Acta Biomater 62:234–245

Jun SW, Kim M-S, Kim J-S, Park HJ, Lee S, Woo J-S, Hwang S-J (2007) Preparation and characterization of simvastatin/hydroxypropyl-β-cyclodextrin inclusion complex using supercritical antisolvent (SAS) process. Eur J Pharm Biopharm 66(3):413–421

Kaur N, Garg T, Goyal AK, Rath G (2016) Formulation, optimization and evaluation of curcumin-β-cyclodextrin-loaded sponge for effective drug delivery in thermal burns chemotherapy. Drug Deliv 23(7):2245–2254

Kellici TF, Ntountaniotis D, Leonis G, Chatziathanasiadou M, Chatzikonstantinou AV, Becker-Baldus J, Glaubitz C, Tzakos AG, Viras K, Chatzigeorgiou P (2015) Investigation of the interactions of Silibinin with 2-Hydroxypropyl-β-cyclodextrin through biophysical techniques and computational methods. Mol Pharm 12(3):954–965

Lee S-H, Kim YH, Yu H-J, Cho N-S, Kim T-H, Kim D-C, Chung C-B, Hwang Y-I, Kim KH (2007) Enhanced bioavailability of soy isoflavones by complexation with β-cyclodextrin in rats. Biosci Biotechnol Biochem 71(12):2927–2933

Li S, Purdy WC (1992) Cyclodextrins and their applications in analytical chemistry. Chem Rev 92(6):1457–1470

Liu RH (2013) Health-promoting components of fruits and vegetables in the diet. Adv Nutr 4(3):384S–392S

Loftsson T, Brewster ME (2012) Cyclodextrins as functional excipients: methods to enhance complexation efficiency. J Pharm Sci 101(9):3019–3032

Loftsson T, Duchêne D (2007) Cyclodextrins and their pharmaceutical applications. Int J Pharm 329:1):1–1)11

Loftsson T, Jarho P, Masson M, Järvinen T (2005) Cyclodextrins in drug delivery. Expert Opin Drug Deliv 2(2):335–351

Lu Z, Cheng B, Hu Y, Zhang Y, Zou G (2009) Complexation of resveratrol with cyclodextrins: solubility and antioxidant activity. Food Chem 113(1):17–20

Manach C, Hubert J, Llorach R, Scalbert A (2009) The complex links between dietary phytochemicals and human health deciphered by metabolomics. Mol Nutr Food Res 53(10):1303–1315

Mangolim CS, Moriwaki C, Nogueira AC, Sato F, Baesso ML, Neto AM, Matioli G (2014) Curcumin–β-cyclodextrin inclusion complex: stability, solubility, characterisation by FT-IR, FT-Raman, X-ray diffraction and photoacoustic spectroscopy, and food application. Food Chem 153:361–370

Marques HMC (2010) A review on cyclodextrin encapsulation of essential oils and volatiles. Flavour Fragr J 25(5):313–326

Martina K, Binello A, Lawson D, Jicsinszky L, Cravotto G (2013) Recent applications of cyclodextrins as food additives and in food processing. Curr Nutr Food Sci 9(3):167–179

McClements DJ, Li F, Xiao H (2015) The nutraceutical bioavailability classification scheme: classifying nutraceuticals according to factors limiting their oral bioavailability. Annu Rev Food Sci Technol 6:299–327

Mellet CO, Fernández JMG, Benito JM (2011) Cyclodextrin-based gene delivery systems. Chem Soc Rev 40(3):1586–1608

Meyer H, Bolarinwa A, Wolfram G, Linseisen J (2006) Bioavailability of apigenin from apiin-rich parsley in humans. Ann Nutr Metab 50(3):167–172

Mohtar N, Taylor KM, Sheikh K, Somavarapu S (2017) Design and development of dry powder sulfobutylether-β-cyclodextrin complex for pulmonary delivery of fisetin. Eur J Pharm Biopharm 113:1–10

Oommen E, Shenoy BD, Udupa N, Kamath R, DEVI P (1999) Antitumour efficacy of Cyclodextrin-complexed and Niosome-encapsulated Plumbagin in mice bearing melanoma B16F1. Pharm Pharmacol Commun 5(4):281–285

Oprean C, Mioc M, Csányi E, Ambrus R, Bojin F, Tatu C, Cristea M, Ivan A, Danciu C, Dehelean C (2016) Improvement of ursolic and oleanolic acids' antitumor activity by complexation with hydrophilic cyclodextrins. Biomed Pharmacother 83:1095–1104

Pápay ZE, Sebestyén Z, Ludányi K, Kállai N, Balogh E, Kósa A, Somavarapu S, Böddi B, Antal I (2016) Comparative evaluation of the effect of cyclodextrins and pH on aqueous solubility of apigenin. J Pharm Biomed Anal 117:210–216

Pinho E, Grootveld M, Soares G, Henriques M (2014) Cyclodextrins as encapsulation agents for plant bioactive compounds. Carbohydr Polym 101:121–135

Pinho E, Soares G, Henriques M (2015) Cyclodextrin modulation of gallic acid in vitro antibacterial activity. J Incl Phenom Macrocycl Chem 81(1–2):205–214

Popat A, Karmakar S, Jambhrunkar S, Xu C, Yu C (2014) Curcumin-cyclodextrin encapsulated chitosan nanoconjugates with enhanced solubility and cell cytotoxicity. Colloids Surf B: Biointerfaces 117:520–527

Popović M, Kaurinović B, Trivić S, Mimica-Dukić N, Bursać M (2006) Effect of celery (Apium graveolens) extracts on some biochemical parameters of oxidative stress in mice treated with carbon tetrachloride. Phytother Res 20(7):531–537

Serri C, Argirò M, Piras L, Mita DG, Saija A, Mita L, Forte M, Giarra S, Biondi M, Crispi S (2017) Nano-precipitated curcumin loaded particles: effect of carrier size and drug complexation with (2-hydroxypropyl)-β-cyclodextrin on their biological performances. Int J Pharm 520(1):21–28

Shulman M, Cohen M, Soto-Gutierrez A, Yagi H, Wang H, Goldwasser J, Lee-Parsons CW, Benny-Ratsaby O, Yarmush ML, Nahmias Y (2011) Enhancement of naringenin bioavailability by complexation with hydroxypropoyl-β-cyclodextrin. PLoS One 6(4):e18033

Silva JC, Almeida JR, Quintans JS, Gopalsamy RG, Shanmugam S, Serafini MR, Oliveira MR, Silva BA, Martins AO, Castro FF (2016) Enhancement of orofacial antinociceptive effect of carvacrol, a monoterpene present in oregano and thyme oils, by β-cyclodextrin inclusion complex in mice. Biomed Pharmacother 84:454–461

Soica C, Danciu C, Savoiu-Balint G, Borcan F, Ambrus R, Zupko I, Bojin F, Coricovac D, Ciurlea S, Avram S (2014) Betulinic acid in complex with a gamma-cyclodextrin derivative decreases proliferation and in vivo tumor development of non-metastatic and metastatic B164A5 cells. Int J Mol Sci 15(5):8235–8255

Stella VJ, He Q (2008) Cyclodextrins. Toxicol Pathol 36(1):30–42

Suzuki R, Inoue Y, Tsunoda Y, Murata I, Isshiki Y, Kondo S, Kanamoto I (2015) Effect of γ-cyclodextrin derivative complexation on the physicochemical properties and antimicrobial activity of hinokitiol. J Incl Phenom Macrocycl Chem 83(1–2):177–186

Wang X, Deng L, Cai L, Zhang X, Zheng H, Deng C, Duan X, Zhao X, Wei Y, Chen L (2011) Preparation, characterization, pharmacokinetics, and bioactivity of honokiol-in-hydroxypropyl-β-cyclodextrin-in-liposome. J Pharm Sci 100(8):3357–3364

Wu H, Liang H, Yuan Q, Wang T, Yan X (2010) Preparation and stability investigation of the inclusion complex of sulforaphane with hydroxypropyl-β-cyclodextrin. Carbohydr Polym 82(3):613–617

Xu X-r, Yu H-t, Hang L, Shao Y, Ding S-h, Yang X-w (2014) Preparation of naringenin/β--cyclodextrin complex and its more potent alleviative effect on choroidal neovascularization in rats. Biomed Res Int 2014

Yang L-J, Ma S-X, Zhou S-Y, Chen W, Yuan M-W, Yin Y-Q, Yang X-D (2013) Preparation and characterization of inclusion complexes of naringenin with β-cyclodextrin or its derivative. Carbohydr Polym 98(1):861–869

Yang Z, Argenziano M, Salamone P, Pirro E, Sprio AE, Di Scipio F, Carere ME, Quaglino E, Cavallo F, Cavalli R (2016) Preclinical pharmacokinetics comparison between resveratrol 2-hydroxypropyl-β-cyclodextrin complex and resveratrol suspension after oral administration. J Incl Phenom Macrocycl Chem 86(3–4):263–271

Yee EM, Hook JM, Bhadbhade MM, Vittorio O, Kuchel RP, Brandl MB, Tilley RD, Black DS, Kumar N (2017) Preparation, characterization and in vitro biological evaluation of (1: 2) phenoxodiol-β-cyclodextrin complex. Carbohydr Polym 165:444–454

Zhang J, Ma PX (2013) Cyclodextrin-based supramolecular systems for drug delivery: recent progress and future perspective. Adv Drug Deliv Rev 65(9):1215–1233

Zhang L, Man S, Qiu H, Liu Z, Zhang M, Ma L, Gao W (2016) Curcumin-cyclodextrin complexes enhanced the anti-cancer effects of curcumin. Environ Toxicol Pharmacol 48:31–38

Zhu Z-Y, Luo Y, Liu Y, Wang X-T, Liu F, Guo M-Z, Wang Z, Liu A-J, Zhang Y-M (2016) Inclusion of chrysin in β-cyclodextrin and its biological activities. J Drug Delivery Sci Technol 31:176–186

Chapter 2
Cyclodextrin-Steroid Interactions and Applications to Pharmaceuticals, Food, Biotechnology and Environment

Éva Fenyvesi, István Puskás, and Lajos Szente

Contents

Abstract Steroids are abundant in living organisms and are responsible for various biological functions. Steroids form inclusion complexes with cyclodextrins, which are used for diverse applications. This chapter reviews the steroid drugs, bile salts, vitamin D and phytosterols in relation to their complexation with cyclodextrins. Emphasis is made on the effect of cyclodextrin–steroid interactions on solubility, stability and bioavailability of entrapped steroids. We discuss the pharmacokinetic changes of cyclodextrin-formulated steroid drugs designed for various routes of administration. We present some cyclodextrins-enabled steroid formulations introduced to the market. Specifically, we discuss sugammadex, the first cyclodextrin derivative approved as active pharmaceutical ingredient, which revolutionized anesthesia. The catalysis or inhibition of biotransformation reactions of steroids by

É. Fenyvesi (✉) · I. Puskás · L. Szente
CycloLab Cyclodextrin R&D Laboratory Co., Budapest, Hungary
e-mail: fenyvesi.e@cyclolab.hu; puskas@cyclolab.hu; szente@cyclolab.hu

S. Fourmentin et al. (eds.), *Cyclodextrin Applications in Medicine, Food, Environment and Liquid Crystals*, Environmental Chemistry for a Sustainable World, https://doi.org/10.1007/978-3-319-76162-6_2

cyclodextrin complexation are examplified. We present laboratory and pilot-scale experiments for the environmental applications such as cyclodextrin-containing sorbents for capturing residual steroid drugs, e.g. contraceptives from purified waste water.

2.1 Introduction

Steroids are a class of organic compounds with gonane (cyclopentaperhydrophenanthrene) structure. The gonane unit consisting of four fused rings with three cyclohexane (A, B and C ring) and a cyclopentane ring (D ring) may have various substituents resulting in hundreds of compounds found in plants, animals and fungi. The example of ergosterol, the first steroid complexed by cyclodextrins is shown in Fig. 2.1.

The diversity of the steroids and complexity of their biological functions are illustrated by the fact that 7 Nobel Prizes were awarded in connection with their structure, isolation, synthesis and biological effects starting in 1927 by H.O. Wieland and finishing in 1975 by V. Prelog (Vardanyan and Hruby 2016).

Steroids have various biological functions:

- building unit of cell membrane (cholesterol, phytosterols),
- digestion of fats (cholic acids),
- controlling the secondary gender features, sexual activity and spermatogenesis (male hormones),
- anabolic activity (muscle mass enhancement),
- controlling the various processes in reproduction, contraception, bone density (estrogen and progesterone).

Corticosteroids are a class of steroid hormones produced in the adrenal cortex of vertebrates. Corticosteroids, which are involved in the carbohydrate, fat and protein metabolism and have anti-inflammatory, immunosuppressive, vasoconstrictive and anti-proliferative effects, belong to the group of glucocorticoids while those involved in the regulation of electrolyte and water balance are mineralocorticoids. Cortisol (hydrocortisone), prednisolone, dexamethasone are examples for the first, and androsterone and progesterone for the second group. Some synthetic corticosteroids, such as prednisone have both effects.

Steroids are ideal guest molecules for complexation by cyclodextrins, the cyclic oligosaccharides prepared by enzymatic conversion of starch (Szejtli 1982; Crini 2014). The specific feature of cyclodextrins that they have a cavity, which is of less hydrophilic character compared to the outer surface is utilized when the high energy water molecules in the cavity are exchanged by hydrophobic guest molecules, such as steroids forming host–guest inclusion complexes. This molecular encapsulation depending on the proper fit between the host and guest molecules often ensures enhanced solubility and improved bioavailability for poorly soluble drugs of low absorption in the living organisms.

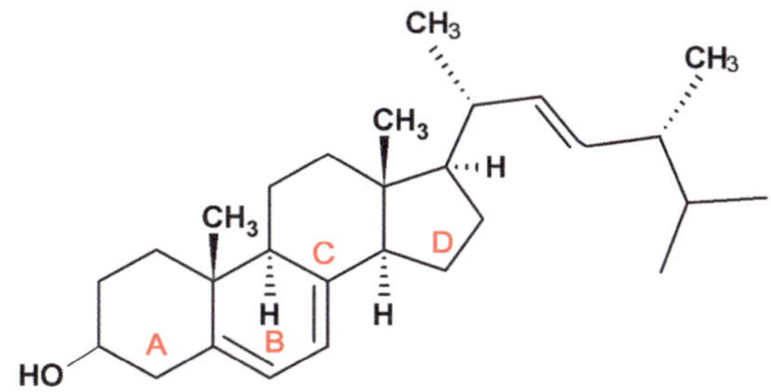

Fig. 2.1 Chemical formula of ergosterol

Steroids are poorly soluble in water, have various polymorphs with different dissolution rates. The release of steroid drugs from different formulations of different manufacturers might be different and the release rate can change also during storage in consequence of recrystallization. The very first publication on cyclodextrin/steroid complexes is most probably by Cramer and Henglein (1957) reporting on the solubility of beta-cyclodextrin/ergosterol complex (0.762 g/100 cm^3) and the ergosterol content in the complex (3.66%) corresponding to molar ratio of guest to host (1:2). The complex was prepared by mixing aqueous beta-cyclodextrin solution (1%) with ergosterol dissolved in diethyl ether and stirred for 5–8 days. The precipitate was filtered and washed with water, acetone and diethyl ether before drying. The guest content was calculated from the cyclodextrin content determined by polarimetry.

Huge number of publications have been devoted to the cyclodextrin–steroid interaction during the 60 years passed after the first paper of Cramer. In the cyclodextrin-related literature the most thoroughly studied steroid is cholesterol (around 2500 publications in Scopus) and steroid drugs (nearly 1000 publications), followed by bile acids and salts (~400). As the structural aspects of cyclodextrin/cholesterol interaction as well as its applications in molecular biology, medicine and food industry have been recently reviewed (Fenyvesi et al. 2016; Szente and Fenyvesi 2017), this Chapter focuses on steroid drugs and bile salts with emphasis on pharmaceutical, biotechnology and environmental use of their complexation. Vitamin D and phytosterols are also discussed. The special cyclodextrin derivative developed for the ultimate entrapment of muscle relaxants is also shown.

2.2 Inclusion Complex Formation of Steroids

2.2.1 Determination of the Association Constants

The affinity of a cyclodextrin to a guest compound can be characterized by the association constant (K) of the equilibrium process of complex association of 1:1 mole/mole host:guest:

$$K = \frac{[SCD]}{[S][CD]}$$

where [S], [CD] and [SCD] stand for the concentrations of the steroid, cyclodextrin and their complex, respectively.

The association constants can be determined by various methods recording the changes of a physical parameter, such as solubility, spectral changes, chromatographic or electrophoretic mobility, etc. as a function of cyclodextrin concentration (Szejtli 1982). *Phase solubility method* is used the most frequently (Uekama et al. 1982a; Habon et al. 1985; Liu et al. 1990). A special variation of this method is the phase distribution method, which is based on the change in octanol/water partition (log P) of a guest molecule with increasing cyclodextrin concentrations in the aqueous phase (Masson et al. 2005). A correction factor should be used for the calculation of K by this method because octanol is also complexed by the cyclodextrins.

Competitive guest binding method is based on the replacement of methyl orange or other dye by steroids in the cavity of various β-cyclodextrin derivatives, such as hydroxypropyl, carboxymethyl, methyl and sulfobutyl β-cyclodextrins (Khomutov et al. 2002). The K values calculated increased with increasing hydrophobicity of the steroids with highest values for carboxymethyl β-cyclodextrin among the β-cyclodextrins studied.

Induced Circular Dichroism titration is based on detection of maximum wavelength and magnitude of the induced circular dichroism bands as a function of cyclodextrin concentration (Chun and Yun 1993). Similarly, the chemical shifts in H^1 NMR spectra (Chun and Yun 1993; Tan et al. 1994; Cameron et al. 2002; Jover et al. 2004) and the fluorescence intensity in the fluorescence spectra (Sadlej-Sosnowska 1997a) as a function of cyclodextrin concentration can be used for the calculation of K.

In addition to the association constants also the thermodynamic parameters of complex formation can be obtained by *calorimetric titrations* (Ollila et al. 2001; Liu et al. 2003; Holm et al. 2009).

Chromatographic techniques, such as thin layer chromatography (TLC, Cserháti and Forgács 1996) and high performance liquid chromatography (HPLC, Sadlej-Sosnowska 1995; Flood et al. 2000; Yanez et al. 2007) utilize the change in chromatographic mobility upon enhancing cyclodextrin concentration in the solution/buffer applied as eluent (Table 2.1). While TLC is useful only for comparison of the binding affinities of various steroids, HPLC offers quantitative results. It should be, however, taken into consideration that the calculated association constants depend also on the ratio of organic solvent in the eluent (Agnus et al. 1994; Sadlej-Sosnowska 1997a; Yanez et al. 2007). Using capillary electrophoresis and monitoring the change in electrophoretic mobility of the steroid as a function of cyclodextrin concentration in the background electrolyte, the K value can be calculated (Shakalisava and Regan 2005; Larsen et al. 2007). This method is useful for ionizable molecules, but in the case of steroid complexes both the guests and the hosts are usually non-ionic molecules what results in uncertainty of the determination.

Electrochemical methods, such as differential pulse voltammetry was used for the determination of association constants for estradiol and estrone with

Table 2.1 Association constants (log K, M^{-1}) of estradiol complexes (1:1) with β-, γ- and hydroxypropyl-β-cyclodextrin determined by various methods at room temperature in water if otherwise not stated

	β-Cyclodextrin	γ-Cyclodextrin	Hydroxypropyl-β-cyclodextrin
Phase solubility	3.9[a] (Albers and Müller 1992)		3.5 (De Paula et al. 2007) 4.4 (Masson et al. 2005)
Octanol/water phase distribution			4.5–4.6 (Masson et al. 2005)
Fluorimetry	3.6 (Sadlej-Sosnowska 1997a) 4.6 (Perez and Escandar 2013) 1.6 (MeOH:Water 20/80, Perez and Escandar 2013)	3.8 (Sadlej-Sosnowska 1997a) 4.2 (Perez and Escandar 2013)	4.8 (Perez and Escandar 2013)
High performance liquid chromatography (HPLC)	3.8 (MeOH:Water 20/80, Sadlej-Sosnowska 1997b) 2.4 (MeCN/buffer[b] 35/65, Yanez et al. 2007)	3.9 (MeOH:Water 20/80, Sadlej-Sosnowska 1997b)	
Capillary electrophoresis	1.9 (Shakalisava and Regan 2005)	3.1 (Shakalisava and Regan 2005)	2.0 (Shakalisava and Regan 2005)
Differential pulse voltammetry	2.0 (MeCN/buffer[b] 35/65, Yanez et al. 2007)		
Chronocoulometry	2.1 (EtOH/buffer[b] 30/70, Yanez and Basquinzay 2008)		2.4 (EtOH/buffer 30/70, Yanez and Basquinzay 2008)

[a]calculated based on the solubility isotherms in Albers and Müller 1992
[b]0.05 N phosphate buffer pH 5.0

β-cyclodextrin in various solvent/buffer systems (Yanez et al. 2007). The inclusion complex formation resulted in decreased peak current when the β-cyclodextrin concentration increased. The K values obtained were different from those determined by HPLC using the same solvent/buffer system (Table 2.1). Chronocoulometry is based on the decrease of the cathodic peak current due to the reduced diffusion of the guest molecule when an inclusion complex is formed (Yanez and Basquinzay 2008).

The K values obtained by different methods might differ in a high extent. For instance, the K values for estradiol complexes listed in Table 2.1 are comparable when determined by phase solubility, octanol-water phase distribution and fluorimetry while much lower values can be obtained by electrochemical methods and capillary electrophoresis. Any series of data determined by the same method seem to be useful for comparison either the steroids or the cyclodextrins, but data created by different methods can be compared with extreme caution if at all.

The complex association constants depend on the structure of both the host and the guest. Increasing hydrophobicity, the presence of a double bond in ring A or of a fluorine substituent in the steroid favors the complexation, as well as the larger cavity size and the methyl substituents on the cyclodextrin ring (Lopata et al. 1985). Larger apparent association constants were obtained for γ-cyclodextrin than for β-cyclodextrin studying 13 steroids by HPLC (Flood et al. 2000). The substituents on the 11- and 17-positions were found to have crucial effect on complexation. γ-cyclodextrin was the preferred host for cardiac steroids with A/B *cis* ring junction, while β-cyclodextrin was slightly more effective in complexation of the molecules with A/B *trans* junction (Shimada et al. 1989). The presence of hydroxyl groups at 3- and 12-positions enhance the interactions.

The constants can be further enhanced by the presence of polymers. For instance the K values for hydrocortisone with hydroxypropyl-, random methyl and maltosyl β-cyclodextrin were enhanced by 15–20% in the presence of 0.25 w/v% polyvinyl-pyrrolidone (Sigurdardottir and Loftsson 1995).

2.2.2 Preparation of Complexes

The properties (solubility, stability and dissolution rate) of the complexed steroid depends not only on the properties of host and guest but also on the method used for the preparation of the inclusion complex. For instance, in the case of hydrocortisone acetate co-precipitation from a common solution resulted in complexes of higher solubility compared to those obtained by co-grinding or freeze drying (Belikov et al. 1991a). On the other hand, β-cyclodextrin/danazol complexes obtained by freeze drying and milling showed enhanced dissolution rate compared to those prepared by kneading and co-precipitation methods (Jadhav et al. 2007). The hydrocortisone acetate γ-cyclodextrin complex prepared by spray drying overperformed that prepared by coprecipitation concerning the entrapped drug content (Becirevic-Lacan and Skalko 1997).

The wettability of the β-cyclodextrin/methylhydroxyprogesterone complex was higher when prepared by mechano-chemical activation (co-grinding) compared to the complex obtained by co-precipitation method (Carli et al. 1987). A more sophisticated way of mechano-chemical activation can be implemented by milling with planetary ball mill leading to fine grinding (Rinaldi et al. 2015).

2.3 Steroid Drugs

2.3.1 Corticosteroids and Other Steroid Drugs

2.3.1.1 Structure of Cyclodextrin/Steroid Complexes

The stoichiometry of the inclusion complexes of steroids with β-cyclodextrin and its derivatives is usually 2:1 (host:guest) (Uekama et al. 1982a). Mass spectrometry showed the presence of mass fragments of both 1:1 and 2:1 complex molecular ions

(Liu et al. 1990). For β-cyclodextrin/spironolactone and hydroxypropyl-β-cyclodextrin/hydrocortisone complexes 3:1 stoichiometry was also published (Moellgaard and Bundgaard 1983; Preiss et al. 1994). The free energy calculations proved that there is a significant decrease in the total hydrophobic surface area during complex formation of steroids with cyclodextrins and the driving force responsible for complex formation is the van der Walls and hydrophobic interactions with additional effect of hydrogen bonding (Jadhav and Vavia 2008; Cai et al. 2009).

The molecular structure of the complexes is determined mainly by the structure of the steroid. Progesterone is fully immersed in the β-cyclodextrin cavity; however, complete inclusion of the hydrocortisone molecule was prevented by the polar hydroxyl groups on its surface (Forgo and Gondos 2002). First, the cyclopentane ring is included into a β-cyclodextrin molecule and as a second step the free A ring is enwrapped by a second β-cyclodextrin forming 2:1 host:guest complex (Liu et al. 1990). The protons of steroidal skeleton of danazol are preferably involved in the complexation with β-cyclodextrin (Jadhav and Vavia 2008). The molecular modeling study indicated inclusion complexation of methylprednisolone with γ-cyclodextrin and hydroxypropyl γ-cyclodextrin by entrance of the A and B rings into the cyclodextrin cavity from its larger rim (Thi et al. 2010). Similarly, the A and B ring of prednisolone were embedded into the cavity of dimethyl β-cyclodextrin from its wider rim (Shi et al. 2014). The inclusion process is enthalpy driven and the primary forces responsible for the inclusion are hydrogen bonding and Van der Waals force.

Inclusion complexation of 2-methoxyestradiol, however, involves insertion of the D-ring from the secondary side of each cyclodextrin molecule independently of the substituents on the cyclodextrin (Caira et al. 2015). The 17-OH group generally forms hydrogen bond with a host glycosidic oxygen atom within the cyclodextrin cavity, while the A-ring and part of the B-ring of the steroid protrude from the secondary side.

Two steroid antibiotics of the fusidane family, sodium fusidate and potassium helvolate enter into the cavity of the γ-cyclodextrin by the side chain, reaching the central region of the steroid (rings C and D), whereas the A and B rings (B only partially) remain outside in the 1:1 complex (Jover et al. 2003). For β-cyclodextrin complexes, rotating-frame nuclear Overhauser effect correlation (ROESY) spectra show a remarkable absence of interactions of the protons of the C and D rings, whereas clear interactions corresponding to the side chain, and A and B rings are observed in accordance with the 2:1 (cyclodextrin:drug) stoichiometry. Repeating the NMR studies using higher sample concentrations strong interaction was found with B, C and D rings of fusidate and the γ-cyclodextrin cavity (Larsen et al. 2007).

Some possible conformations of cyclodextrin/steroid complexes can be seen in Fig. 2.2.

2.3.1.2 Solubilizing Effect

Most of the steroids are poorly soluble in water and their solubility can be enhanced by complexation with hydrophilic cyclodextrins. Several examples are listed in Table 2.2.

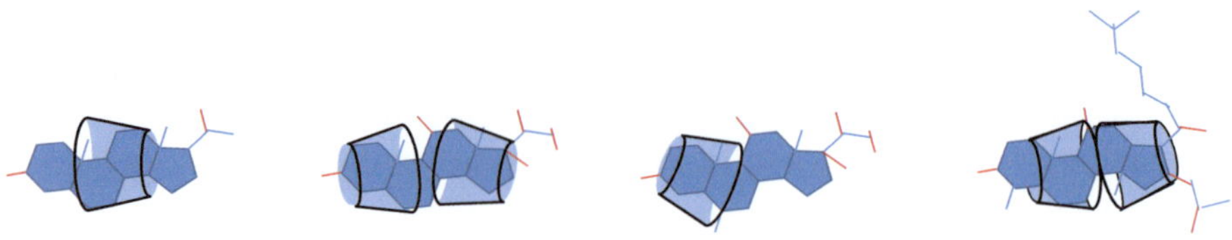

Fig. 2.2 Schemes of some possible structures of cyclodextrin/steroid complexes: β-cyclodextrin/progesterone (**A**), β-cyclodextrin/hydrocortisone (**B**), γ-cyclodextrin/methylprednisolone (**C**) and β-cyclodextrin/sodium fusidate (**D**). Various parts of the steroid skeleton are entrapped depending on the substituents in the steroid molecule and the cavity size of the cyclodextrin

While the inclusion complexes of various steroids with β- and γ-cyclodextrin can precipitate from aqueous solution (show Bs type solubility isotherm according to the classification of Higuchi and Connors 1965, which means that the complex has limited solubility in water), most of the cyclodextrin derivatives form soluble inclusion complexes (A type isotherms, which means that the solubility of steroid increases with increasing cyclodextrin concentration). Even the complexes with the native cyclodextrins have enhanced solubility, wettability and dissolution rate compared to the steroid alone. Alkylation of β-cyclodextrin increases not only the solubility of β-cyclodextrin but also the solubility of the complexes with steroids (Szejtli 1984; Müller and Brauns 1986). Condensation products of β-cyclodextrin with propylene oxide or epichlorohydrin (hydroxypropyl β-cyclodextrin and dihydroxypropyl β-cyclodextrin, and its polymers, respectively) have high water solubility and form water soluble complexes with various steroids (Pitha et al. 1986). Enhanced complexation of steroid drugs was observed with 6-O-acyl β-cyclodextrins of increasing chain length, while the stoichiometric ratio of 2:1 cyclodextrin:drug remained unaffected (Liu et al. 1992). Both the length of the alkyl chain and the degree of alkylation had an influence on the binding of steroids, such as 6-methylprednisolone, prednisolone, triamcinolone, D(−)norgestrel and hydrocortisone by sulfobutylether-alkylether β-cyclodextrin derivatives (Tongiani et al. 2009).

It is not easy to find general rules on how the various substituents on the cyclodextrin molecules affect the complexation with steroids because the various research groups used different sets of cyclodextrin derivatives. For instance, the solubilizing effect on 2-methoxyestradiol (anticancer drug) followed the order (Caira et al. 2015):

dimethyl β- > random methylated β- > hydroxypropyl β- ~ γ- > trimethyl β- > β- > triacetyl β- ~ α- > trimethyl α-cyclodextrin.

While hardly any difference in the solubilizing effect on prednisolone was found among the various cyclodextrins studied, the solubilizing effect on methyl prednisolone followed the order (Larsen et al. 2005):

γ- > β- > maltosyl β- > glucuronyl glucosyl β- ~ sulfobutyl ether γ- > hydroxypropyl β- > α-cyclodextrin.

Table 2.2 Examples of cyclodextrin/steroid complexes and the effects obtained by complexation

Steroid	Cyclodextrin	Effect	References
Alfaxalone	Hydroxypropyl β-cyclodextrin	Solubility enhancement	Brewster et al. 1989
Allopregnanolone	Hydroxypropyl β-cyclodextrin	Solubility enhancement	Ford et al. 2005
Androstenediol	Hydroxypropyl β-cyclodextrin	Solubility enhancement, improved stability	Pitha et al. 1992
Androstenedione	Hydroxypropyl β-cyclodextrin	Solubility enhancement, improved stability	Pitha et al. 1992
Beclomethasone	γ-cyclodextrin	Improved aerodynamic performance in inhalant formulations	Cabral-Marques and Almeida 2009
Beclomethasone dipropionate	α-, β-, γ-cyclodextrin	Solubility enhancement; Improved release from ointment base and percutaneous absorption	Uekama et al. 1982a; Uekama et al. 1985
Betamethasone	α-, β-, γ-cyclodextrin	Solubility enhancement	Uekama et al. 1982a
Betamethasone valerate	α-, β-, γ-cyclodextrin	Solubility enhancement	Uekama et al. 1982a
Budesonide	Sulfobutyl ether β-cyclodextrin	Fast, long lasting relief for allergic rhinitis symptoms, faster onset of action	Salapatek et al. 2011
Ciclesonide	Crystalline methyl β-cyclodextrin (CRYSMEB)	Enhanced solubility	Arriagas and Cabral-Marques 2013
Cortexolone	β-cyclodextrin, soluble β-cyclodextrin polymer	Solubility enhancement, improved bioconversion to hydrocortisone	Kompantseva et al. 1990; Volkova et al. 1999
Corticosterone	β-, hydroxypropyl β- and sulfobutyl ether β-cyclodextrin	Solubility enhancement, skin permeation enhancement	Kralova and Mitterhauszerova 1989; Hashem et al. 2003; Shaker et al. 2003
Cortisone	α-, β-, γ-cyclodextrin	Solubility enhancement	Uekama et al. 1982a
Cortisone acetate	α-, β-, γ-cyclodextrin	Solubility enhancement	Uekama et al. 1982a
Cyproterone acetate	γ-, hydroxypropyl β- and dimethyl β-cyclodextrin	Solubility enhancement	De Hassonville et al. 2003
Danazol	β-, hydroxypropyl β- and sulfobutyl ether β-cyclodextrin	Enhanced dissolution rate	Badawy et al. 1996; Jadhav and Vavia 2008

(continued)

Table 2.2 (continued)

Steroid	Cyclodextrin	Effect	References
Dexamethasone	Hydroxypropyl β- and glucuronyl glucosyl β-cyclodextrin	Prolonged suppression of stress-induced release of adrenocorticotropic hormone and corticosterone in rats; Increased water solubility, enhanced drug absorption into the eye, improved aqueous stability and reduced local irritation; Enhanced bioavailability in the first hour	Uekama et al. 1982a; Anderson et al. 1989; Shinoda et al. 1999; Dietzel et al. 1990; Loftsson and Stefansson 2002; Stefansson and Loftsson 2003; Jansook et al. 2010
Dexamethasone acetate	β-, γ-, hydroxypropyl β-cyclodextrin	Increased stability	Uekama et al. 1982a; Vianna et al. 1998
Digoxin	γ-, hydroxypropyl β-cyclodextrin	Solubility enhancement and improved bioavailability in dogs and humans; Stabilization against acidic hydrolysis	Uekama et al. 1981; Uekama et al. 1982b; Müller and Brauns 1986; Seo and Uekama 1989; Helm et al. 1994
Digitoxin	Hydroxypropyl β-cyclodextrin	Solubility enhancement; Stabilizing effect	Müller and Brauns 1986; Helm et al. 1994
Estradiol	Hydroxypropyl β-, random methylated β-cyclodextrin and water-soluble β-cyclodextrin polymer	Solubility enhancement; Improved buccal and nasal absorption	Pitha et al. 1986; Pike et al. 2004
2-Methoxy estradiol	Di- and trimethyl β-, random methylated β-, hydroxypropyl β-cyclodextrin	Solubility enhancement	Caira et al. 2015
Estriol	Glucuronyl glucosyl β-cyclodextrin	Solubility enhancement	Koizumi et al. 1987
Fluocinolone	α-, β-, γ-cyclodextrin	Solubility enhancement	Uekama et al. 1982a
Fluticasone propionate	Hydroxypropyl β-cyclodextrin	Improved dissolution rate	Lohade et al. 2007
Fusidic acid/ sodium fusidate	γ-cyclodextrin polymer	Solubility enhancement	Wintgens and Amiel 2010
Hydrocortisone	γ-, hydroxypropyl β-, hydroxypropyl γ-, dimethyl β- and glucuronyl glucosyl β-cyclodextrin	Increasing transdermal permeation; Increased solubility	Uekama et al. 1982a; Szejtli 1984; Müller and Brauns 1986; Okada and Koizumi 1998; Kear et al. 2008; Jansook et al. 2010
Hydrocortisone acetate	α-, β-, γ-cyclodextrin	Improved release from hydrogels (α > β > γ)	Uekama et al. 1982a; Ceschel et al. 1993

(continued)

Table 2.2 (continued)

Steroid	Cyclodextrin	Effect	References
Loteprednol etabonate	γ-, hydroxypropyl β-cyclodextrin, dimethyl and maltosyl β-cyclodextrin	Solubility and stability enhancement, enhanced permeability through hamster cheek pouch, improved tissue partitioning in rats p.o.	Alberth et al. 1991; Bodor et al. 1992; Bodor et al. 2000
Methyl testosterone	Dimethyl β-cyclodextrin	Solubility enhancement	Szejtli 1984
Methyl prednisolone	α-, β- and γ-, hydroxypropyl β-, hydroxypropyl γ-, sulfobutyl ether β-, sulfobutyl ether γ-, glucuronyl glucosyl β-cyclodextrin	Solubility enhancement	Larsen et al. 2005; Thi et al. 2010
Norprogesterone	β-, hydroxypropyl β- and hydroxyethyl β-cyclodextrin	Solubility enhancement	Ahmed 1998
Paramethasone	α-, β-, γ-cyclodextrin	Solubility enhancement	Uekama et al. 1982a
Prednisolone	β-, dimethyl β-cyclodextrin	Increased solubility, enhanced bioavalability, decreased particle size	Uekama et al. 1982a; Uekama et al. 1983; Arimori et al. 1984; Arimori and Uekama 1987; Belikov et al. 1991b; Larsen et al. 2005; Shi et al. 2014
Prednisolone acetate	α-, β-, γ-cyclodextrin	Solubility enhancement	Uekama et al. 1982a
Pregnanolone	Hydroxypropyl β-cyclodextrin	Enhanced solubility and anesthetic activity	Brewster et al. 1995
Pregnenolone	Hydroxypropyl β-cyclodextrin	Enhanced solubility and reduced anesthetic activity	Brewster et al. 1995
Progesterone	α-, β-, γ-, dimethyl β-, hydroxypropyl β-cyclodextrin, β-cyclodextrin polymer	Solubility enhancement; Enhanced buccal absorption	Uekama et al. 1982a; Szejtli 1984; Pitha et al. 1986
Progestin	Hydroxypropyl β-and hydroxyethyl β-cyclodextrin	Stabilization	Ahmed 1997
Testosterone	γ-, hydroxypropyl β-cyclodextrin, β-cyclodextrin polymer	Solubility enhancement; Improved buccal absorption	Pitha and Pitha 1985; Pitha et al. 1986; Pitha et al. 1987

(continued)

Table 2.2 (continued)

Steroid	Cyclodextrin	Effect	References
Spironolactone	β-cyclodextrin	Enhanced solubility and bioavailability	Kata and Haragh 1981; Moellgaard and Bundgaard 1983; Wouessidjewe et al. 1989
Triamcinolone	α-, β-, γ-cyclodextrin	Solubility enhancement	Uekama et al. 1982a
Triamcinolone acetate	α-, β-, γ-cyclodextrin	Solubility enhancement	Uekama et al. 1982a
Triamcinolone diacetate	α-, β-, γ-cyclodextrin	Solubility enhancement	Uekama et al. 1982a
Triamcinolone acetonide	Hydroxypropyl β-cyclodextrin	Solubility enhancement and increased transdermal permeation	Habon et al. 1985; Kear et al. 2008

In solubilization of progesterone hydroxypropyl- and trimethyl β-cyclodextrin were more efficient than hydroxypropyl γ- and sulfobutyl β-cyclodextrin (Lahiani-Skiba et al. 2006), on the contrary the order of γ- > hydroxypropyl γ- > hydroxypropyl β-cyclodextrin was observed for methylprednisolone (Thi et al. 2010).

The degree of the solubility enhancement depends on the structure of the steroids, too. Hydroxypropyl β-cyclodextrin solubilized testosterone derivatives more effectively than estradiol esters (Albers and Müller 1992). Within a homologous series of steroid hormones, the steepest linear solubility isotherms were found for 17-methyl and 3-methyl derivatives. The interaction decreased with increasing length of the ester side chain suggesting that the unsubstituted steroids are preferred for complexation.

Most often hydrocortisone was the model drug in the studies concerning the aggregation of cyclodextrins and their complexes (Loftsson et al. 2002). The self-assembly of cyclodextrin/hydrocortisone complexes into smaller or larger aggregates explains the precipitation observed at higher β- or γ-cyclodextrin concentrations. Even the well soluble complexes with hydroxypropyl β-cyclodextrin form aggregates, which do not permeate through synthetic membranes at the expected rate (Loftsson 2014). Water soluble polymers stabilize the cyclodextrin/drug complexes and their aggregates and hence improve the complexation efficiency of γ-cyclodextrin and hydroxypropyl γ-cyclodextrin (Jansook et al. 2010). Polymers as ternary components can further enhance the solubility of β-cyclodextrin/steroid complexes. For instance, polyethylene glycol (PEG 6000) was advantageous for progesterone (Lahiani-Skiba et al. 2006), hydroxypropyl methylcellulose for hydrocortisone, dexamethasone and pregnenolone (Brewster et al. 1995; Jansook and Loftsson 2009; Loftsson 2014).

2.3.1.3 Stabilizing Effect

The cyclodextrin complexation can either increase or decrease the chemical stability of the included guest molecules depending on the structure of the complex. If the sensitive part of the guest is buried in the cavity a stabilizing effect can be observed.

If the sensitive part is located near to the hydroxyl groups at the rims enhanced hydrolysis may occur. Both catalysis and inhibition of the degradation can be detected for the steroid drugs as well. For instance, the degradation of hydrocortisone was accelerated by β-cyclodextrin, which shifted the keto-enol equilibrium of the dihydroxyacetone side chain to the more reactive enol form (Andersen and Bundgaard 1983). hydroxypropyl β-cyclodextrin reduced the degradation of the testosterone derivatives and estradiol esters and the stabilizing effect changed parallel to the solubilizing effect (Albers and Müller 1992). The degradation of contraceptive progestin was also accelerated by β-cyclodextrin, while decelerated by hydroxypropyl and hydroxyethyl β-cyclodextrin (Ahmed 1997).

Betamethasone 17-valerate undergoes a facile hydroxide ion-catalyzed rearrangement to the less active 21-valerate ester in basic aqueous solution. Whereas α-cyclodextrin did not affect the rate of rearrangement, β-cyclodextrin caused a rate acceleration of up to three-fold and γ- and dimethyl β-cyclodextrin resulted in a marked retardation (Andersen and Bundgaard 1984). The hydroxide ion-catalyzed rearrangement of hydrocortisone 17-butyrate into the 21-butyrate of lower biological activity was catalyzed by β-cyclodextrin and inhibited by α-, γ-cyclodextrin and dimethyl β-cyclodextrin at pH 2–8 (Chun and Kim 1992).

Complexation with dimethyl β-cyclodextrin made 9α-hydroxyandrost-4-ene-3,17-dione resistant against bacterial degradation by Mycobacterium sp. VKM Ac-1817D (Khomutov et al. 2007).

2.3.1.4 Changes in Pharmacokinetics

The advantages of using steroid drugs in complexed form related to the uncomplexed ones depend on the route of administration.

Using *intravenous* application, the aqueous hydroxypropyl β-cyclodextrin formulation showed identical anesthetic profile in dogs compared to alfaxalone, a neuroactive steroid, general anesthetic dissolved in Chremofor as solubilizer (Estes et al. 1990). The adverse side effects of Chremofor could be avoided. Compared to the pure drug prolonged effect of hydroxypropyl β-cyclodextrin complex of dexamethasone in suppressing stress-induced elevations of plasma adrenocorticotropic hormone (ACTH) was observed when administered to rats by a single tail vein injection (Anderson et al. 1989).

Testosterone dissolved in hydroxypropyl β-cyclodextrin solution and administered *subcutaneously* to senescent rats intensified androgen-sensitive behavior and physiology, stimulated spermatogenesis and increased muscle weight without substantial enlargement of the prostate (Taylor et al. 1989).

The bioavailability of danazol complexed by hydroxypropyl β-cyclodextrin was 2.5 times higher than that of the pure drug administered *orally* to rats (Badawy et al. 1996). β-Cyclodextrin/danazol complex showed complete inhibition of implantation when given orally postcoitally to mice (Jadhav et al. 2007). This complex was safe up to 2000 mg/kg as a single oral dose. The complexation did not affect the pharmacokinetics and relative bioavailability of ethinyl estradiol in an open-labeled, crossover study, when 18 healthy postmenopausal women received ethinyl estradiol uncomplexed or complexed with β-cyclodextrin as a single oral dose (Blode et al. 2008).

Complexes with hydroxypropyl, dihydroxypropyl β-cyclodextrin and β-cyclodextrin polymer ensured enhanced *sublingual/buccal* absorption of steroids to avoid first-pass loss while reduced the absorption from the gastrointestinal tract (Pitha et al. 1986). Sublingual administration of hydroxypropyl β-cyclodextrin/testosterone complex in 5 mg dose of the steroid was found to be useful in testosterone replacement therapy for hypogonadal states because of avoiding first-pass metabolism, rapid absorption and quick metabolism mimicking the endogenous episodic secretion of testosterone (Stünkel et al. 1991). All patients showed significant improvements in sexual motivation and performance. The sublingual formulation was suggested for treatment of boys with delayed puberty and elder men with androgen deficiency (Salehian et al. 1995). The sublingual administration of precursor steroids, such as androstenediol solubilized by hydroxypropyl β-cyclodextrin is used as nutritional supplement causing rapid increase in serum testosterone concentration (Brown et al. 2002). It reduces the individual variability characteristic after oral administration.

In a clinical study, 78% of 50 patients with *oral disease* showed improvement and 48% showed considerable clinical improvement without harmful side effects when hydrocortisone solubilized in hydroxypropyl β-cyclodextrin aqueous solution was applied (Kristmundsdottir et al. 1996). Based on these results a mouthwash solution containing hydrocortisone (0.3% w/v) and hydroxypropyl β-cyclodextrin (4.5 w/v%) and hydroxypropylmethylcellulose (0.5% w/v) was developed and evaluated positively in an open clinical efficacy study as topical steroid therapy of oral mucosal lesions (102 patients with aphthous ulceration, lichen planus, and other mucosal conditions) (Holbrook et al. 1998).

Solubilizing budesonide with sulfobutyl ether β-cyclodextrin a formulation with improved nebulization properties compared to those of the conventional budesonide suspension was developed for *pulmonary* administration (Basu et al. 2009). Albumin microspheres containing fluticasone propionate complexed by hydroxypropyl β-cyclodextrin were proposed for the treatment of asthma (Lohade et al. 2007).

The *nasal* absorption of estradiol and progesterone enhanced from around 20% for the suspension to ~60% for aqueous solution with dimethyl β-cyclodextrin (Hermens et al. 1990; Schipper et al. 1990). Based on these results a nasal formulation of estradiol was marketed for hormone replacement therapy of postmenopausal disorders by Servier till it was withdrawn in 2006 for commercial reasons.

The permeability of 17β-estradiol, hydrocortisone and testosterone through hairless mouth skin was enhanced by both hydroxypropyl and dimethyl β-cyclodextrin suggesting improved *transdermal* delivery (Loftsson et al. 1991). The enhancing effect was dependent on the cyclodextrin to drug molar ratio: using hydroxypropyl β-cyclodextrin in excess resulted in a reduced flux, which can be slightly improved by small amounts of polymers, such as of poly(vinylpyrrolidone) or hydroxypropyl methyl cellulose (Loftsson and Siguroardottir 1994; Loftsson et al. 1994a). Similarly, the skin penetration of hydrocortisone enhanced by iontophoresis showed maximum delivery from 9% hydroxypropyl β-cyclodextrin solution (Chang and Banga 1998). Both increasing and decreasing hydroxypropyl β-cyclodextrin concentration (3% and 15%) resulted in lower drug delivery. The rise in flux with increased cyclodextrin concentration and fall with excess cyclodextrin was described with the diffusion model based on the carrier function of cyclodextrins from bulk solution through the undisturbed aqueous diffusion layer to the lipophilic surface of biological membranes (Masson et al. 1999). On the other hand Shaker et al. (2003) found no accelerating effect of hydroxypropyl β-cyclodextrin on the skin permeation of corticosterone through the hairless mouth skin.

The in vivo *ocular* bioavailability of hydrocortisone in the rabbit was determined following topical administration of solutions containing hydrocortisone (1%) with hydroxypropyl β-cyclodextrin alone, or containing the mucoadhesive, viscosity-enhancing polymers sodium hyaluronate (0.2 and 0.5% w/v) or Carbopol 934P (0.1% w/ v). A 1% hydrocortisone suspension was used as control. Formulation of hydrocortisone as a solution with hydroxypropyl β-cyclodextrin in the absence of polymer increased the bioavailability of hydrocortisone in the aqueous humour by approximately 55% and cornea by 75% when compared to suspension. Inclusion of either polymer did not result in any further increase in ocular bioavailability over that noted for the polymer-free solution. The in vitro corneal permeability of hydrocortisone was also evaluated. Permeability was the greatest when formulated either as a suspension, or as an hydroxypropyl β-cyclodextrin solution in which the concentration of free (uncomplexed) hydrocortisone is equivalent to that of a saturated solution (Bary et al. 2000). Enhanced and prolonged dexamethasone concentration in aqueous humour of rabbits was reported after applying eye drops containing the hydroxypropyl β-cyclodextrin/drug complex (Loftsson et al. 1994b). Similar results were obtained in human experiment (Kristinsson et al. 1996).

Table 2.3 gives some examples on the steroid drugs marketed in cyclodextrin-enhanced formulations. Formerly several anabolic steroids were available in hydroxypropyl β-cyclodextrin-complexed form as nutritional supplements for oral intake: the enhanced solubility promised higher absorption and increased androgenic effects. These formulations have been banned after US Congress passed the Designer Anabolic Steroid Act to protect consumers from dangerous anabolic steroids marketed as dietary supplements causing liver injury, shrinkage of testicles, infertility, etc.

Water soluble estradiol (encapsulated by hydroxypropyl β-cyclodextrin) suitable for cell cultures is sold by Sigma Alldrich (2017).

2.3.1.5 Complexes in Novel Drug Delivery Systems

Recent developments of novel drug formulations including nanoparticles, sheets, hydrogels, etc. provided benefits also for steroid administration. Some examples are introduced as follows:

Table 2.3 Examples for cyclodextrin/steroid formulations marketed

Steroid	Cyclodextrin	Route of administration	Trade name	Company
Alfaxalone	Sulfobutyl ether β-cyclodextrin	i.v.	Phaxan	Drawbridge pharm., chemic labs, US
Alfaxalone	Hydroxypropyl β-cyclodextrin	i.v. (veterinary)	Alfaxan	Jurox, CTD
Dexamethasone	β-cyclodextrin	Dermal	Glymesason	Fujinaga
Ethinyl estradiol	β-cyclodextrin	Oral	Safyral/Beyaz/Lorina	Bayer healthcare/Sandoz
17β-Estradiol	Random methylated β-cyclodextrin	Nasal	Aerodiol[a]	Servier
Hydrocortisone	Hydroxypropyl β-cyclodextrin	Buccal	Dexocort	Actavis
Progesterone	Hydroxypropyl β-cyclodextrin	Injection	Lubion	Hikma (Portugal)

[a]Withdrawn from the market from economic reasons

Nanospheres and nanocapsules of amphiphilic β-cyclodextrin can be used to obtain the desired loading and steroid releasing properties (Bilensoy et al. 2006). Poly(isobutylcyanoacrylate) *nanoparticles* combined with hydroxypropyl β-cyclodextrin with high steroid-loading capacity was reported by daSilveira et al. (2000). Chitosan *microspheres* with hydroxypropyl β-cyclodextrin/hydrocortisone complex showed enhanced drug dissolution (Filipovic-Grcic et al. 2000). Both hydrocortisone and progesterone were released from solid lipid nanoparticles at a lower extent when incorporated as complexed with β- and hydroxypropyl β-cyclodextrin than as free molecules (Cavalli et al. 1999). Cyclodextrin-modified dextran self-assemble to form nanoparticles with improved solubilizing effect on hydrocortisone (Fulop et al. 2013).

Retarded dissolution of hydrocortisone from poly(vinyl alcohol) (PVA) *films* was recorded when it was incorporated as an hydroxypropyl β-cyclodextrin complex (Becirevic-Lacan and Filipovic-Grcic 2000). Similarly hydroxypropyl β-cyclodextrin/ethinyl estradiol complex was formulated as PVA film for transmucosal or buccal administration of the hormone (Podhaisky and Bracht 2007). The complexation ensured homogeneous distribution and prolonged dissolution of spironolactone acetonide in poly(ethylene oxide) films developed for buccal delivery (Miro et al. 2013). In situ gelling nasal inserts were developed by adding of methyl β-cyclodextrin to carrageenan (sulfated polysaccharides from seaweeds) to ensure slower absorption of eastradiol with lower peak serum level compared to the nasal spray (Aerodiol) (Werner et al. 2004).

Polyethyleneimine-silk fibroin *hydrogel* developed for the treatment of pressure sores and containing herbal extract and hydrocortisone acetate provided faster healing compared to the control (Lee et al. 2012). Hydrogels of crosslinked cyclodextrin polymer ensured high estradiol load and sustained release (Rodriguez-Tenreiro et al. 2007).

Cyclodextrin *immobilized to cotton textile* with monochlorothiazinyl β-cyclodextrin was loaded with various drugs including hydrocortisone to obtain antiallergic therapeutic pajamas (Radu et al. 2013).

Nanodimer cyclodextrin ligands obtained by crosslinking β-cyclodextrin with toluene 2,4-diisocyanate were prepared by *imprinting* technique: coupling the cyclodextrin/steroid complex with the crosslinking agent, then removing the steroid resulted in hosts of extremely high affinity toward the specific steroid with association constant of 10^7 M^{-1}(Khomutov and Donova 2011).

2.3.1.6 Muscle Relaxants

Muscle relaxants are an important group of anesthetics. The most commonly used ones belong to the vecuronium group having steroid structure. Rocuronium is of special interest because of its rapid onset of action and it is free from unwanted side effects. Neuromuscular reversal agents are drugs that prevent the action of neuromuscular blocking agents after surgery and restore normal muscle function. At present, acetylcholinesterase inhibitors, such as physostigmine and neostigmine are commonly used as reversal agents, however, they are not selective for the neuromuscular junction, but increase the amount of acetylcholine in all parts of the body. This can result in many serious unwanted side-effects causing clinical difficulties such as cardiac arrythmia, bronchoconstriction, increased salivation, nausea, vomiting etc.

A recently developed cyclodextrin, Sugammadex (octakis-(6-deoxy-6-S-mercaptopropionyl-γ-cyclodextrin sodium salt) is a neuromuscular reversal agent with practically no side effects (Bom et al. 2002). This cyclodextrin derivative has been designed to specifically encapsulate the steroidal blocking drugs like rocuronium by forming stable inclusion complexes of extremely high binding constant of about 10^7 M^{-1} as determined by isothermal calorimetry (Cameron et al. 2002). For comparison, the K values for β- and γ-cyclodextrin are 3.3 x 10^3 M^{-1} and 1.8 x 10^4 M^{-1}. Once the steroidal agent have entered the cavity of Sugammadex molecule, its action is prevented and almost an instantaneous recovery of neuromuscular function occurs. The entrapped agent is then excreted by the kidneys. Sugammadex is selective and efficient, representing a new class of agents for the reversal of neuromuscular block.

The X-ray crystallographic investigations of the solid-state structure of the Sugammadex/rocuronium bromide 1:1 mol/mol inclusion complex indicated that all four steroidal rings of rocuronium were in a close contact with the lipophilic wall of the Sugammadex cavity, and the positively charged quaternary ammonium moiety occurring at ring D of the guest is loosely surrounded by the carboxylic function of the thiopropanoic acid substituents in Sugammadex. (Fig. 2.3). In order to induce encapsulation both rocuronium bromide and the modified cyclodextrin undergo conformational changes: ring A of rocuronium bromide 'switches' from the more

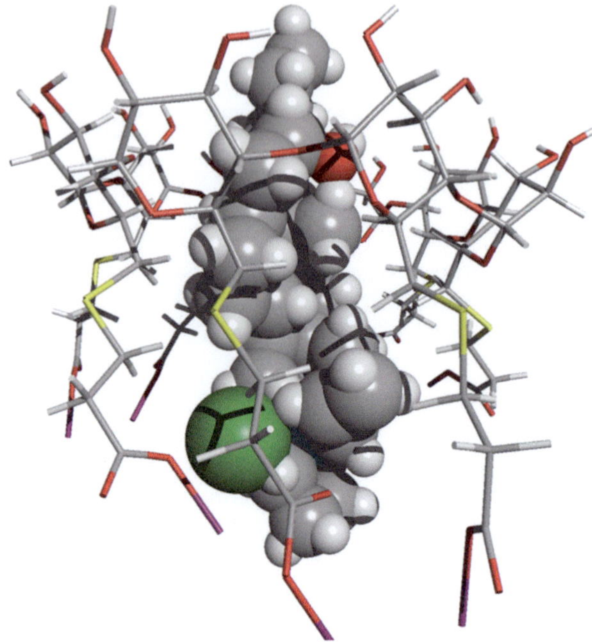

Fig. 2.3 Structure of Sugammadex/rocuronium 1:1 complex, a tight host-guest packing, where Sugammadex surrounds the entire rocuronium by inclusion. (Szente 2006)

sterically encumbered chair to the sterically less demanding twist-boat, whilst the modified cyclodextrin "opens" its cavity to allow the steroid to enter (Cooper et al. 2005).

The structure/activity relationship was nicely demonstrated when cavity geometry (size and depth) and reversal potency on neuromuscular block were compared. Even parent γ-cyclodextrin provided a good reversal activity against induced neuromuscular block. However, the specifically modified γ-cyclodextrin tailored for entire entrapment of the bulky steroidal agent showed a pharmacological potency of at least an order of magnitude higher, than did the parent γ-cyclodextrin itself (Bom et al. 2002). The pharmacological efficacy of Sugammadex at a 1.0 mg/kg dose was excellent: it reversed neuromuscular block in 95% depth within 3 min, while the standard treatment of animals with neostigmine at 40 μg/kg dose produced the same level of recovery within 6 min only (Cameron et al. 2002; de Boer et al. 2007). This means that the action of Sugammadex is twice as fast as is the standard treatment. Furthermore, no remarkable changes in hemodynamic parameters, heart rate and blood pressure were observed upon treatment with Sugammadex.

Dose finding and safety studies indicated that Sugammadex was well tolerated with practically no side effect. In this study twenty seven male surgical patients received placebo or Sugammadex treatment (at 0.5; 1.0; 2.0; 3.0 and 4.0 mg/kg doses) for reversal of 0.6 mg/kg rocuronium bromide-induced neuromuscular block. All treated patients recovered without any clinical consequences (Sorgenfrei et al. 2006). An average of 59–77% of the administered Sugammadex was excreted unchanged in the urine within 16 h, mostly in the first 8 h. Sugammadex also increased the proportion of the rocuronium dose excreted unchanged in the urine within 16 h. From the results of this safety study it has been concluded that at doses of 2.0 mg/kg or higher, Sugammadex safely and effectively reversed 0.6 mg/kg rocuronium-induced neuromuscular block in a dose dependent manner (Hunter and Flockton 2006; de Boer et al. 2007).

The potential for displacement of the captured rocuronium or vecuronium by any drug commonly used in or shortly after anesthesia, commonly prescribed drugs such as antidepressants and cardiovascular drugs, drugs (both steroidal and nonsteroidal) acting on steroidal receptors, such as the corticosteroids, and the selective estrogen receptor modulator toremifene was studied (Zwiers et al. 2011). Out of 300 drugs only three (flucloxacillin, fusidic acid and toremifene) showed high association values (>10^4) thus the potential for displacement, but the clinical study found no evidence, no reoccurance of blockade occurred.

The recent reviews state that Sugammadex not only overperforms the traditional treatment with neostigmine in faster recovery from deep neuromuscular blockade but also demonstrated efficacy in various special patient populations, including patients with hepatic dysfunction, pulmonary disease, cardiac disease, and morbidly obese patients (Schaller and Fink 2013; Keating 2016). Although further data are needed, the six randomized controlled trials with 253 patients showed that Sugammadex was fast and effective in the reversal of rocuronium-induced neuromuscular blockade during surgery in general anesthesia in pediatric patients, too (Won et al. 2016). Somewhat slower recovery was observed in elderly patients and also the clearance decreased (Lee et al. 2016). Intravenous Sugammadex was generally well tolerated with lower risk of adverse effects compared to neostigmine (Carron et al. 2016). Sugammadex revolutionized anesthesia. It has been approved as active pharmaceutical agent: the first drug among the cyclodextrin derivatives otherwise used as excipients!

2.3.2 *Bile Acids/Salts*

Bile acids are polar derivatives of cholesterol synthesized by the liver. They are surfactant-like molecules and have a role in the fat digestion. In addition to their physiological role, some of them are also therapeutic agents for bile reflux gastritis, cholesterol gallstone diseases, etc.

Bile salt molecules have a rigid steroidal skeleton with an OH group on one side (hydrophilic part) and a hydrocarbon chain on the other side (hydrophobic part). Bile salts form micelles above a concentration characteristic for the given bile salt, the critical micelle concentration (cmc). While β-cyclodextrin interacts mainly with the hydrophilic side, the larger γ-cyclodextrin with the hydrophobic side (Tan et al. 1994). The association constant of 1:1 complexes with β-cyclodextrin are in the range of 10^3–10^5 M^{-1} (Tan and Lindenbaum 1991). At higher cyclodextrin concentration 2:1 cyclodextrin/bile salt complexes are also formed. The best fitting is observed with β-cyclodextrin as the α-cyclodextrin ring is too small for complex-

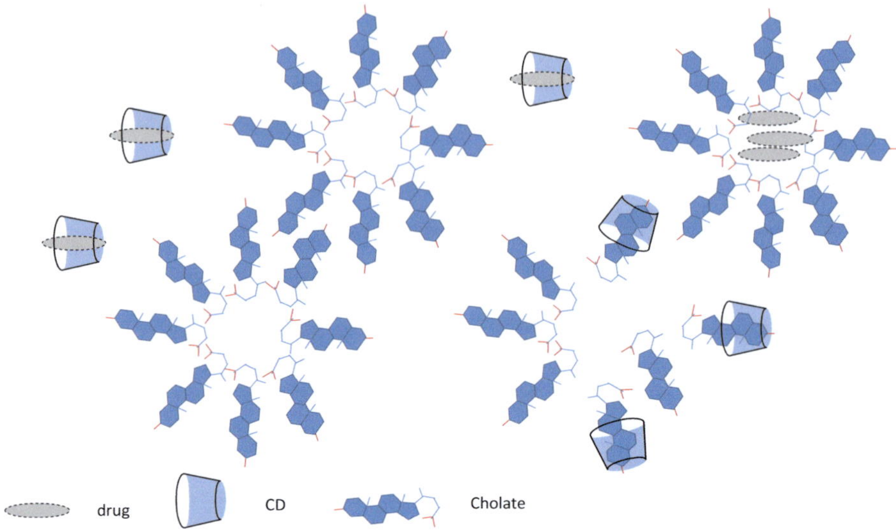

Fig. 2.4 Simultaneous partitioning of the drug between cyclodextrin and the micelles formed by bile salt. Cyclodextrin form complexes both with drug and with bile salt

ation of bile acids and their salts, while γ-cyclodextrin is too large (Manabe et al. 1994).

Trihydroxy bile salts form 1:1 complexes with β- and γ-cyclodextrin, while dihydroxy bile salts form 2:1 complexes with β-cyclodextrin and 1: 1 complexes with γ-cyclodextrin. Rotating-frame nuclear Overhauser effect correlation spectroscopy (ROESY) experiments stated that the side chain, ring D, and part of ring C of the steroid body of the bile salts are included into the cavities of β- and γ-cyclodextrin, in 1:1 complexes. The A ring of the steroid body is included into the cavity of the second β-cyclodextrin in the 2:1 complexes (Ramos Cabrer et al. 2003).

There are large variations in the association constants depending on the structure of bile salts (Holm et al. 2009). The log K values based on microcalorimetry titrations fall in the range of 3.3–5.0 M^{-1} with glycocholate and taurochenodeoxycholate representing the lowest and highest values in a series of 10 bile salts (Tan and Lindenbaum 1991).The log K values for glycocholate complex with hydroxypropyl and methyl β-cyclodextrin are somewhat higher: 3.6 and 3.8 M^{-1}, resp. (Ollila et al. 2001). In the case of methyl β-cyclodextrins the methylation at O3 hydroxyls hinders complexation by partially blocking the cavity entrance, while methyl groups at O2 positions promote complexation by extending the hydrophobic cavity (Schonbeck et al. 2011).

The bitter taste of ursodeoxycholic acid can be masked by complexing its arginine salt (Szejtli and Szente 2005).

The fate of cyclodextrin-complexed drug in the gastrointestinal system might be influenced by the presence of bile salts, which compete for the cavity of the cyclodextrins (Fig. 2.4). On the other hand, the drugs partition between the cyclodextrin

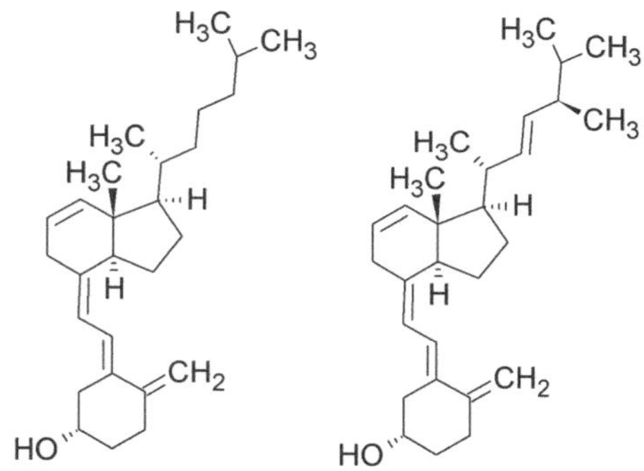

Fig. 2.5 Structures of vitamin D2 (ergocalciferon) and D3 (cholecalciferon)

cavities and the micelles formed by the bile salts (Miyajima et al. 1986; Frijlink et al. 1990). These micelles are disrupted by cyclodextrin complexation (Horsky and Pitha 1996).

Animal experiments showed that β-cyclodextrin stimulated the bile acid synthesis. The bile acid contents of the cecum and colon on treated hamsters were 2.7-fold higher than those of control animals in experiments when hamsters and Rico rats were fed with β-cyclodextrin (Riottot et al. 1993).

2.3.3 Vitamin D

Chemically, the various forms of vitamin D are secosteroids, i.e., steroids in which one of the bonds in the steroid rings is broken. In humans, the most important compound of Vitamin D group is vitamin D3 (cholecalciferol) and vitamin D2 (ergocalciferol) (Fig. 2.5). They are recognized as prohormone regulating the calcium and phosphate homeostasis, cell proliferation and cell differentiation. The main source is the dermal synthesis by sunlight, but can be ingested by diet or supplements.

Complexation of vitamin D3 with β-cyclodextrin results in enhanced aqueous solubility and better stability against heat, light and oxidation (Szejtli and Bolla 1980). The molar ratio of cholecalciferol to both β- and γ-cyclodextrin is 1:2 (Delaurent et al. 1998). Improved bioavailability of the β-cyclodextrin complex was observed in albino rats (Szejtli et al. 1980). The reduced plasma phosphate and calcium levels of hypovitaminotic animals were normalized faster compared to the animals, which obtained vitamin D3 uncomplexed.

The isomerization of previtamin D3 to vitamin D3 happens much faster in aqueous β-cyclodextrin solution compared to that in organic solvent, such as n-hexane (Tian and Holick 1995).

The solubilizing effect of the methylated cyclodextrin derivatives is higher, but the glucosylated-, maltosylated or galactosylated (branched) derivatives is weaker than that of the unmodified cyclodextrins (Okada et al. 2011). The complex with dimethyl β-cyclodextrin is prepared by adding portion wise vitamin D3 to the dimethyl β-cyclodextrin solution always waiting for complete dissolution then evaporating to dryness by freeze drying (Szejtli 1988).

It is very important that in the presence of cholic acids no complex formation of β-cyclodextrin with vitamins A and D3 was observed proving that no sequestration of these vitamins occur upon cyclodextrin application in food as it was previously assumed (Comini et al. 1994).

Vitamin D3 complex was used to improve the quality of rice. Vitamin D3 remained stable during storage, increased rehydration rate and degree of gelatinization of rice and inhibited the retrogradation (Tian et al. 2007).

The product of vitamin D3 metabolism, 25-hydroxyvitamin D3, is the precursor to the biologically active hormone, 1α,25-dihydroxyvitamin D3. The improved solubility of vitamin D3 by complexation can be applied for intensified production of various derivatives, such as mono-, di and trihydroxyvitamin D2 and D3 mimicking the metabolism of these vitamins by cytochrome P450scc (Tuckey et al. 2008; Nguyen et al. 2009). The presence of cyclodextrin is advantageous for the production of 25-hydroxy-vitamin D2 or D3 in the recombinant yeast cells expressing human cytochrome P450 2R1 enzyme (Yasuda et al. 2013).

2.3.4 Plant Sterols

Some vegetable oils and nuts contain phytosterols, such as β-sitosterol, stigmasterol and ergosterol and phytostanols such as sitostanol and campestanol. Both phytosterols and phytostanols have sterane skeleton. Phytosterols have similar role in the plant cells as cholesterol in mammalian cells: they regulate membrane fluidity and permeability (Hartmann 1998). They inhibit the absorption of cholesterol, can also modulate the activity of membrane-bound enzymes and are beneficial in various cancers.

Unlike cholesterol, which fits only into the cavity of β-cyclodextrin, β-sitosterol forms complex with all the natural cyclodextrins with the highest encapsulation efficiency (93%) with β-cyclodextrin (Rinaldi et al. 2015). The solubility of hydroxypropyl β-cyclodextrin/phytosterols complex was higher than that of the β-cyclodextrin complex (Meng et al. 2012).

The health effects of these nutraceutical compounds were studied using cyclodextrin complexation for solubilisation of the otherwise poorly soluble compounds. Hydroxypropyl β-cyclodextrin-complexed β-sitosterol was effective at cell growth inhibition of tumor cells (Awad et al. 1996). Comparing the effect of the most common dietary sterols: β-sitosterol in vegetarian diet and cholesterol of Western diet, β-sitosterol was found to inhibit the growth of prostate cancer cells and induce apoptosis in a higher extent than cholesterol (Von Holtz et al. 1998). Both sterols were delivered by a cyclodextrin vehicle. Similar results were obtained with breast cancer cells (Awad et al. 2001). The positive effects of phytosterols, β-sitosterol and campesterol over cholesterol were demonstrated in the treatment of macrophages involved in the inflammatory process in cardiovascular diseases: the prostaglandin release was reduced and the atheroma development slowed down on the effect of phytosterols while cholesterol was ineffective (Awad et al. 2004). One of the reasons might be that the efflux of β-sitosterol and sitostanol from macrophages was more efficient than the efflux of campesterol and cholesterol, while there was not significant difference in their influx (Hovenkamp et al. 2007).

The uptake of β-sitosterol by phospholipid membranes was improved by complexation (Castelli et al. 2006). Using hydroxypropyl β-cyclodextrin as carrier, β-sitosterol increased the fluidity of the inner mitochondrial membrane and mitochondrial adenosine triphosphate (ATP) content (Shi et al. 2012). Further advantages of incorporating β-sitosterol into the membrane, beneficial also for neurodegenerative diseases such as Alzheimer's disease, are increasing the resistance to oxidative stress and lipid peroxidation (Shi et al. 2013), and promoting nonamyloidogenic cleavage of endogenous amyloid precursor protein (Wang et al. 2013).

Plant growth regulator 24-epibrassinolide (brassinosteroid) showed improved activity when complexed with β-cyclodextrin in rice lamina inclination assay (De Azevedo et al. 2002). Cultivating tomato cells in the presence of cyclodextrins an extracellular accumulation of two sterols (isofucosterol and β-sitosterol) and taraxasterol, a common tomato fruit cuticular triterpene, was observed (Briceno et al. 2012). Similarly the addition of β-cyclodextrin, especially methyl β-cyclodextrin to the cell suspension cultures of carrot not only induced the biosynthesis of phytosterols but also promoted their secretion into the culture medium where they could be isolated (Sabater-Jara and Pedreno 2013).

There are several patents on cyclodextrin/phytosterol formulations and their application in healthy food, beverages, pharmaceuticals and nutraceuticals (some examples: Stewart et al. 1998; Hamano and Suzuki 1999; Tripp et al. 2004; Hashimoto et al. 2008; Tadashi and Masaki 2012). In addition to the numerous products in Japan there is at least one European product containing phytosterols (cranberry seed oil) formulated with γ-cyclodextrin (Wacker Fine Chemicals 2005).

2.4 Applications in Biotechnology

The first disclosure describing the use of cyclodextrins for chemical manipulation of steroids was a patent application filed in Hungary in 1981, wherein the cyclodextrins were used to intensify steroid biotransformation (Udvardy Nagy et al. 1981). Later, this patent protection was extended to various countries worldwide. It was known prior to the elaboration of the cyclodextrin-intensified process that some steps of the synthesis of steroid compounds may be economically carried out by means of microorganisms, according to the classical semisynthetic route (Leigh et al. 1952). Without added cyclodextrin, the rate of bioconversion employing microorganism cells (or isolated enzymes) was limited due to the poor solubility of steroids in water. In addition, often "end product inhibition" hindered the progress

Table 2.4 Cyclodextrin-assisted steroid bioconversion reactions (I) (Udvardy Nagy et al. 1981)

Substrate	Microorganism	Cyclodextrin	End product
Hydrocortisone	*Arthrobacter simplex*	α-cyclodextrin	Prednisolone, β-hydroxy-prednisolone
17α-methyl-testosterone	*Arthrobacter simplex*	β-cyclodextrin	δ-1-methyl-testosterone
5-Pregnene-3β,17α,21-triol-20-on-21-acetate	*Flavobacterium lucecoloratum*	α-cyclodextrin	Reichstein's compound S (4-pregnene-17α,21-diol-3,20-dione)
4-Pregnene-17α,21-dihydroxy-3,20-dione-17-acetate	*Curvularia prasadii*	β-cyclodextrin	Hydrocortisone, hydrocortisone-17-acetate, Reichstein's compound S-17-acetate, Reichstein's compound S
Hydrocortisone	*Arthrobacter simplex*	β-cyclodextrin	Prednisolone
Sitosterine	*Mycobacterium sp.*	β-cyclodextrin	Androst-4-diene-3,17-dione
Progesterone	*Ophiobolus herpotrichus*	β-cyclodextrin	21-Dihydroxy-progesterone
Reichstein's compound S (4-pregnene-17α,21-diol-3,20-dione)	*Curvularia lunata*	β-cyclodextrin	Hydrocortisone
16α-Methyl-Reichstein's compound S	*Curvularia lunata*	β-cyclodextrin	11β,17α,21-Trihydroxy-16α-methyl-pregn-4-ene-3,20-dione
3β,17α,21-Trihydroxy-pregn-5-en-21-acetate	*Flavobacterium dehydrogenans*	β-cyclodextrin	Reichstein's compound S
Lanatoside-A	*Streptomyces purpurascens*	β-cyclodextrin	Digoxin

of the reaction and in certain cases side reactions could also occur. It was found that application of cyclodextrins resulted in manifold advantageous consequences. The precursor steroid compounds (primarily as substrates of the enzymatic conversion reactions) also behaved as guest molecules in concurrent cyclodextrin inclusion complex formation. Consequently, the intrinsic solubility of the substrates increased whilst their potential toxic/inhibition effect could be masked being entrapped in inclusion complex form. In Table 2.4 the examples of this pioneering work are briefly summarized.

The disclosed technology was in fact industrially applied at Kőbánya Pharmaceutical Company (Gedeon Richter Chemical Works) in Budapest, Hungary (Szejtli 1998). This application of cyclodextrins for steroid bioconversion was included in the first review authored by Szejtli in the mid-80's dealing with the use of cyclodextrins in the field of biotechnology amongst a number of other applications (Szejtli 1986). He reviewed the topic again a few years later (Szejtli 1990).

Hesselink et al. (1989) systematically studied the bioconversion of easily available, cheap steroid precursors, such as cholesterol, sitosterol and 4-cholestenone into androst-4-ene-3,17-dione and androsta-1,4-dien-3,17-dione by using mycobacteria. The effect of α-, β- and γ-cyclodextrins were studied – each cyclodextrin was applied in identical molar ratio. Even though β-cyclodextrin was postulated as optimal host for these sterols due to its cavity size, it was shown that all three cyclodextrins could enhance the rate of bioconversion, however, to different extent:

Cholesterol: without cyclodextrin $<$ α- $<<$ β- $\approx$ γ-cyclodextrin,
Sitosterol: without cyclodextrin $\approx$ α- $<$ β- $<$ γ-cyclodextrin,
4-cholestenone: without cyclodextrin $<$ α- $<$ β- $<$ γ-cyclodextrin.

It was also found that the mycobacteria were not affected by the presence of cyclodextrins.

According to studies performed by Wördenbag et al. (1990), utilizing the complexation performance of β-cyclodextrin, the phenolic steroid 17β-estradiol could be ortho-hydroxylated into mainly 4-hydroxyestradiol, by a phenoloxidase isolated from *Mucuna pruriens*. Mushroom tyrosinase converted 17β-estradiol in the presence of β-cyclodextrin solely into 2-hydroxyestradiol, with a maximal yield of 30% after 6–8 hrs. Uncomplexed 17β-estradiol could not be converted in the same conditions.

To date, various similar approaches were utilized applying a steroid precursor and a microbial culture (or enzyme derived/extracted from microbial culture) to obtain valuable steroid compounds with cyclodextrin-assisted procedure. Some illustrative examples are summarized in Table 2.5.

Khomutov et al. utilized host-guest molecular complexation with cyclodextrin for the control of steroid bioconversions (Khomutov et al. 2007). In contrast to the examples shown in Table 2.4 and 2.5, opposite, enzyme inhibitory effect was observed in case the cyclodextrin has exceptionally high affinity to the substrate. Inhibitory effect of methylated β-cyclodextrin on steroid degradation was studied using the degradation of 9α-hydroxyandrost-4-ene-3,17-dione (9-OH-AD) by mycobacteria. The formation of the methyl β-cyclodextrin/9-OH-AD complex was evidenced by 1H NMR-spectroscopy. This inclusion complex was shown to be resistant to enzymatic degradation.

Jadoun et al. (1993) investigated the oxidation of cholesterol to cholest-4-en-3-one by *Rhodococcus erythropolis*. As shown in Table 2.5, alkylated cyclodextrins behaved as process enhancers. On the contrary, natural β- and γ-cyclodextrin totally inhibited the reaction of cholesterol due to the very low solubility of the corresponding cyclodextrin complexes.

Table 2.5 Cyclodextrin-assisted steroid bioconversion reactions (II)

Substrate	Microorganism	Cyclodextrin	End product	Reference
6α-Fluorocortexolone; 16α-,17α-acetonide	*Tieghemella orchidisCurvularia lunata*	β-cyclodextrin	11β-Hydroxy metabolites	Alekhina et al. 1993
Hydrocortisone	*Corynebacterium simplex*	β-cyclodextrin	Prednisolone	Alekhina et al. 1993
6α-Methylhydrocortisone	*Corynebacterium simplex*	β-cyclodextrin	6α-Methylprednisolone	Alekhina et al. 1993
Cortisone acetate	*Arthrobacter simplex*	β-, random methylated β-, sulfobutyl ether β-, hydroxypropyl β-cyclodextrin	Prednisone acetate	Ma et al. 2009
17α-Hydroxypregna-4-ene-3,20-dione-21-acetate (RSA) 17α-Hydroxypregna-4-ene-3,20-dione	*Absidia coerulea*	β-cyclodextrin	11β-Hydroxy metabolites	Shen et al. 2010.
Progesterone	*Rhizopus nigricans*	β-cyclodextrin	11α-Hydroxyprogesterone	Roglič et al. 2005; Znidarsic-Plazl and Plazl 2010
4-Androstene-3,17-dione	*Nocardioides simplex*	β-cyclodextrin	Androst-1,4-diene-3,17-dione (ADD)	Luthra et al. 2015
Androstenedione	*Saccharomyces cerevisiae*	β-, γ-, hydroxypropyl β-cyclodextrin (effective) α-, di- and trimethyl β-cyclodextrin (slight effect).	Testosterone	Singer et al. 1991
Cholesterol	*Rhodococcus erythropolis*	Hydroxypropyl β-, di- and trimethyl β-cyclodextrin	Cholest-4-en-3-one	Jadoun and Bar 1993

2.5 Removal of Steroid Hormones from Wastewater

The complexation of cyclodextrins with steroids can be used for their removal from water. The appearance of hormones and hormone like compounds in surface waters and eventually also in tap water is an emerging concern nowadays. The waste water treatment plants are not required to remove the drug residues including also the contraceptives. The sophisticated analytical techniques made possible to detect these compounds and initiated the development of additional purification steps aiming the removal of residual drugs and other xenobiotics of estrogenic activity.

Cyclodextrins are widely used for environmental problems enhancing the soil remediation technologies as well as waste water treatments (Gruiz et al. 2011; Landy et al. 2012).

Cyclodextrin polymers prepared by crosslinking of cyclodextrins with epichlorohydrin are selective sorbents for the removal of hormones and other endocrine disrupting chemicals at very low (0.1–0.01 nanomole) concentrations (Oishi and Moriuchi 2010). The removal capacity followed the order of β-cyclodextrin polymer ≥ γ-cyclodextrin polymer ≫ α-cyclodextrin polymer. Over 90% removal rate was obtained for 17β-estradiol in municipal sewage plant effluent spiked with 17β-estradiol and cholesterol in 1:100 molar ratio (the concentration of 17β-estradiol was 10^{-11} mol/L). The estrogenic activity was measured by yeast two-hybrid assay.

In small scale laboratory experiments the filters containing cyclodextrin polymers used for purification of drinking water were able to absorb more than 90% of the bisphenol-A and of the estrogenic hormones. Both the analytical chemistry and toxicity results showed efficient elimination of these pollutants (Nagy et al. 2014). Especially, the toxicity of the filtrate decreased considerably. In the laboratory experiment modeling post-purification of waste water by fluidization technology, hormones, such as β-estradiol, ethinylestradiol, estriol were removed efficiently (87–99%) similarly the bisphenol-A (94%) from the spiked test solution.

Silica coated with β-cyclodextrin using hexamethyl diisocyanate as crosslinking agent was used for the removal of emerging contaminants, such as estrogen hormones, perfluoro compounds, bisphenol A and 1,4-dioxane from water (Bhattarai et al. 2012). More than 95% of 17-β-estradiol was removed from single component solution and >90% of most of the estrogens in multicomponent system even after 4 regeneration cycles. Magnetic graphene oxide was modified with β-cyclodextrin/poly(l-glutamic acid) to get a sorbent with excellent binding capacity for 17β-estradiol (Jiang et al. 2016).

After the several laboratory scale experiments with only a few mL of model solutions a pilot scale experiment was also performed by applying 1 kg of the β-cyclodextrin polymer to 300 L purified waste water spiked with 9 emerging pollutants (drugs including steroid hormones) at ~2×10^{-8}mol/L concentration (Fenyvesi and Szente 2016). The β-cyclodextrin polymer obtained in the form of tiny beads by crosslinking β-cyclodextrin with epichlorohydrin was found to effectively (>95%) remove hormones, such as estradiol (>99.9%), ethinylestradiol (99.9%) and estriol (96.1%) together with bisphenol A (99.4%) of estrogenic activity.

2.6 Conclusion

Steroids are well-studied guest molecules in cyclodextrin research due to (i) abundant synthetic and natural compounds belonging to this group, (ii) their physiological and pharmaceutical importance and (iii) their proper fitting into the cavities of β- and γ-cyclodextrin cavities. The most abundant representative of steroids in the living organisms is cholesterol, but this is not in the scope of the present review focusing on steroid drugs, cholates, vitamin D and phytosterols. Steroid drugs are typical examples for demonstrating the advantages of complexation: enhanced solubility of the usually poorly soluble compounds, enhanced stability against light-induced degradation, hydrolysis and oxidation, enhanced or sustained release from cyclodextrin-enabled formulations, improved bioavailability at various administration routes, catalysis of biotransformations, etc. No wonder that several steroid drugs have been marketed in their complexed form. Based on the high affinity of cyclodextrins to steroids a γ-cyclodextrin derivative (Sugammadex) able to selectively entrap rocuronium and other steroid anesthetics with similar structure, has been approved for the reversal of neuromuscular blockade. Sugammadex is the first cyclodextrin, which is used not for drug delivery but for drug removal from the circulation, revolutionizing anesthesia with this new approach. Several examples are shown on the catalytic effect of complexation in steroid bioconversions. As Sugammadex can remove curare drugs from the body, cyclodextrin polymers can efficiently remove steroid drugs from purified wastewater reducing the risk these micropollutants pose to both the humans and the environment.

References

Agnus B, Gosselet NM, Sebille B (1994) Indirect photodetection of pregnanolone on a Cyclobond column by high-performance liquid chromatography. J Chromatogr 663(1):27–33. https://doi.org/10.1016/0021-9673(94)80492-3

Ahmed SM (1997) Effect of cyclodextrins on the chemical stability of ST1435, a contraceptive steroid progestin, in aqueous solution. J Incl Phenom Mol Recognit Chem 27(1):85–96. https://doi.org/10.1023/A:1007946525783

Ahmed SM (1998) Improvement of solubility and dissolution of 19-norprogesterone via inclusion complexation. J Incl Phenom Mol Recognit Chem 30(2):111–125. https://doi.org/10.1023/A:1007979430842

Albers E, Müller BW (1992) Complexation of steroid hormones with cyclodextrin derivatives: substituent effects of the guest molecule on solubility and stability in aqueous solution. J Pharm Sci 81(8):756–761. https://doi.org/10.1002/jps.2600810808

Alberth M, Wu WM, Winwood D, Bodor N (1991) Lipophilicity, solubility and permeability of loteprednol etabonate: a novel, soft anti-inflammatory corticosteroid. Aust J Biol Sci 2(2):115–125

Alekhina TM, Ryzhkova VM, Gusarova TI, Kurakov VV, Klabnichkina GA (1993) Microbiological transformation of steroid-β-cyclodextrin inclusion compounds. Khim Farm Zh 27(4):59–62

Sigma Alldrich (2017) http://www.sigmaaldrich.com/catalog/product/sigma/e4389 (Accessed on 13 Mar 2017)

Andersen FM, Bundgaard H (1983) Influence of β-cyclodextrin on the stability of hydrocortisone in aqueous solution. Arch Pharm Chem, Sci Ed 11(2):61–66. https://doi.org/10.1007/BF02974095
Andersen FM, Bundgaard H (1984) The influence of cyclodextrin complexation on the stability of betamethasone 17-valerate. Int J Pharm 20(1–2):155–162. https://doi.org/10.1016/0378-5173(84)90226-6
Anderson WR, Simpkins JW, Brewster ME, Bodor NS (1989) Evidence for prolonged suppression of stress-induced release of adrenocorticotropic hormone and corticosterone with a brain-enhanced dexamethasone-redox delivery system. Neuroendocrinology 50(1):9–16. https://doi.org/10.1159/000125196
Arimori K, Uekama K (1987) Effects of beta- and gamma-cyclodextrins on the pharmacokinetic behavior of prednisolone after intravenous and intramuscular administrations to rabbits. J Pharmacobiodyn 10:390–395. https://doi.org/10.1248/bpb1978.10.390
Arimori K, Sakai A, Otagiri M, Uekama K (1984) Improvement in biopharmaceutics of prednisolone by β- and γ-cyclodextrins. J Incl Phenom 1(4):387–394. https://doi.org/10.1007/BF00665481
Arriagas N, Cabral-Marques H (2013) Obtention and characterization of a ciclesonide/methyl-beta cyclodextrin complex. Abstract book of 3rd European Conference on Cyclodextrins, October 2–4, 2013, Antalya, Turkey, P-30
Awad AB, Chen YC, Fink CS, Hennessey T (1996) β-Sitosterol inhibits HT-29 human colon cancer cell growth and alters membrane lipids. Anticancer Res 16(5A):2797–2804
Awad AB, Williams H, Fink CS (2001) Phytosterols reduce in vitro metastatic ability of MDA-MB-231 human breast cancer cells. Nutr Cancer 40(2):157–164. https://doi.org/10.1207/s15327914nc402_12
Awad AB, Toczek J, Fink CS (2004) Phytosterols decrease prostaglandin release in cultured P388D1/MAB macrophages. Prostaglandins Leukot Essent Fatty Acids 70(6):511–520. https://doi.org/10.1016/j.plefa.2003.11.005
Badawy FSI, Marshall AL, Ghorab MM, Adeyeye MC (1996) A study of the complexation between danazol and hydrophilic cyclodextrin derivatives. Drug Dev Ind Pharm 22(9 & 10):959–966. https://doi.org/10.3109/03639049609065926
Bary AR, Tucker IG, Davies NM (2000) Considerations in the use of hydroxypropyl-beta-cyclodextrin in the formulation of aqueous ophthalmic solutions of hydrocortisone. Eur J Pharm Biopharm 50:237–244. https://doi.org/10.1016/s0939-6411(00)00108-9
Basu K, Nair A, Williamson PA, Mukhopadhyay S, Lipworth BJ (2009) Airway and systemic effects of soluble and suspension formulations of nebulized budesonide in asthmatic children. Ann Allergy Asthma Immunol 103(5):436–441. https://doi.org/10.1016/s1081-1206(10)60365-1
Becirevic-Lacan M, Filipovic-Grcic J (2000) Effect of hydroxypropyl-β-cyclodextrin on hydrocortisone dissolution from films intended for ocular drug delivery. Pharmazie 55(7):518–520
Becirevic-Lacan M, Skalko N (1997) Hydrocortisone/cyclodextrin complex. Complexation methods and complex incorporation in liposomes. STP Pharma Sci 7(5):343–347
Belikov VG, Kompantseva EV, Gavrilin MV, Dranik LI (1991a) Effects of β-cyclodextrin on water solubility of some steroidal hormones. Farmatsiya (Moscow) 2:35–37
Belikov VG, Kompantseva EV, Gavrilin MV, Umnova EF (1991b) Potential use of β-cyclodextrin to improve the procedure for obtaining prednisolone. Khim-Farm Zh 25(2):48–49
Bhattarai B, Manickavachagam M, Suri R (2012) Development of novel cyclodextrin adsorbents for the removal of emerging contaminants from water. Abstracts of papers, 244th ACS National meeting and exposition, Philadelphia, PA, United States, August 19–23, 2012, ENVR-5
Bilensoy E, Sen M, Hincal AA (2006) Effect of drug physicochemical properties on in vitro characteristics of amphiphilic cyclodextrin nanospheres and nanocapsules. J Microencapsul 23(1):59–68. https://doi.org/10.1080/02652040500286227
Blode H, Schuermann R, Benda N (2008) Novel ethinyl estradiol-beta-cyclodextrin clathrate formulation does not influence the relative bioavailability of ethinyl estradiol or coadministered drospirenone. Contraception 77(3):171–176. https://doi.org/10.1016/j.contraception.2007.10.009

Bodor N, Murakami T, Wu WM (1992) Soft drugs. 18. Oral and rectal delivery of loteprednol etabonate, a novel soft corticosteroid, in rats-for safer treatment of gastrointestinal inflammation. Pharm Res 12(6):869–674

Bodor N, Drustrup J, Wu W (2000) Effect of cyclodextrins on the solubility and stability of a novel soft corticosteroid, loteprednol etabonate. Pharmazie 55(3):206–209

Bom A, Bradley M, Cameron K, Clark JK, Van Egmond J, Feilden H, MacLean EJ, Muir AW, Palin R, Rees DC, Zhang MQ (2002) A novel concept of reversing neuromuscular block: chemical encapsulation of rocuronium bromide by a cyclodextrin-based synthetic host. Angew Chem Int Ed Engl 41:266–270. https://doi.org/10.1002/1521-3757(20020118)114:2<275::AID-ANGE275>3.0.CO;2-A

Brewster ME, Estes KS, Bodor N (1989) Development of a non-surfactant formulation for alfaxalone through the use of chemically-modified cyclodextrins. J Parenter Sci Technol 43(6):262–265

Brewster ME, Anderson WR, Loftsson T, Huang MJ, Bodor N, Pop E (1995) Preparation, characterization and anesthetic properties of 2-hydroxypropyl-β-cyclodextrin complexes of pregnanolone and pregnenolone in rat and mouse. J Pharm Sci 84(10):1154–1159. https://doi.org/10.1002/jps.2600841004

Briceno MA, Almagro Z, Sabater-Jara L, Calderon AB, Pedreno AA, Ferrer MA (2012) Enhancement of phytosterols, taraxasterol and induction of extracellular pathogenesis-related proteins in cell cultures of *Solanum lycopersicum* cv micro-tom elicited with cyclodextrins and methyl jasmonate. J Plant Physiol 169(11):1050–1058. https://doi.org/10.1016/j.jplph.2012.03.008

Brown GA, Martini ER, Roberts BS, Vukovich MD, King DS (2002) Acute hormonal response to sublingual androstenediol intake in young men. J Appl Physiol 92:142–146

Cabral-Marques H, Almeida R (2009) Optimisation of spray-drying process variables for dry powder inhalation (DPI) formulations of corticosteroid/cyclodextrin inclusion complexes. Eur J Pharm Biopharm 73(1):121–129. https://doi.org/10.1016/j.ejpb.2009.05.002

Cai W, Sun T, Liu P, Chipot C, Shao X (2009) Inclusion mechanism of steroid drugs into β-cyclodextrins. Insights from free energy calculations. J Phys Chem B 113(22):7836–7843. https://doi.org/10.1021/jp901825w

Caira MR, Bourne SA, Samsodien H, Smith VJ (2015) Inclusion complexes of 2-methoxyestradiol with dimethylated and permethylated β-cyclodextrins: models for cyclodextrin–steroid interaction. Beilstein J Org Chem 11:2616–2630

Cameron KS, Fletcher D, Fielding L (2002) An NMR study of cyclodextrin complexes of the steroidal neuromuscular blocker drug rocuronium bromide. Magn Reson Chem 40:251–260. https://doi.org/10.3762/bjoc.11.281

Carli F, Colombo I, Torricelli C (1987) Drug beta-cyclodextrin systems by mechanochemical activation. Chim Oggi 5:61

Carron M, Zarantonello F, Tellaroli P, Ori C (2016) Efficacy and safety of sugammadex compared to neostigmine for reversal of neuromuscular blockade: a meta-analysis of randomized controlled trials. J Clin Anesth 35:1–12. https://doi.org/10.1016/j.jclinane.2016.06.018

Castelli F, Sarpietro MG, Micieli D, Trombetta D, Saija A (2006) Differential scanning calorimetry evidence of the enhancement of beta-sitosterol absorption across biological membranes mediated by beta-cyclodextrins. J Agricult Food Chem 54(26):10228–10233. https://doi.org/10.1021/jf062228x

Cavalli R, Peira E, Caputo O, Gasco MR (1999) Solid lipid nanoparticles as carriers of hydrocortisone and progesterone complexes with β-cyclodextrins. Int J Pharm 182(1):59–69. https://doi.org/10.1016/s0378-5173(99)00066-6

Ceschel GC, Berlati F, Ronchi C, Porziotta E (1993) Determination of the release of corticosteroids in topical formulations. Boll Chim Farm 132(2):49–50

Chang SL, Banga AK (1998) Transdermal iontophoretic delivery of hydrocortisone from cyclodextrin solutions. J Pharm Pharmacol 50(6):635–640. https://doi.org/10.1111/j.2042-7158.1998.tb06897.x

Chun IK, Kim BY (1992) Influence of various cyclodextrins on the stability of hydrocortisone 17-butyrate in aqueous solution. Arch Pharm Res 15(2):176–183. https://doi.org/10.1007/bf02974095

Chun IK, Yun DS (1993) Inclusion complexation of hydrocortisone butyrate with cyclodextrins and dimethyl-β-cyclodextrin in aqueous solution and in solid state. Int J Pharm 96:91–103. https://doi.org/10.1016/0378-5173(93)90216-3

Comini S, Olivier P, Riottot M, Duhamel D (1994) Interaction of β-cyclodextrin with bile acids and their competition with vitamins A and D3 as determined by 1H-NMR spectrometry. Clin Chim Acta 228(2):181–194. https://doi.org/10.1016/0009-8981(94)90288-7

Cooper A, Nutley M, MacLean EJ, Cameron K, Fielding L, Mestres J, Palin R (2005) Mutual induced fit in cyclodextrin-rocuronium complexes. Org Biomol Chem 3(10):1863–1871. https://doi.org/10.1039/B415903A

Cramer F, Henglein FM (1957) Regularities of formation of cyclodextrin inclusion complexes. Chem Ber 90:2561–2571

Crini G (2014) Review: a history of cyclodextrins. Chem Rev 114(21):10940–10975. https://doi.org/10.1021/cr500081p

Cserháti T, Forgács E (1996) Effect of β-cyclodextrin derivatives on the retention of steroidal drugs. J Chromatog B Biomed Appl 681(1):205–211. https://doi.org/10.1016/0378-4347(95)00469-6

daSilveira AM, Duchene D, Ponchel G (2000) Influence of solubility and partition coefficient on the loading of combined poly(isobutylcyanoacrylate) and hydroxypropyl-beta-cyclodextrin nanoparticles by steroids. STP Pharma Sci 10(4):309–314

De Azevedo MBM, Zullo MAT, Alderete JB, De Azevedo MMM, Salva TJG, Duran N (2002) Characterisation and properties of the inclusion complex of 24-epibrassinolide with beta-cyclodextrin. Plant Growth Regul 37(3):233–240

de Boer HD, van Egmond J, Driessen JJ, Booij LHD (2007) Update on the management of neuromuscular block: focus on sugammadex. Neuropsychiatr Dis Treat 3(5):539–544

De Hassonville SH, Perly B, Piel G, Van Hees T, Barillaro V, Bertholet P, Delattre L, Evrard B (2003) Inclusion complexes of cyproterone acetate with cyclodextrins in aqueous solution. J Incl Phenom Macrocycl Chem 44(1–4):289–292

De Paula D, Rodrigues Oliveira DC, Tedesco AC, Bentley MVLB (2007) Enhancing effect of modified beta-cyclodextrins on in vitro skin permeation of estradiol. Brazilian. J Pharm Sci 43:111–120. https://doi.org/10.1590/S1516-93322007000100014

Delaurent C, Siouffi AM, Pepe G (1998) Cyclodextrin inclusion complexes with vitamin D3: investigations of the solid complex characterization. Chem Anal 43(4):601–616

Dietzel K, Estes KS, Brewster ME, Bodor NS, Derendorf H (1990) The use of 2-hydroxypropyl-β-cyclodextrin as a vehicle for intravenous administration of dexamethasone in dogs. Int J Pharm 59(3):225–230. https://doi.org/10.1016/0378-5173(90)90113-i

Estes KS, Brewster ME, Webb AI, Bodor N (1990) A non-surfactant formulation for alfaxalone based on an amorphous cyclodextrin: activity studies in rats and dogs. Int J Pharm 65(1–2):101–107. https://doi.org/10.1016/0378-5173(90)90014-U

Fenyvesi E, Szente L (2016) Cyclodextrin-enabled novel waste water treatment technologies: from the test tube to the waste water treatment plant. Magyar Kem Foly 122(2–4):83–89. http://www.mkf.mke.org.hu/images/stories/docs/2016_234a/MKF_2016_83.pdf (Accessed on 20 Mar 2017)

Fenyvesi É, Vikmon MA, Szente L (2016) Cyclodextrins in food technology and human nutrition: benefits and limitations. Crit Rev Food Sci Nutr 56:1981–2004. https://doi.org/10.1080/10408398.2013.809513

Filipovic-Grcic J, Voinovich D, Moneghini M, Becirevic-Lacan M, Magarotto L, Jalsenjak I (2000) Chitosan microspheres with hydrocortisone and hydrocortisone-hydroxypropyl-β-cyclodextrin inclusion complex. Eur J Pharm Sci 9(4):373–379. https://doi.org/10.1016/S0928-0987(99)00078-0

Flood KG, Reynolds ER, Snow NH (2000) Characterization of inclusion complexes of betamethasone-related steroids with cyclodextrins using high-performance liquid chromatography. J Chromatog A 903(1–2):49–65. https://doi.org/10.1016/S0021-9673(00)00867-0

Ford MM, Nickel JD, Finn DA (2005) Treatment with and withdrawal from finasteride alter ethanol intake patterns in male C57BL/6J mice: potential role of endogenous neurosteroids. Alcohol (New York) 37(1):23–33. https://doi.org/10.1016/j.alcohol.2005.11.002

Forgo P, Gondos G (2002) A study of β-cyclodextrin inclusion complexes with progesterone and hydrocortisone using rotating frame Overhauser spectroscopy. Monatsh Chem 133(1):101–106. https://doi.org/10.1007/s007060270011

Frijlink HW, Eissens AC, Schoonen AJM, Lerk CF (1990) The effects of cyclodextrins on drug absorption. II. In vivo observations. Int J Pharm 64(2–3):195–205. https://doi.org/10.1016/0378-5173(90)90269-A

Fulop Z, Nielsen TT, Larsen KL, Loftsson T (2013) Dextran-based cyclodextrin polymers: their solubilizing effect and self-association. Carbohydr Polym 97(2):635–642. https://doi.org/10.1016/j.carbpol.2013.05.053

Gruiz K, Molnár M, Fenyvesi E, Hajdu C, Atkári Á, Barkács K (2011) Cyclodextrins in innovative engineering tools for risk-based environmental management. J Incl Phenom Macrocycl Chem 70(3–4):299–306. https://doi.org/10.1007/s10847-010-9909-y

Habon I, Stadler-Szoke A, Szejtli J (1985) Cyclodextrin inclusion complexes of steroids. Magy Kem Lapja 40(5):231–236. (in Hungarian)

Hamano Y, Suzuki K (1999) Moisture-retaining composition, and composition for external use, containing the same. Patent JP19990266200

Hartmann MA (1998) Plant sterols and the membrane environment. Trends Plant Sci 3:170–175. https://doi.org/10.1016/S1360-1385(98)01233-3

Hashem FM, Halim A, Ghanem E, El-Leithy S, Shaker DS (2003) Effect of cyclodextrins on the solubility and in-vitro permeation of corticosterone through hairless mouse skin. Bull Faculty Pharm (Cairo Univ) 41(2):117–129

Hashimoto Y, Shono J, Kamoi A, Obuaki T, Nakamura M, Hamajima T, Aoyagi M, Nakagawa N (2008) Method for preventing decomposition/deterioration of lipophilic component in the presence of water. Patent US2011256283

Helm H, Andersen A, Müller BW, Waaler T (1994) Stabilization of digitalis glycosides through 2-hydroxypropyl-β-cyclodextrin complexation in aqueous solution at pH 7.3. Pharmazie 49(7):494–496

Hermens WAJJ, Deurloo MJM, Romeyn SG, Verhoef JC, Merkus FWHM (1990) Nasal absorption enhancement of 17β-estradiol by dimethyl-β-cyclodextrin in rabbits and rats. Pharm Res 7:500–503

Hesselink PGM, van Vliet S, De Vries H, Witholt B (1989) Optimization of steroid side chain cleavage by mycobacterium sp. in the presence of cyclodextrins. Enzym Microb Technol 11:398–404

Higuchi T, Connors KA (1965) Phase-solubility techniques. Adv Anal Chem Instrument 4:117–212

Holbrook WP, Kristmundsdottir T, Loftsson T (1998) Aqueous hydrocortisone mouthwash solution: clinical evaluation. Acta Odontol Scand 56(3):157–160

Holm R, Shi W, Hartvig RA, Askjaer S, Christian Madsen J, Westh P (2009) Thermodynamics and structure of inclusion compounds of tauro- and glyco-conjugated bile salts and beta-cyclodextrin. Phys Chem Chem Phys 11(25):5070–5078. https://doi.org/10.1039/b820487j

Horsky J, Pitha J (1996) Hydroxypropyl cyclodextrins: potential synergism with carcinogens. J Pharm Sci 85(1):96–100. https://doi.org/10.1021/js9501324

Hovenkamp E, Lourbakos A, Duchateau GSMJE, Tareilus EW, Trautwein EA (2007) Preferential efflux of phytosterols over cholesterol from macrophages. Lipids 42(12):1125–1132. https://doi.org/10.1007/s11745-007-3119-4

Hunter JM, Flockton E (2006) The doughnut and the hole: a new pharmacological concept for anaesthetists. Brit J Anaesth 97:123–126. https://doi.org/10.1093/bja/ael158

Jadhav GS, Vavia PR (2008) Physicochemical, in silico and in vivo evaluation of a danazol-beta-cyclodextrin complex. Int J Pharm 352(1–2):5–16. https://doi.org/10.1016/j.ijpharm.2007.10.005

Jadhav GS, Vavia PR, Nandedkar TD (2007) Danazol-beta-cyclodextrin binary system: a potential application in emergency contraception by the oral route. AAPS PharmSciTech 8(2):35. https://doi.org/10.1208/pt0802035

Jadoun J, Bar R (1993) Microbial transformations in a cyclodextrin medium. Part 3. Cholesterol oxidation by *Rhodococcus Erythropolis*. Appl Microbiol Biotechnol 40(2):230–240. https://doi.org/10.1007/BF00170372

Jansook P, Loftsson T (2009) CDs as solubilizers: effects of excipients and competing drugs. Int J Pharm 379(1):32–40. https://doi.org/10.1016/j.ijpharm.2009.06.005

Jansook P, Moya-Ortega MD, Loftsson T (2010) Effect of self-aggregation of gamma-cyclodextrin on drug solubilization. J Incl Phenom Macrocycl Chem 68(1–2):229–236. https://doi.org/10.1007/s10847-010-9779-3

Jiang L, Liu Y, Liu S, Li M, Hu X, Zeng G, Hu X, Liu S, Liu S, Huang B, Li M (2016) Fabrication of β-cyclodextrin/poly (L-glutamic acid) supported magnetic graphene oxide and its adsorption behavior for 17β-estradiol. Chem Eng J 308:597–605. https://doi.org/10.1016/j.cej.2016.09.067

Jover A, Budal RM, Al-Soufi W, Meijide F, Vazquez Tato J, Yunes RA (2003) Spectra and structure of complexes formed by sodium fusidate and potassium helvolate with beta- and gamma-cyclodextrin. Steroids 68(1):55–64. https://doi.org/10.1016/s0039-128x(02)00115-0

Jover A, Budal RM, Meijide F, Soto VH, Tato JV (2004) Determination of microscopic equilibrium constants for the complexation of ditopic guests by cyclodextrins from NMR experiments. J Phys Chem B 108(49):18850–18859. https://doi.org/10.1021/jp0370573

Kata M, Haragh L (1981) Spray embedding of spironolactone with β-cyclodextrin. Pharmazie 36(11):784–785

Kear CL, Yang J, Godwin DA, Felton LA (2008) Investigation into the mechanism by which cyclodextrins influence transdermal drug delivery. Drug Dev Ind Pharm 34(7):692–697. https://doi.org/10.1080/03639040701842428

Keating GM (2016) Sugammadex: a review of neuromuscular blockade reversal. Drugs 76(10):1041–1052. https://doi.org/10.1007/s4026510016-0604-1

Khomutov SM, Donova MV (2011) Nanodimer cyclodextrin ligands with high affinity to steroids. J Incl Phenom Macrocycl Chem 70(3):353–357. https://doi.org/10.1007/s10847-010-9896-z

Khomutov SM, Sidorov IA, Dovbnya DV, Donova MV (2002) Estimation of cyclodextrin affinity to steroids. J Pharm Pharmacol 54(5):617–622. https://doi.org/10.1211/0022357021778925

Khomutov SM, Sukhodolskaya GV, Donova MV (2007) The inhibitory effect of cyclodextrin on the degradation of 9alpha-hydroxyandrost-4-ene-3,17-dione by *Mycobacterium sp*. VKM ac-1817D. Biocatal Biotransformation 25(5):386–392. https://doi.org/10.1080/10242420701510510

Koizumi K, Okada Y, Kubota Y, Utamura T (1987) Inclusion complexes of poorly water-soluble drugs with glucosyl-cyclodextrins. Chem Pharm Bull 35:3413–3418. https://doi.org/10.1248/cpb.35.3413

Kompantseva EV, Gavrilin MV, Botezat-Belyi YK, Umnova EF, IYa A (1990) Investigation of β-cyclodextrin-cortexolone interaction. Khim Farm Zh 24(10):81–82

Kralova K, Mitterhauszerova L (1989) Interactions of β-cyclodextrin with steroid compounds in aqueous solutions. Pharmazie 44(9):623–625

Kristinsson JK, Fridriksdottir H, Thorisdottir S, Sigurdardottir AM, Stefansson E, Loftsson T (1996) Dexamethasone-cyclodextrin-polymer co-complexes in aqueous eye drops. Invest Ophthalmol Vis Sci 37:1199–1203

Kristmundsdottir T, Loftsson T, Holbrook WP (1996) Formulation and clinical evaluation of a hydrocortisone solution for the treatment of oral disease. Int J Pharm 139(1, 2):63–68. https://doi.org/10.1016/0378-5173(96)04565-6

Lahiani-Skiba M, Barbot C, Bounoure F, Joudieh S, Skiba M (2006) Solubility and dissolution rate of progesterone-cyclodextrin-polymer systems. Drug Dev Ind Pharm 32(9):1043–1058. https://doi.org/10.1080/03639040600897093

Landy D, Mallard I, Ponchel A, Monflier E, Fourmentin S (2012) Remediation technologies using cyclodextrins: an overview. Environ Chem Lett 10:225–237. https://doi.org/10.1007/s10311-011-0351-1

Larsen KL, Aachmann FL, Wimmer R, Stella VJ, Kjolner UM (2005) Phase solubility and structure of the inclusion complexes of prednisolone and 6-alpha-methyl prednisolone with various cyclodextrins. J Pharm Sci 94(3):507–515

Larsen KL, Andersen SB, Morbak AL, Wimmer R (2007) Inclusion complexes of fusidic acid and three structurally related compounds with cyclodextrins. J Incl Phenom Macrocycl Chem 57:185–190. https://doi.org/10.1007/s10847-006-9198-7

Lee JC, Seo MS, Kim SR (2012) A beta-cyclodextrin, polyethyleneimine and silk fibroin hydrogel containing *Centella asiatica* extract and hydrocortisone acetate: releasing properties and in vivo efficacy for healing of pressure sores. Clin Exp Dermatol 37(7):762–771. https://doi.org/10.1111/j.1365-2230.2011.04331.x

Lee LL, Athanassoglou V, Pandit JJ (2016) Neuromuscular blockade in the elderly patient. J Pain Res 9:437–444. https://doi.org/10.2147/JPR.S85183

Leigh HM, Meister PD, Weintraub A, Reineke LM, Eppstein SH, Murray HC, Peterson DH (1952) Microbiological transformations of steroids. I. Introduction of oxygen at carbon-11 of progesterone. J Am Chem Soc 73(23):5933–5936. https://doi.org/10.1021/ja01143a033

Liu FY, Kildsig DO, Mitra AK (1990) Beta-cyclodextrin/steroid complexation: effect of steroid structure on association equilibria. Pharm Res 7(8):869–873

Liu FY, Kildsig DO, Mitra AK (1992) Complexation of 6-acyl-O-β-cyclodextrin derivatives with steroids - effects of chain length and substitution degree. Drug Dev Ind Pharm 18(15):1599–1612. https://doi.org/10.3109/03639049209040889

Liu Y, Yang YW, Cao R, Song SH, Zhang HY, Wang LH (2003) Thermodynamic origin of molecular selective binding of bile salts by aminated beta-cyclodextrins. J Phys Chem B 107(50):14130–14139. https://doi.org/10.1021/jp036366q

Loftsson T (2014) Self-assembled cyclodextrin nanoparticles and drug delivery. J Incl Phenom Macrocycl Chem 80:1–7

Loftsson T, Siguroardottir AM (1994) The effect of poly(vinylpyrrolidone) and hydroxypropyl methyl cellulose on HPβCD complexation of hydrocortisone and its permeability through hairless mouse skin. Eur J Pharm Sci 2(4):297–230. https://doi.org/10.1016/0928-0987(94)90013-2

Loftsson T, Stefansson E (2002) Cyclodextrins in eye drop formulations: enhanced topical delivery of corticosteroids to the eye. Acta Ophthalmol Scand 80(2):144–150. https://doi.org/10.1034/j.1600-0420.2002.800205.x

Loftsson T, Olafsdottir BJ, Bodor N (1991) The effects of cyclodextrins on transdermal delivery of drugs. Eur J Pharm Biopharm 37:30–33

Loftsson T, Frioriksdottir H, Ingvarsdottir G, Jonsdottir B, Siguroardottir AM (1994a) The influence of 2-hydroxypropyl-β-cyclodextrin on diffusion rates and transdermal delivery of hydrocortisone. Drug Dev Ind Pharm 20(9):1699–1708. https://doi.org/10.3109/03639049409050210

Loftsson T, Fridriksdottir H, Thorisdottir S, Stefansson E (1994b) The effect of hydroxypropyl methylcellulose on release of dexamethazone from aqueous 2-hydroxypropyl-β-cyclodextrin formulations. Int J Pharm 104:181–184

Loftsson T, Masson M, Sigurdsson HH (2002) Cyclodextrins and drug permeability through semi-permeable cellophane membranes. Int J Pharm 232(1–2):35–43. https://doi.org/10.1016/S0378-5173(01)00895-X

Lohade AA, Singh DJ, Parmar JJ, Hegde DD, Menon MD, Soni PS, Samad A, Gaikwad RV (2007) Albumin microspheres of fluticasone propionate inclusion complexes for pulmonary delivery. Indian J Pharm Sci 69(5):707–709. https://doi.org/10.4103/0250-474X.38486

Lopata A, Darvas F, Stadler-Szoke A, Szejtli J (1985) Quantitative structure-stability relationships among inclusion complexes of cyclodextrins. Part 2. Steroid hormones. QSAR strategies Des Bioact Compd, Proc Eur Symp Quant Struct-Act Relat, 5th, Meeting Date 1984, 353–6. Seydel JK (ed) VCH: Weinheim, Fed Rep Ger

Luthra U, Khadpekar S, Trivedi A, Shetty A, Kumar H (2015) Biotransformation of 4-androstene-3, 17-dione to androst-1,4-diene-3, 17-dione by *Nocardioides Simplex*. WJPPS 4(11):1935–1943

Ma YH, Wang M, Fan Z, Shen YB, Zhang LT (2009) The influence of host–guest inclusion complex formation on the biotransformation of cortisone acetate 1-dehydrogenation. J Steroid Biochem Mol Biol 117(4–5):146–151. https://doi.org/10.1016/j.jsbmb.2009.08.007

Manabe M, Kawamura H, Morita H, Ikushima K (1994) Determination of binding constant for cyclodextrin with bile salt by the differential conductivity method. Proc 7th Int Symp Cyclodextrins, Tokyo, 25–28 April 1994, pp 275–278

Masson M, Loftsson T, Masson G, Stefansson E (1999) Cyclodextrins as permeation enhancers: some theoretical evaluations and in vitro testing. J Control Release 59(1):107–118. https://doi.org/10.1016/S0168-3659(98)00182-5

Masson M, Sigurdardottir BV, Matthiasson K, Loftsson T (2005) Investigation of drug-cyclodextrin complexes by a phase-distribution method: some theoretical and practical considerations. Chem Pharm Bull 53(8):958–964

Meng X, Pan Q, Liu Y (2012) Preparation and properties of phytosterols with hydroxypropyl β-cyclodextrin inclusion complexes. Eur Food Res Technol 235:1039–1047. https://doi.org/10.1007/s00217-012-1833-5

Miro A, d'Angelo I, Nappi A, La Manna P, Biondi M, Mayol L, Musto P, Russo R, Rotonda MIL, Ungaro F, Quaglia F (2013) Engineering poly(ethylene oxide) buccal films with cyclodextrin: a novel role for an old excipient? Int J Pharm 452:283–291. https://doi.org/10.1016/j.ijpharm.2013.05.030

Miyajima K, Yokoi M, Komatsu H, Nakagaki M (1986) Interaction of β-cyclodextrin with bile salts in aqueous solutions. Chem Pharm Bull 34(3):1395–1398. https://doi.org/10.1248/cpb.34.1395

Moellgaard AF, Bundgaard H (1983) Inclusion complexation of spironolactone with cyclodextrins. Arch Pharm Chem Sci Ed 11(1):7–14

Müller BW, Brauns U (1986) Hydroxypropyl-β-cyclodextrin derivatives: influence of average degree on substitution on complexing ability and surface activity. J Pharm Sci 75(6):571–572. https://doi.org/10.1002/jps.2600750609

Nagy ZM, Molnár M, Fekete-Kertész I, Molnár-Perl I, Fenyvesi É, Gruiz K (2014) Removal of emerging micropollutants from water using cyclodextrin. Sci Total Environ 485–486:711–719. https://doi.org/10.1016/j.scitotenv.04.003

Nguyen MN, Slominski A, Li W, Ng YR, Tuckey RC (2009) Metabolism of vitamin D2 to 17,20,24-trihydroxyvitamin D2 by cytochrome P450scc (CYP11A1). Drug Metab Dispos 37(4):761–767. https://doi.org/10.1124/dmd.108.025619

Oishi K, Moriuchi A (2010) Removal of dissolved estrogen in sewage effluents by beta-cyclodextrin polymer. Sci Total Environ 409(1):112–115. https://doi.org/10.1016/j.scitotenv.2010.09.031

Okada Y, Koizumi K (1998) Some properties and the inclusion behavior of three positional isomers of 61,6nDi-O-alpha-D-glucosyl-cyclomaltoheptaoses (β-cyclodextrins). Chem Pharm Bull 46(2):319–323. https://doi.org/10.1248/cpb.46.319

Okada Y, Semma M, Ichikawa A (2011) Physicochemical and biological properties of 2-O-α-D-galactosyl-cyclomaltohexaose (α-cyclodextrin) and -cyclomaltoheptaose (β-cyclodextrin). Carbohydr Res 346(8):1018–1022. https://doi.org/10.1016/j.carres.2011.02.023

Ollila F, Pentikaeinen OT, Forss S, Johnson MS, Slotte JP (2001) Characterization of bile salt/cyclodextrin interactions using isothermal titration calorimetry. Langmuir 17(22):7107–7111. https://doi.org/10.1021/la0109258

Perez RL, Escandar GM (2013) Spectrofluorimetric study of estrogen-cyclodextrin inclusion complexes in aqueous systems. Analyst (Cambridge UK) 138(4):1239–1248. https://doi.org/10.1039/c2an36395j

Pike MC, Spicer DV, Daniels A, Daniels JR (2004) Nasal spray steroid formulation and method. US Pat Appl. Publ. US 2004022738

Pitha J, Pitha J (1985) Amorphous water-soluble derivatives of cyclodextrins: nontoxic dissolution enhancing excipients. J Pharm Sci 74(9):987–990. https://doi.org/10.1002/jps.2600740916

Pitha J, Harman SM, Michel ME (1986) Hydrophilic cyclodextrin derivatives enable effective oral administration of steroidal hormones. J Pharm Sci 75(2):165–167. https://doi.org/10.1002/jps.2600750213

Pitha J, Anaissie EJ, Uekama K (1987) Gamma-cyclodextrin:testosterone complex suitable for sublingual administration. J Pharm Sci 76:788–790. https://doi.org/10.1002/jps.2600761007

Pitha J, Hoshino T, Torres-Labandeira J, Irie T (1992) Preparation of drug-hydroxypropyl cyclodextrin complexes by a method using ethanol or aqueous ammonium hydroxide as cosolubilizers. Int J Pharm 80(2–3):253–258. https://doi.org/10.1016/0378-5173(92)90282-7

Podhaisky HP, Bracht S (2007) Pharmaceutical films for oral use. PCT Int. Appl. WO 2007085498
Preiss A, Mehnert W, Froemming KH (1994) Complexation of hydrocortisone with β-cyclodextrin and hydroxypropyl-β-cyclodextrin. Arch Pharm (Weinheim, Ger) 327(11):729–734
Radu CD, Salariu M, Avadanei M, Ghiciuc C, Foia L, Lupusoru EC, Ferri A, Ulea E, Lipsa F (2013) Cotton-made cellulose support for anti-allergic pajamas. Carbohydr Polym 95:479–486. https://doi.org/10.1016/j.carbpol.2013.03.037
Ramos Cabrer P, Alvarez-Parrilla E, Al-Soufi W, Meijide F, Rodriguez Nunez E, Vazquez Tato J (2003) Complexation of bile salts by natural cyclodextrins. Supramol Chem 15(1):33–43. https://doi.org/10.1080/1061027031000073207
Rinaldi L, Binello A, Stolle A, Curini M, Cravotto G (2015) Efficient mechanochemical complexation of various steroid compounds with α-, β- and γ-cyclodextrin. Steroids 98:58–62. https://doi.org/10.1016/j.steroids.2015.02.016
Riottot M, Olivier P, Huet A, Caboche JJ, Parquet M, Khallou J, Lutton C (1993) Hypolipidemic effects of β-cyclodextrin in the hamster and in the genetically hypercholesterolemic Rico rat. Lipids 28(3):181–188
Rodriguez-Tenreiro C, Alvarez-Lorenzo C, Rodriguez-Perez A, Concheiro A, Torres-Labandeira JJ (2007) Estradiol sustained release from high affinity cyclodextrin hydrogels. Eur J Pharm Biopharm 66(1):55–62. https://doi.org/10.1016/j.ejpb.2006.09.003
Roglič U, Žnidaršič-Plazl P, Plazl I (2005) The influence of β-cyclodextrin on the kinetics of progesterone transformation by *Rhizopus Nigricans*. Biocatal Biotransformation 23(5):299–305. https://doi.org/10.1080/10242420500175929
Sabater-Jara A, Pedreno M (2013) Use of β-cyclodextrins to enhance phytosterol production in cell suspension cultures of carrot (*Daucus carota* L.). Plant Cell Tissue Organ Cult 114:249–258. https://doi.org/10.1007/s11240-013-0320-4
Sadlej-Sosnowska N (1995) Molecular complexation: β-cyclodextrin and steroid hormones inclusion complexes studied by high-performance liquid chromatography. Eur J Pharm Sci 3(1):1–5. https://doi.org/10.1016/0928-0987(94)00006-L
Sadlej-Sosnowska N (1997a) Fluorometric determination of association constants of three estrogens with cyclodextrins. J Fluoresc 7(3):195–200. https://doi.org/10.1007/BF02758219
Sadlej-Sosnowska N (1997b) Influence of the structure of steroid hormones on their association with cyclodextrins: a high-performance liquid chromatography study. J Incl Phenom Mol Recognit Chem 27(1):31–40. https://doi.org/10.1023/A:1007964116987
Salapatek AM, Lee J, Patel D, D'Angelo P, Liu JH, Zimmerer RO, Pipkin JD (2011) Solubilized nasal steroid (CDX-947) when combined in the same solution nasal spray with an antihistamine (CDX-313) provides improved, fast-acting symptom relief in patients with allergic rhinitis. Allergy Asthma Proc 32(3):221–229. https://doi.org/10.2500/aap.2011.32.3444
Salehian B, Wang C, Alexander G, Davidson T, McDonald V, Berman N, Dudley RE, Ziel F, Swerdloff RS (1995) Pharmacokinetics, bioefficacy, and safety of sublingual testosterone cyclodextrin in hypogonadal men: comparison to testosterone enanthate–A clinical research center study. J Clin Endocrinol Metab 80(12):3567–3575. https://doi.org/10.1210/jcem.80.12.8530600
Schaller SJ, Fink H (2013) Sugammadex as a reversal agent for neuromuscular block: an evidence-based review. Core Evid 8:57–67. https://doi.org/10.2147/CE.S35675
Schipper NGM, Hermens WAJJ, Romeyn SG, Verhoef J, Merkus FWHM (1990) Nasal absorption of 17β-estradiol and progesterone from a dimethyl cyclodextrin inclusion formulation in rats. Int J Pharm 64(1):61–66. https://doi.org/10.1016/0378-5173(90)90179-8
Schonbeck C, Westh P, Madsen JC, Larsen KL, Stade LW, Holm R (2011) Methylated β-cyclodextrins: influence of degree and pattern of substitution on the thermodynamics of complexation with tauro- and glyco-conjugated bile salts. Langmuir 27(10):5832–5841. https://doi.org/10.1021/la200381f
Seo H, Uekama K (1989) Enhanced bioavailability of digoxin by gamma-cyclodextrin complexation: evaluation for sublingual and oral administrations in humans. J Pharm Soc Jpn 109(10):778–782

Shakalisava Y, Regan F (2005) Determination of association constants of inclusion complexes of steroid hormones and cyclodextrins from their electrophoretic mobility. Electrophoresis 27(15):3048–3056. https://doi.org/10.1002/elps.200500842

Shaker DS, Ghanem AH, Li SK, Warner KS, Hashem FM, Higuchi WI (2003) Mechanistic studies of the effect of hydroxypropyl-beta-cyclodextrin on in vitro transdermal permeation of corticosterone through hairless mouse skin. Int J Pharm 253(1–2):1–11. https://doi.org/10.1016/S0378-5173(02)00625-7

Shen Y, Ma B, Zheng Y, Luo J, Zhi Y, Wang M (2010) The mechanism of β-cyclodextrin on the 11β-hydroxylation biotransformation of steroid. Proc 3rd Int Conf Biomed Engin Inform, BMEI 2010, Yantai, China, 5, 5639462, pp. 1964–1967

Shi C, Wu F, Xu J (2012) Incorporation of beta-sitosterol into mitochondrial membrane enhances mitochondrial function by promoting inner mitochondrial membrane fluidity. J Bioenerg Biomembr 45(3):301–305. https://doi.org/10.1007/s10863-012-9495-3

Shi C, Wu F, Zhu X, Xu J (2013) Incorporation of beta-sitosterol into the membrane increases resistance to oxidative stress and lipid peroxidation via estrogen receptor-mediated PI3K/GSK3beta signaling. Biochim Biophys Acta 1830:2538–2544. https://doi.org/10.1016/j.bbagen.2012.12.012

Shi JH, Chen K, Xu J (2014) Characterization of the inclusion interaction between prednisolone and di-O-methyl-beta-cyclodextrin: spectroscopic methods and molecular modeling. J Mol Liquids 194:172–178. https://doi.org/10.1016/j.molliq.2014.01.023

Shimada K, Oe T, Hirose Y, Komine Y (1989) Retention behavior of cardiac steroids using cyclodextrin in the mobile phase in high-performance liquid chromatography. J Chromatogr 478(2):339–347

Shinoda T, Kagatani S, Maeda A, Konno Y, Hashimoto H, Hara K, Fujita K, Sonobe T (1999) Sugar-branched-cyclodextrins as injectable drug carriers in mice. Drug Dev Ind Pharm 25(11):1185–1192. https://doi.org/10.1081/DDC-100102286

Sigurdardottir AM, Loftsson T (1995) The effect of polyvinylpyrrolidone on cyclodextrin complexation of hydrocortisone and its diffusion through hairless mouse skin. Int J Pharm 126(1, 2):73–78. https://doi.org/10.1016/0378-5173(95)04095-1

Singer Y, Shity H, Bar R (1991) Microbial transformations in a cyclodextrin medium. Part 2. Reduction of androstenedione to testosterone by *Saccharomyces Cerevisiae*. Appl Microbiol Biotechnol 35(6):731–737. https://doi.org/10.1007/BF00169886

Sorgenfrei IF, Norrild K, Larsen PB, Stensballe J, Ostergaard D, Prins ME, Viby-Mogensen J (2006) Reversal of rocuronium-induced neuromuscular block by the selective relaxant binding agent sugammadex: a dose-finding and safety study. Survey Anesthesiol 50(6):286–287. https://doi.org/10.1097/01.sa.0000238953.66744.e7

Stefansson E, Loftsson T (2003) Cyclodextrins in eye drop formulations. J Incl Phenom Macrocycl Chemi 44(1–4):23–27. https://doi.org/10.1034/j.1600-0420.2002.800205.x

Stewart DJ, Milanova RK, Zawistowski J, Wallis SH (1998) Compositions comprising phytosterol and/or phytostanol having enhanced solubility and dispersability and the use of these compositions to lower serum cholesterol. Patent US19980092497

Stünkel CA, Dudley RE, Yen SS (1991) Sublingual administration of testosterone-hydroxypropyl-beta-cyclodextrin inclusion complex simulates episodic androgen release in hypogonadal men. J Clin Endocrinol Metab 72(5):1054–1059. https://doi.org/10.1210/jcem-72-5-1054

Szejtli J (1982) Cyclodextrins and their inclusion complexes. Akadémiai Kiadó, Budapest

Szejtli J (1984) Industrial applications of cyclodextrins. In: Atwood JL, Davies JED, MacNicol DD (eds) Inclusion compdounds, vol 3. Academic, London, pp 331–390

Szejtli J (1986) Cyclodextrins in biotechnology. Starch/Staerke 38:388–390. https://doi.org/10.1002/star.19860381107

Szejtli J (1988) Cyclodextrin technology. Kluwer

Szejtli J (1990) The cyclodextrins and their applications in biotechnology. Carbohydr Polym 12(4):375–392. https://doi.org/10.1016/0144-8617(90)90088-A

Szejtli J (1998) Ciklodextrinek alkalmazása vegyipari termékekben és eljárásokban. (Application of cyclodextrins in chemical industrial products and technologies, *in Hungarian*). Magy Kém Lapja 53(6):271–282

Szejtli J, Bolla E (1980) Stabilization of fat-soluble vitamins with β-cyclodextrin. Starch 32(11):386–391

Szejtli J, Szente L (2005) Elimination of bitter, disgusting tastes of drugs and foods by cyclodextrins. Eur J Pharm Biopharm 61:115–125. https://doi.org/10.1016/j.ejpb.2005.05.006

Szejtli J, Bolla-Pusztai E, Szabo P, Ferenczy T (1980) Enhancement of stability and biological effect of cholecalciferol by β-cyclodextrin complexation. Pharmazie 35(12):779–787

Szente L (2006) Cyclodextrins in anesthesiology. Cyclodextrin News 20(10):265–269

Szente L, Fenyvesi É (2017) Cyclodextrin-lipid complexes: cavity size matters. Struct Chem 28(2):479–492. https://doi.org/10.1007/s11224-016-0884-9

Tadashi H, Masaki N (2012) Composition containing complex and production method thereof. Patent WO2012JP60575

Tan X, Lindenbaum S (1991) Studies on complexation between β-cyclodextrin and bile salts. Int J Pharm 74(2–3):127–135. https://doi.org/10.1016/0378-5173(91)90229-H

Tan ZJ, Zhu XX, Brown GR (1994) Formation of inclusion complexes of cyclodextrins with bile salt anions as determined by NMR titration studies. Langmuir 10(4):1034–1039. https://doi.org/10.1021/la00016a014

Taylor GT, Weiss J, Pitha J (1989) Testosterone in a cyclodextrin-containing formulation: behavioral and physiological effects of episode-like pulses in rats. Pharm Res 6:641–646

Thi TD, Nauwelaerts K, Froeyen M, Baudemprez L, Van Speybroeck M, Augustijns P, Annaert P, Martens J, Van Humbeeck J, Van den Mooter G (2010) Comparison of the complexation between methylprednisolone and different cyclodextrins in solution by 1H-NMR and molecular modeling studies. J Pharm Sci 99(9):3863–3873. https://doi.org/10.1002/jps.22227

Tian XQ, Holick MF (1995) Catalyzed thermal isomerization between previtamin D3 and vitamin D3 via beta-cyclodextrin complexation. J Biol Chem 270(15):8706–8711. https://doi.org/10.1074/jbc.270.15.8706

Tian Y, Jin Z, Deng L, Xu X, Zhao J (2007) Effect of vitamin D3-beta-cyclodextrin inclusion compound on the quality improvement in extruded nutrition-fortified rice. Shipin Yu Fajiao Gongye 33(12):52–55. (Chinese) (Chem Abstr 150:471236)

Tongiani S, Ozeki T, Stella VJ (2009) Sulfobutyl ether-alkyl ether mixed cyclodextrin derivatives with enhanced inclusion ability. J Pharm Sci 98(12):4769–4780. https://doi.org/10.1002/jps.21791

Tripp ML, Minich D, Irving T, Ramachandran R, Bland J, Nelson P (2004) Encapsulated fractions isolated or derived from hops. Patent Appl US2009087512

Tuckey RC, Nguyen MN, Slominski A (2008) Kinetics of vitamin D3 metabolism by cytochrome P450scc (CYP11A1) in phospholipid vesicles and cyclodextrin. Int J Biochem Cell Biol 40(11):2619–2626. https://doi.org/10.1016/j.biocel.2008.05.006

Udvardy Nagy I, Bartho I, Hantos G, Trinn M, Vida Z, Szejtli J, Stadler I, Habon I, Balázs Á (1981). Process for the intensification of the microbiological transformation of steroid compounds. Hungarian Patent Application HU19810002785

Uekama K, Fujinaga T, Otagiri M, Seo H, Tsuruoka M (1981) Enhanced bioavailability of digoxin by γ-cyclodextrin complexation. J Pharmacobiodyn 4(9):735–737

Uekama K, Fujinaga T, Hirayama F, Otagiri M, Yamasaki M (1982a) Inclusion complexations of steroid hormones with cyclodextrins in water and in solid phase. Int J Pharm 10(1):1–15. https://doi.org/10.1016/0378-5173(82)90057-6

Uekama K, Fujinaga T, Hirayama F, Otagiri M, Kurono Y, Ikeda K (1982b) Effects of cyclodextrins on the acid hydrolysis of digoxin. J Pharm Pharmacol 34(10):627–630. https://doi.org/10.1111/j.2042-7158.1982.tb04690.x

Uekama K, Otagiri M, Uemura Y, Fujinaga T, Arimori K, Matsuo N, Tasaki K, Sugii A (1983) Improvement of oral bioavailability of prednisolone by β-cyclodextrin complexation in humans. J Pharmacobiodyn 6(2):124–127

Uekama K, Otagiri M, Sakai A, Irie T, Matsuo N, Matsuoka Y (1985) Improvement in the percutaneous absorption of beclomethasone dipropionate by γ-cyclodextrin complexation. J Pharm Pharmacol 37(8):532–535. https://doi.org/10.1111/j.2042-7158.1985.tb03061.x

Vardanyan R, Hruby V (2016) Synthesis of best-seller drugs. Academic press, Cambridge, US

Vianna RFL, Bentley MVLB, Ribeiro G, Carvalho FS, Neto AF, de Oliveira DCR, Collett JH (1998) Formation of cyclodextrin inclusion complexes with corticosteroids: their characterization and stability. Int J Pharm 167(1–2):205–213. https://doi.org/10.1016/S0378-5173(98)00068-4
Volkova DA, Bannikova GE, Lopatin SA, Il'in MM, Gabinskaya KN, Andryushina VA, Gracheva IM, Varlamov VP (1999) Effect of a β-cyclodextrin-epichlorohydrin copolymer on the solubility of cortexolone. Pharm Chem J 33(11):601–603
Von Holtz RL, Fink CS, Awad AB (1998) β-Sitosterol activates the sphingomyelin cycle and induces apoptosis in LNCaP human prostate cancer cells. Nutr Cancer 32(1):8–12. https://doi.org/10.1080/01635589809514709
Wacker Fine Chemicals (2005). OmegaDry® Cranberry. http://www.cyclodextrin.org/CDPDF/PT_OmegaDry_Cranberry.pdf (Accessed on 12 Mar 2017)
Wang J, Wu F, Shi C (2013) Substitution of membrane cholesterol with beta-sitosterol promotes nonamyloidogenic cleavage of endogenous amyloid precursor protein. Neurosci 247:227–233. https://doi.org/10.1016/j.neuroscience.2013.05.022
Werner U, Damge C, Maincent P, Bodmeier R (2004) Properties of in situ gelling nasal inserts containing estradiol/ methyl beta-cyclodextrin. J Drug Deliv Sci Technol 14(4):275–284. https://doi.org/10.1016/S1773-2247(04)50048-5
Wintgens V, Amiel C (2010) Water-soluble γ-cyclodextrin polymers with high molecular weight and their complex forming properties. Eur Polym J 46(9):1915–1922. https://doi.org/10.1016/j.eurpolymj.2010.06.014
Won YJ, Lim BG, Lee DK, Kim H, Kong MH, Lee IO (2016) Sugammadex for reversal of rocuronium-induced neuromuscular blockade in pediatric patients: a systematic review and meta-analysis. Medicine (Baltimore) 95(34):e4678. https://doi.org/10.1097/MD.0000000000004678
Wördenbag HJ, Pras N, Frijlink HW, Lerk CF, Malingre TM (1990) Cyclodextrin-facilitated bioconversion of 17 β-estradiol by a phenoloxidase from Mucuna pruriens cell cultures. Phytochemistry 29(5):1551–1554. https://doi.org/10.1016/0031-9422(90)80119-2
Wouessidjewe D, Crassous A, Duchene D, Coleman A, Rysanek N, Tsoucaris G, Perly B, Djedaïni F (1989) Inclusion of spironolactone in cyclomaltoheptaose: a guest affected by the hospitality of the host. Carbohydr Res 192:313–321. https://doi.org/10.1016/0008-6215(89)85189-4
Yanez C, Basquinzay R (2008) Chronocoulometry diffusion coefficients as a measure of cyclodextrin-estradiol complex association. J Electroanal Chem 622(2):242–245. https://doi.org/10.1016/j.jelechem.2008.06.005
Yanez C, Basualdo J, Jara-Ulloa P, Squella JA (2007) Inclusion complexes of estrone and estradiol with β-cyclodextrin: voltammetric and HPLC studies. J Phys Org Chem 20:499–555. https://doi.org/10.1002/poc.1186
Yasuda K, Endo M, Ikushiro S, Kamakura M, Ohta M, Sakaki T (2013) UV-dependent production of 25-hydroxyvitamin D2 in the recombinant yeast cells expressing human CYP2R1. Biochem Biophys Res Commun 434:311–315. https://doi.org/10.1016/j.bbrc.2013.02.124
Znidarsic-Plazl P, Plazl I (2010) Development of a continuous steroid biotransformation process and product extraction within microchannel system. Catal Today 157(1–4):315–320. https://doi.org/10.1016/j.cattod.2010.01.042
Zwiers A, van den Heuvel M, Smeets J, Rutherford S (2011) Assessment of the potential for displacement interactions with sugammadex. A pharmacokinetic-pharmacodynamic modelling approach. Clin Drug Investig 31(2):101–111. https://doi.org/10.2165/11584730-000000000-00000

Chapter 3
Cyclodextrin-Based Nanosystems in Targeted Cancer Therapy

Nazlı Erdoğar and Erem Bilensoy

Contents

Abstract Cancer is a disease that shows uncontrolled cell division and invasion in other tissues, resulting in high incidence and mortality worldwide. Classical chemotherapeutic agents have poor water solubility and lack of specificity, resulting in systemic toxicity and limitations of the maximum drug dose. Alternatively, the use of nanosized drug carriers overcomes these drawbacks and increase *therapeutic* efficacy of many chemotherapy drugs. Nanocarriers for drug delivery allows systems to combine properties for multiple functions. Here, cyclodextrins are of particular interest due to their good inclusion capability, excellent biocompatibility, and ability to self-assemble and form various stable nanoscale systems such as micellar aggregates, nanoreservoirs and nanoparticles for biomedical applications. These cyclodextrin-based nanosystems show several advantages in terms of stability, safety, the ability to encapsulate hydrophobic drugs and good *in vivo* tolerance.

Here we review cyclodextrin-based nanocarriers such as liposomes, nanoparticles, micelles for delivery of anticancer drugs in cancer therapy. We discuss preparation and characterization. We compare cyclodextrin-based nanocarrier groups for their potential in targeted delivery of anticancer agents such as chemotherapeutic

N. Erdoğar · E. Bilensoy (✉)
Department of Pharmaceutical Technology, Hacettepe University Faculty of Pharmacy, Ankara, Turkey
e-mail: eremino@hacettepe.edu.tr

S. Fourmentin et al. (eds.), *Cyclodextrin Applications in Medicine, Food, Environment and Liquid Crystals*, Environmental Chemistry for a Sustainable World, https://doi.org/10.1007/978-3-319-76162-6_3

drugs, and siRNA for cancer therapy or diagnosis. The major benefits are recognition of cancer cells and delivery of drugs to the target tissue, high antitumor efficacy and low adverse effects, prolonged blood circulation, higher drug loading, and controlled release. *In vitro* and *in vivo* studies performed in cell cultures and animal models have given promising results.

3.1 Introduction

The World Health Organization attributed 8.8 million deaths to cancer in 2015, which is nearly 1 in 6 of all global deaths. New global cancer incidences are expected to increase from 14.1 million in 2012 to approximately 23.6 million in 2030. One of the main reasons is the lack of selective delivery of antitumor drugs to tumor tissue. High systemic exposure to anticancer agents induces dose-limiting toxicity. Therefore, targeted delivery is very important to overcome current limitations in cancer therapy. Recent developments in nanomedicine are expected to improve drug delivery, thereby increasing efficacy and safety while decreasing the unwanted effects of anticancer drugs in cancer therapy (Wicki et al. 2015; Xu et al. 2015).

Researchers have studied to improve on current standards in drug delivery relating to distribution, intracellular uptake, and dose efficacy by taking an advantage of nanosystems to encapsulate drugs and target sites of disease (Farokhzad and Langer 2009). Most clinically available anticancer nanomedicine are passively targeted first generation drugs that basically based on controlling the pharmacokinetics and biodistribution of a compound by modulating its physicochemical properties (Golden et al. 1998). Even though (enhanced permeability and retention EPR) effect provides the accumulation of nanodrugs in tumor tissues with passive targeting, it is inadequate to control adverse effects of cytotoxic drugs and fully exploit the benefits of targeted delivery (Wicki et al. 2015). Second generation nanomedicines are relying on drug delivery technologies with an active targeting vector or nanocarriers with stimuli-responsive properties. Hence, they possess the promise of improved targeting and increased efficacy (Farokhzad and Langer 2006). Nanoparticulate delivery with active targeting can increase drug retention in tumor tissue due to enhanced cellular binding, reduce nonspecific uptake, and resistance mechanisms (van der Meel et al. 2013).

Cyclodextrins are cyclic oligosaccharides produced from enzymatic degradation of starch with three-dimensional truncated cone or torus shape because of the chair conformation of glucopyranose unit that comprises the native cyclodextrin (Fig. 3.1) (Merisko-Liversidge et al. 2003). These cyclic oligosaccharides, containing six (α-cyclodextrin), seven (β-cyclodextrin), eight (γ-cyclodextrin), or more (α-1,4-)-linked D-glucopyranose units, consist of an apolar inner cavity and a polar external face. This property makes them soluble in aqueous medium to bind hydrophobic drugs in their cavity with noncovalent bonds (Simões et al. 2015). The physico-

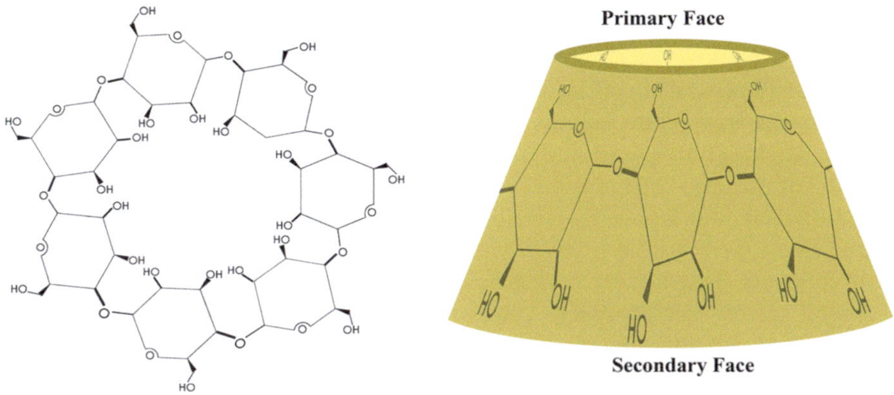

Chemical Structure of β-Cyclodextrin **3D Structure of β-Cyclodextrin**

Fig. 3.1 Chemical structure and 3 dimensional structure of β-cyclodextrin

chemical properties of cyclodextrins presented important advantages for pharmaceutical and biomedical applications.

Cyclodextrins have been extensively utilized to form inclusion complexes with drugs through host–guest interactions in pharmaceutical science and industry. The other consequential applications include enhancement of drug solubility and stability (Brewster and Loftsson 2007), enhancing drug absorption (Carrier et al. 2007), masking taste or odour (Uekama et al. 1998), controlling drug release profiles (Hirayama and Uekama 1999), reducing unwanted side effects (Davis and Brewster 2004), and improving drug permeability across biological barriers (Loftsson and Brewster 2011). Pharmaceutical formulations using cyclodextrins can be administered with in oral, nasal, ocular, rectal, parenteral and dermal applications (Loftssona and Jarvinen 1999; Matsuda and Arima 1999; Merkus et al. 1999; Carrier et al. 2007). To improve the pharmaceutical features of natural cyclodextrins relating to solubility, inclusion capability, controlled drug delivery capacity, and toxicity, chemically modified cyclodextrins have been synthesized, including highly soluble, amphiphilic, or hydrophobic derivatives (Duchene et al. 1999; Szente and Szejtli 1999). Recently, there are 30–40 marketed pharmaceutical products based on drug/cyclodextrin complexes. For instance, α-CD/Alprostadil (iv solution), SBE-β-CD/voriconazole (iv solution), β-CD/cephalosporin (tablet), 2-HPβCD/cisapride (suppository), M-β-CD/chloramphenicol (eye drop solution) and 2-HPγCD/diclofenac sodium (eye drop solution) take place in the market list in Japan, Europe and USA (Loftsson and Duchene 2007). Different cyclodextrin-containing nanopharmaceuticals have been used for patients with solid tumors in clinical trials.

3.2 Cyclodextrin-Based Nanocarriers for Cancer Treatment

The major drawback in cancer chemotherapy is the severe adverse effects of anticancer drugs on healthy tissues (Kobayashi et al. 2002; Ross and Small 2002; Sehouli et al. 2002). These side effects cause dose reduction, treatment delay, or tapering of therapy (Dharap et al. 2005). The adverse effects on healthy tissues can be significantly diminished by developing drug delivery systems with different targeting techniques for cancer cells (Minko 2004; Minko et al. 2004). As a result of this, uptake of drug was decreased by healthy cells and enhanced the influx and retention of the drug in cancer cell. Nanoparticulate drug delivery systems are promising due to their unique accumulation property at tumor site. Targeted anticancer drug delivery has severe advantages such as avoiding reticuloendothelial system, utilizing EPR effect and tumour specific targeting. Nanocarriers encapsulate, complex, or conjugate drug molecules to deliver them to the desired site (Kwon and Okano 1996; Brigger et al. 2002; Brannon-Peppas and Blanchette 2004).

New drug delivery systems overcome limitations related to conventional methods and offer effective cancer therapy with minimized adverse effects (Brigger et al. 2002; Chidambaram et al. 2011). Targeted therapy is to target antitumor drugs into cancer cells in safer and more efficient manner. In passive targeting, nanoparticles have accumulated by abnormal vascular structure and defective lymphatic function called 'Enhanced Permeability and Retention Effect'. By the help of nanosized drug delivery systems, most of the chemotherapeutic drugs that have a nonspecific distribution in both cancer and normal cells reach in cancer cells due to increased vascular permeability by the EPR effect. For this reason, active targeting (ligand-mediated targeting) was developed using targeting ligands that selectively bind surface molecules into surface of nanoparticles, they uptake by targeted tumour cells which receptors overexpressed in resulting specific internalization and uptake by targeted diseased cells while eliminating adverse effects in normal tissues (Bertrand et al. 2014; Xu et al. 2015).

Cyclodextrin-based nanocarriers include two concepts in a single delivery system that prepared in two approaches; (i) anticancer drug has a complexation with cyclodextrin and encapsulation of drug-cyclodextrin complex into carrier, (ii) drug encapsulated designed cyclodextrin-based nanocarrier (Gidwani and Vyas 2015). These system has many important advantages such as higher drug loading, prolonged blood circulation, reduced adverse effects, controlled/sustained release, increased residence time at mucosal membranes and better pharmacokinetic properties (Maeda et al. 2009). In this section, we will give synthesis, preparation and characterization of recent cyclodextrin-based nanocarriers used in cancer therapy. In addition, their applicability to deliver anticancer drugs will discuss with literature analysis.

3.2.1 Cyclodextrin Polymers

The use of cyclodextrin polymers in pharmaceutical area has been under investigation for last 30 years due to their versatility, molecular structure and molecular weight (Davis and Brewster 2004; Simões et al. 2015). Several types of cyclodextrin polymers can be synthesized in several different ways:

- Crosslinking with reagents such as epichlorohydrin, diepoxides, diisocyanates.
- Copolymerization of vinyl- or methacryoyl-modified-cyclodextrin monomers with other monomers
- Click chemistry
- Schiff-base forming (Simões et al. 2015).

The use of cyclodextrin polymers offers some advantages compared with native cyclodextrins. First, some of fabricated cyclodextrin polymers which have the capability to complex with poorly water soluble drugs due to dramatically increased inclusion formation constant when cyclodextrins are organized into supramolecular entities have been worked by the scientists (Ma and DeQuan 1999). Due to extensive use of these water-soluble cyclodextrin polymers for pharmaceutical area, Gidwani *et al.* have synthesized epichlorohydrin β-cyclodextrin (EPI-βCD) by single step condensation and polymerization reaction. EPI-βCD has been found safer and superior to native cyclodextrins in terms of complexation and solubility properties. These properties make EPI-βCD a favourable pharmaceutical excipient or drug carrier to improve physicochemical and biological properties of water-insoluble drugs (Gidwani and Vyas 2014). Cyclodextrin polymers can form self-assembling nanoparticles/gels that can encapsulate active agents resulting in higher drug loading and controlled drug release. Supramolecular gels of α-cyclodextrin-polymer (poly-αCD) with various poly(ethylene oxide)-based copolymers was reported by Simoes *et al.* to entrap vancomycine in both cyclodextrin cavity and supramolecular structure and sustained the release of vancomycine (Simões et al. 2014). In another study by di Cagno *et al.*, new synthesized βCD-dextran polymers were found to positively affect the diffusion of drug molecules in drug release studies leading to increased encapsulation efficiency and these cyclodextrin polymers have found to be suitable for both local and systemic drug delivery. Due to presence of dextran backbone and the absence of substituents that may lead to steric hindrance at the CD opening, drug-cyclodextrin complex showed higher solubilization with higher extent and higher stability compared to natural β-cyclodextrin (di Cagno et al. 2014).

Different β-cyclodextrin polyrotaxanes were synthesized and evaluated in terms of antitumor efficacy by Yu *et al.* Between hydroxy groups on polyrotaxanes and a hydrophilic spacer, a succinate-based paclitaxel ester derivative was attached covalently. These conjugate has an excellent drug loading approximately 29% and drug release can be rendered by esterase catalysis. Cellular uptake studies established that conjugate can be internalized by tumour cells and showed pharmacological activity of paclitaxel. In vivo antitumor performance, biodistribution and tumour

Table 3.1 Physicochemical properties of cyclodextrin polymer strands and Insert Therapeutics-101 (IT-101)

Parameter	Cyclodextrin polymer strand	IT-101
Particle size	8 nm	36 nm
PDI	0.323	0.277
Zeta potential	Not determined	−1.81 +/− −0.79 mV

penetration of paclitaxel conjugate proved that it has significantly superior efficacy to inhibit tumour growth and prolong survival of tumour-bearing mice compared to Taxol (Yu et al. 2013).

In another study, a new cyclodextrin polymer carrier, Poly(β-cyclodextrin triazine) (PCDT) complex can be obtained to can enhance cytotoxicity of curcumin and paclitaxel that are frequently used anticancer drugs with low solubility, poor stability and side effects in various cancer cell lines. After complexation of the curcumin and paclitaxel with cyclodextrin carrier, solubility increased approximately 3500 and 941 times higher than free paclitaxel, respectively. Cell culture studies proved both paclitaxel and its complexes have shown dose dependent cytotoxic effects, while PCDT did not show any effects on cell growth. (Boztas et al. 2013).

IT-101 (Insert Therapeutics-101), a cyclodextrin-containing polymer conjugate, is investigated for the treatment of cancer in clinic. Linear β-cyclodextrin copolymer and polyethylen glycol was synthesized following camptothecin conjugation to obtain IT-101 that has low toxicity and high solubility in water. As can be seen in Table 3.1, cyclodextrin polymer can self assemble into nanoparticles with 30–40 nm diameter and approximately neutral surface charge that is appropriate for in vivo use.

In vivo studies with tumour-bearing mice models showed longer circulation time, enhanced accumulation and higher antitumor efficacy due to extended release of anticancer drug camptothecin from cyclodextrin containing polymer (Schluep et al. 2006; Davis 2009). Based on obtained successful results of animal studies, phase I and II studies are finished and results will be published in near future.

CRLX101, another polymer conjugate of camptothecin under clinical investigation, assemblies into nanoparticles in size 30–40 nm diameter showed high antitumor efficacy and safety profile in tumour-bearing mice model in preclinical studies. Following this study (CRLX101) has also shown increased antitumor activity as compared with other camptothecin derivatives in first-in-human phase 1/2a study (Gaur et al. 2012; Weiss et al. 2013).

3.2.2 Cyclodextrin-Based/Grafted Polymeric Nanoparticles

Several studies have been carried out in order to develop cyclodextrin-based nanoparticles for cancer therapy in the literature. Nanoparticles can be classified as nanospheres or nanocapsules based on manufacturing process. To obtain nanoparticles, main preparation methods can be divided as nanoprecipitation, emulsion/solvent evaporation and detergent removal techniques (Bilensoy and Hincal 2009).

Among them, nanoprecipitation technique is the best choice due to some advantages such as simple and rapid production with mild solvents, unimodal size distribution, good reproducibility (Lemos-Senna et al. 1998).

The presence of cyclodextrins in nanoparticles can modify the physicochemical properties of drugs therefore increasing drug solubility and bioavailability, improving stability, or masking adverse side effects (Lakkakula and Macedo Krause 2014). For instance, paclitaxel is one of the most common anticancer drugs used in the treatment of breast, lung and ovarian cancer (Zhao et al. 2010; Yao et al. 2011). Despite its unique mechanism, paclitaxel exhibit several major drawbacks such as low water solubility, organic stabilizers used such as Cremophor EL, which induces serious toxicities including hypersensitivity, neurotoxicity and neuropathy (Wang et al. 2011). In a study, 6-O-capro-β-cyclodextrin and 6-N-capro-β-cyclodextrin were developed to obtain nanospheres and nanocapsules directly from the inclusion complex with a diameter of 150 nm using nanoprecipitation technique resulting in high entrapment efficiency (approximately 65%) and controlled release profile. According to cell culture studies, blank nanoparticles were non-toxic than Cremophor EL in L929 fibroblast cells. In addition, Paclitaxel-loaded cyclodextrin nanoparticles has higher anticancer activity in comparison with commercial product in MCF-7 cells (Bilensoy et al. 2008a; Bilensoy et al. 2008b).

On the other hand, oral administration of paclitaxel can be thought alternative route against intravenous administration due to many advantages like better patient compliance, low cost and development of chronic treatment regimens (Malingre et al. 2001; Peltier et al. 2006). Agueros *et al.* studied oral bioavailability of paclitaxel by encapsulated cyclodextrin complex in poly(anhydride) nanoparticles. The sustained release of drug was found to be approximately 27–33-fold higher with an oral bioavailability of more than 80% depending on bioadhesive properties of cyclodextrin nanoparticles (Agueros et al. 2010). Recently, Ye *et al.* developed novel paclitaxel/heptakis (2,6-di-*O*-methyl)-β-cyclodextrin (DM-β-CD) inclusion complex-loaded chitosan nanoparticles using ionic gelation method to decrease side effects and enhance therapeutic efficacy of drug. Paclitaxel/DM-β-CD inclusion complex dramatically enhanced the solubility of paclitaxel in water. They carried out pharmacokinetic studies in rats by intravenous injection of commercial formulation Taxol and Paclitaxel/DM-β-CD inclusion complex-loaded chitosan nanoparticles comparatively. Paclitaxel/DM-β-CD inclusion complex-loaded chitosan nanoparticles exhibited a significant increase in $AUC_{0\rightarrow 24h}$ and mean residence time by 2.7-fold and 1.4-fold, respectively. Therefore, the novel drug/DM-β-CD inclusion complex-loaded chitosan nanoparticles have given any promising results for the significantly improved delivery and controlled release of paclitaxel (Ye et al. 2015).

Although passive targeting of nanomedicines has been studied in different animal models over the last 30 years, researchers have been focused to design actively targeted nanoparticulate drug carriers based on recognition of the ligand by its target substrate (Bertrand et al. 2014). Erdogar *et al.* designed and optimized paclitaxel loaded actively targeted nanoparticles using two new folate-conjugated cyclodextrin derivatives (FCD-1 and FCD-2) synthesized by own research group. Physicochemical

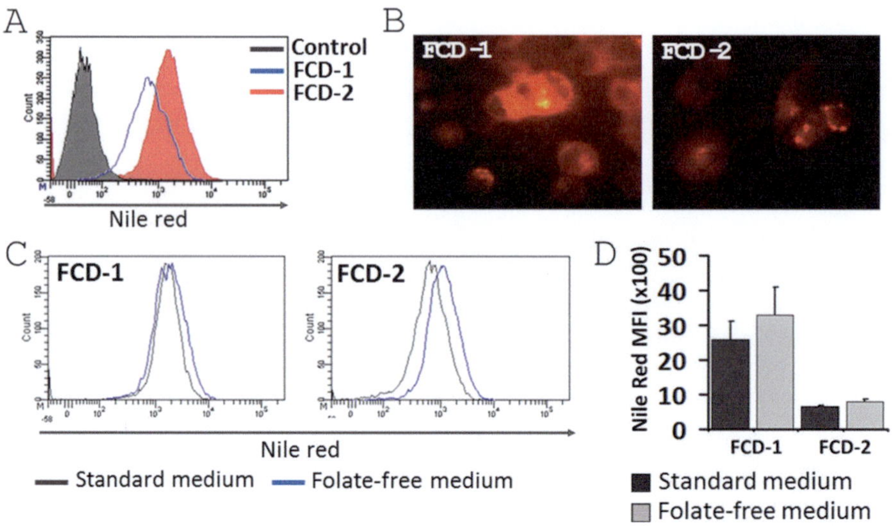

Fig. 3.2 Uptake assays of Nile red-loaded folate-conjugated cyclodextrin (FCD) nanoparticles by ZR-75-1 and T-47D breast cancer cells

characterization studies were performed using scanning electron microscopy and differential scanning calorimetric analysis. Drug release studies showed initial burst release followed by a longer sustained release. Blank nanoparticles had no cytotoxicity against L929 cells. Cytotoxicity studies demonstrated breast cancer cells became more sensitive to cytotoxic effects of paclitaxel delivered by FCD-1 and FCD-2. In cellular uptake studies, folate expression is shown by T-47D and ZR-75-1 human breast cancer cell lines at different surface levels (Fig. 3.2a). Cellular uptake of FCD-1 was much more efficient than that of FCD-2 in both cell lines (Fig. 3.2b and c). Compared to FCD-2, FCD-1 nanoparticles were more homogenously distributed through the cells with cytoplasmic and nuclear localization. (Fig. 3.2c). Due to weak competition between folate and nanoparticles by using folate free media, cellular uptake of nanoparticles was largely increased (Fig. 3.2d and e). These novel folate-conjugated cyclodextrin nanoparticles can be found as promising alternative systems for safer and effective delivery of paclitaxel with a folate-dependent mechanism (Erdogar et al. 2016).

Following that, antitumor effect of paclitaxel loaded folate-conjugated cyclodextrin nanoparticles were evaluated *in vivo* in comparison to the Cremophor EL-based paclitaxel solution by the same research group. Tumour size, survival rate, weight change and metastasis of Balb-c mice were studied after cell injection. As can be seen in Fig. 3.3A, increase of tumour diameter was the highest for mice treated with paclitaxel in Cremophor EL solution. While the longest survival rate was seen in mice treated with paclitaxel-loaded FCD-1 nanoparticles which indicates the lower toxicity of these nanoparticles, mice treated with paclitaxel-loaded FCD-2 nanoparticles has the shortest survival rate in Fig. 3.3B. Mice treated with paclitaxel (PCX)-

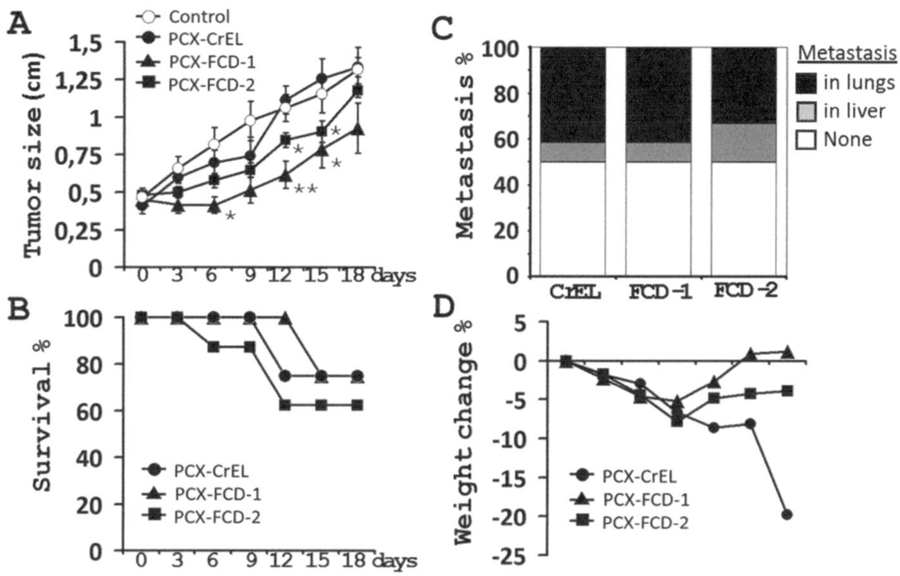

Fig. 3.3 (**A**) Geometric mean diameter (**B**) Survival rate % (**C**) % of metastasis (**D**) Weight change % of mice treated with paclitaxel in Cremophor EL solution and paclitaxel (PCX) loaded nanoparticles (n = 8/group). *Cr* Cremophor, *FCD* folate conjugated cyclodextrin

loaded FCD nanoparticles displayed a similar inhibitory effect on tumour metastasis at the in vivo level. For paclitaxel–Cremophor EL, FCD-1 nanoparticles and FCD-2 nanoparticles group, the lung metastasis foci was desirably reduced with 17%, 17% and 20% metastatic nodules on lungs, respectively. Mice treated with all treatment groups showed excellent anti-metastasis efficacy, with almost no metastatic nodules at livers as healthy liver appearance (Fig. 3.3C). Minimum weight loss was observed for mice treated with paclitaxel loaded FCD-1 nanoparticles, while mice given Cremophor showed significant toxicity throughout the study with maximum weight loss. There was a slight increase in body mass, which was a result of tumour growth in control group (Fig. 3.3D). Due to delayed release profile, prolonged circulation time and enhanced intracellular uptake, selectively localized FCD-1 nanoparticles in animal tumour tissue showed higher antitumor efficacy compared to free paclitaxel, as shown by changes such as longer survival rate, minimum weight loss and normal histopathological appearance (Erdoğar et al. 2017).

Another wide spectrum anticancer drug, camptothecin, initially isolated from the tree Camptotheca acuminata has still not used clinically due to low water solubility, high toxicity and poor stability under physiological conditions (Lorence and Nessler 2004). Camptothecin was complexed with 2-hydroxypropyl-β-cyclodextrin and nanoparticles were prepared directly by desolvation between poly(anhydride) and camptothecin:cyclodextrin inclusion complex for oral administration by Huarte *et al.* On the other hand, camptothecin was encapsulated into β-cyclodextrin-grafted α, β-poly(aspartic acid) (β-CD-*graft*-PAsp) hollow nanospheres to overcome

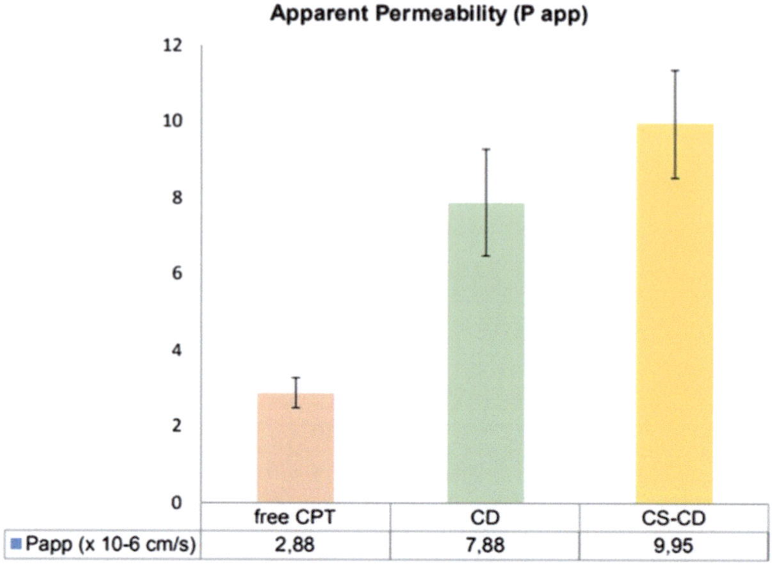

Fig. 3.4 Apparent permeability coefficient (P_{aap}) of different formulations: Camptothecin in DMSO solution, Camptothecin-loaded 6-O-CAPRO and chitosan-coated camptothecin loaded 6-O-CAPRO nanocapsules (n = 4)

problems associated with its solubility and poor physiological stability (Zeng et al. 2013; Huarte et al. 2016).

Nanocapsules are core-shell particles that have an inner liquid core surrounded by a polymeric shell (Anton et al. 2008). This carrier has many advantages such as improved loading capacity for lipophilic drugs and drug stability, control drug release and improve pharmacokinetic properties of drug (Mora-Huertas et al. 2010). Ünal *et al.* developed a novel hybrid cyclodextrin nanocapsules for oral delivery of camptothecin. An amphiphilic cyclodextrin derivative per modified on the primary face 6-O-CAPRO was used to obtain drug stability in the body and improve bioavailability. Camptothecin loaded cyclodextrin nanocapsules have been found stable in gastric fluids. Following this, permeation of drug both in solution form and in nanocapsule formulations across the Caco-2 cell monolayer is reported in Fig. 3.4 as apparent permeability coefficient. Results indicate that in terms of permeability coefficients, there is a significant difference between the formulations and free drug. Encapsulation of camptothecine in nanocapsules increased drug transport three-fold compared with free camptothecine. Intestinal uptake of cyclodextrin nanocapsules was significantly higher in vivo animal model as shown in Fig. 3.4 (Ünal et al. 2015a; Ünal et al. 2015b).

Core-shell nanocapsules were prepared to enhance targeted delivery of cyclodextrin-based supramolecules for magneto-chemotherapy by Lin *et al.* Supermolecule loaded nanocapsules accumulate at tumour sites with passive targeting and permeability of supermolecules increased resulting in effective cancer therapy (Lin et al. 2016). Novel lipid nanocapsules comprising of β-cyclodextrin-poly(4-acryloylmorpholine) conjugate as shell component were evaluated to improve poor low solubility and stability of anticancer drug 4-hydroxynonenal (HNE). Enhancement of antitumoral activity was observed in different cell lines due to increased stability of HNE loaded nanocapsules. Furthermore, promising results were found for lipid nanocapsules on a 3D human reconstructed model of skin melanoma to encourage topical application on the epidermal surface for melanoma (Pizzimenti et al. 2015).

3.2.3 *Cyclodextrin-Based Micelles*

Polymeric micelles can be used as promising carriers to increase the solubility and improve bioavailability of anticancer drugs (Deng et al. 2012). On the other hand, some micellar systems have shown detrimental effects on solubilizing ability after cyclodextrin incorporation. To overcome this problem, three different strategies were developed for the incorporation of cyclodextrins into micellar system following;

- Cyclodextrins reversibly form complexes with hydrophobic groups
- Cyclodextrins form host-guest interaction with macromolecules leading supramolecular polymer micelles
- Cyclodextrins form star-shaped unimolecular micelles or block copolymer micelles (Simões et al. 2015).

Zhang *et al.* studied intracellular pH-sensitive supramolecular block amphiphiles that self-assemble into supramolecular micelles and present pH-sensitive behaviour in acidic environment. This amphiphiles comprised of benzimidazole modified poly(ε-caprolactone) and β-cyclodextrin terminated dextran. Doxorubicin was efficaciously loaded into the supramolecular micelles through hydrophobic interactions. The doxorubicin release from micelles was increased in HepG2 human liver cancer cell line in acid conditions. Doxorubicin-loaded pH-sensitive supramolecular micelles showed higher cellular proliferation inhibition towards HepG2 human liver cancer cell line in vitro (Zhang et al. 2013).

In another study, a well-defined, cyclodextrin-conjugated amphiphilic copolymers were synthesized by combining poly(lactic acid) and monomethoxy poly(ethylene glycol) to form micelles self assembly. Transmission electron microscopy images confirmed the size of blank micelles and indomethacin loaded micelles by dynamic light scattering with mean diameter 173.4 nm and 159.2 nm, respectively. High drug loading, controlled drug release and prolonged circulation time were observed in micelles depending on inclusion complexes with β-cyclodextrin. In addition, indomethacin loaded micelles showed low cytotoxicity due to indomethacin was loaded in the core (He et al. 2013).

A new micellar vector has been synthesized using folate-conjugated poly(ethylene glycol)-poly(D,L-lactide)-β-cyclodextrin polymer. Polyethylene glycol was conjugated with β-cyclodextrin following folate capping to form an amphiphilic complex. Removal of chloroform resulted in an oil/water method in which the supramolecular amphiphilic polymer formed micelles with a mean diameter of 100–200 nm. Depending on folate mechanism, doxorubicin loaded cyclodextrin micelles showed 86% of tumour growth inhibition and reduce the cardiotoxicity of doxorubicin (Zhang et al. 2014).

As a conclusion, cyclodextrin conjugated micellar systems are seen as promising nanocarriers for various hydrophobic drugs in targeted cancer therapy with many advantages such as improved drug release, loading efficiency, toxicities in normal tissues.

3.2.4 Cyclodextrin-Based Liposomes

Liposomes are defined as spherical vesicles consisting of one or more phospholipidic bilayers organized around an aqueous inner phase. Liposomes have been used to encapsulate hydrophilic and lipophilic drugs within the aqueous and lipid layers resulting in high drug loading. In addition, encapsulation in liposomes makes better pharmacological properties and therapeutic effect of drugs. This nanosystem was also used to increase permeability and bioavailability due to structural similarity with biological membranes. However, fast release of drug from liposomes in systemic circulation and retention of lipophilic drugs in the lipid bilayer are the main problems for this nanocarrier (Lim et al. 2008; Sun et al. 2015).

Cyclodextrin liposomal drug delivery system was first used by Mc Cormack and Gregoriadis in 1994 (Mccormack and Gregoriadis 1994). This combine system has offered many advantages such as enhance drug solubility and targeting. Due to encapsulation of drug/CD inclusion complex in internal aqueous phase of liposomes, drug/lipid ratio increased and systemic toxicity reduced because of drug targeting (Maestrelli et al. 2006; Gillet et al. 2009; Cavalcanti et al. 2011). Further, dissociation and renal excretion of cyclodextrin inclusion complexes has been prevented by this system (Frank et al. 1976; McCormack and Gregoriadis 1996).

γ-cyclodextrins has shown favourable drug delivery properties such as big internal cavity, higher aqueous solubility, higher permeability and in vivo biocompatibility and thus Hydroxypropyl-γ-cyclodextrin has been used to increase higher amount of drug in liposomes (Laza-Knoerr et al. 2010). For example, incorporation of 2-hydroxypropyl-γ-cyclodextrin (2-HPγCD) in the aqueous phase during liposome formation increased docetaxel aqueous solubility and antitumor efficacy. In formulation development, burst/fast release of drug from liposomes in the systemic circulation is a critical drawback. In order to control the release rate of docetaxel, 2-HPγCD was applied to load drug and then obtained complex was incorporated in the liposome assembly. As a result of this, higher drug concentration was observed in tumour cells (Sun et al. 2015).

Chen *et al.* prepared and characterized hydroxycamptothecin loaded 2-hydroxypropyl-β-cyclodextrin inclusion liposomes. The inclusions were shown to have a sustained release compared to commercially available hydroxycamptothecin. The hydroxycamptothecin inclusion liposomes had better stability at 4 °C. These liposomes also exhibited better inhibition effect for different cancer cell lines at a dose-dependent manner (Chen et al. 2015).

In another study, Ağardan *et al.* evaluated the effectiveness of raloxifene loaded liposomes and cochleates in breast cancer therapy. Raloxifene liposomes showed higher transport in the presence of dimethyl-β-CD (DMβCD) that open tight junctions of the membranes and increase drug permeability in Caco-2 cells. Higher antitumor activity was observed with raloxifene loaded liposomes containing DMβCD (Agardan et al. 2016).

3.2.5 Cyclodextrin-Based Magnetic Nanoparticles

Photodynamic therapy (PDT) is a promising approach for improved cancer treatment. Clinical protocol comprise of a standard dose of the drug, a dose from a light source and a drug-light interval (Calixto et al. 2016). Phthalocyanins (Pc) are widely used as photosensitizers in photodynamic therapy for cancer treatment due to excellent photophysical properties (Garland et al. 2009; Mitsunaga et al. 2011). The conjugation of phtalocyanins with cyclodextrins can improve their amphiphilicity, biocompatibility and availability at the surface of cancer cell membranes (Mazzaglia et al. 2003; Sortino et al. 2006). Making use of this, Conte *et al.* developed a novel biodegradable nanoassemblies based on heptakis (2-oligo(ethyleneoxide)-6-hexadecylthio-)-β-CD (SC16OH) and zinc-phthalocyanine (ZnPc) to deliver highly lipophilic phthalocyanines. Due to amphiphilic properties, this cyclodextrin increase the poor solubility of phtalocyanines. They found potential photodynamic anticancer effects in HeLa cancer cells, suggesting a promising Zn-Pc delivery system (Conte et al. 2014). Similarly, Lourenço *et al.* prepared phthalocyanines conjugated with α-, β-, and γ-cyclodextrins. The results evidenced that Pc-α-CD and Pc-γ-CD showed a potential photoactivity against UM-UC-3 human bladder cancer cells because of cyclodextrins can damage the membrane barrier resulting in higher uptake and efficacy in cancer cells (Lourenço et al. 2014).

Mesoporous silica nanoparticles have been applied for stimuli-responsive controlled delivery for unique structural characteristics like environmentally stable mesoporous structures, large surface area, tunable pore size, and well-defined surface properties. Recently, cyclodextrins in different forms are found to keep guest molecules for the pore of silica nanoparticles until they are removed by external stimuli like pH, temperature and enzyme (Kim et al. 2010). Badruddoza *et al.* developed β-cyclodextrin conjugated magnetic, fluorescent silica core-shell nanoparticles to deliver hydrophobic drugs in cancer therapy. In many studies, these magnetic silica nanoparticles can only deliver water-soluble drugs so development of inclusion complex between carboxymethyl-β-cyclodextrin and hydrophobic drugs was thought in this study. Dynamic light scattering analyses confirmed the long-term stability and homogenous particle size distribution of nanoparticles that are necessary in biological environment. To evaluate carboxymethyl-β-cyclodextrin role in silica nanoparticles for drug loading capacity, drug inclusion and release studies were realized with retinoic acid, hydrophobic anticancer drug. β-cyclodextrin conjugated mesoporous silica nanoparticles showed significant retinoic acid adsorption compared to nanoparticles without cyclodextrin. The initial burst effect following controlled release was observed indicating more stable cyclodextrin inclusion complex. Increased cellular uptake with retinoic acid loaded β-cyclodextrin conjugated mesoporous silica nanoparticles established efficient drug delivery into tumour cells (Badruddoza et al. 2013).

On the other hand, magnetic nanoparticles (MNPs) are seen as attractive tools for biomedical applications due to have unique features such as nontoxicity, injectability, biocompatibility (Ito et al. 2005). The purpose of magnetic drug targeting in cancer therapy is to increase chemotherapeutics level in tumour tissue while the overall dose is reduced at the same time. This can be carried out with cyclodextrin conjugated magnetic nanoparticles bound to a chemotherapeutic agent. The interaction of the particles with the field gradient leads to an accumulation in tumour tissue and reduces adverse effects in healthy tissues (Alexiou et al. 2011). Otherwise, cyclodextrin conjugated magnetic nanoparticles facilitate to increase drug solubility and permeability, stability, controlled drug release (Hirayama and Uekama 1999; Brewster and Loftsson 2007; Carrier et al. 2007; Loftsson and Brewster 2011). On the other hand, iron oxide magnetic nanoparticles showed agglomeration and increased particle size because of high surface area and dipol-dipol bonds. To overcome this problem, β-cyclodextrin was conjugated to magnetic nanoparticles by Tarasi et al. FTIR and TGA studies results showed that cyclodextrin was successfully conjugated to surface of magnetic nanoparticles. Blank cyclodextrin-conjugated nanoparticles were non-toxic in HEK 293 normal kidney cell line. Docetaxel loaded β-cyclodextrin conjugated nanoparticles showed higher cytotoxicity in HeLa and MDA-MB-231 cell lines than free docetaxel depending on increased cellular uptake (Tarasi et al. 2016).

3.2.6 Cyclodextrin-Based si-RNA (Short Interfering RNA) Delivery System

siRNA based delivery systems have provided as potential for the treatment of a wide range of incurable diseases such as cancer. However, the main problem is the lack of suitable delivery vector for siRNA because of naked siRNA has a short half-life, poor cellular uptake and poor pharmacokinetic properties (Guo et al. 2011). Thus, delivery systems need to be developed that can improve siRNA stability, protect from degradation, avoid immune response and delivery siRNA in target cells (Elsabahy et al. 2011).

Polymer-based siRNA delivery systems are defined as cyclodextrin-containing polycation nanoparticles that self-assemble with siRNA to form colloidal particles with a mean diameter of 50 nm. For in vivo application, these particles are stabilized by polyethylen glycol (PEG) coating that occurs via inclusion complex formation between the terminal adamantanes and the cyclodextrins. For non-targeted in vivo delivery, cyclodextrin containing polycation nanoparticles and the conjugates of adamantane-polyethylen glycol were combined to generate stable but nontargeted polyplexes. For targeted in vivo delivery, some of the polyethylen glycol chains contain targeting ligands such as transferrin for specific interactions with cell-surface receptors (Hu-Lieskovan et al. 2005).

The star-shaped copolymers that contain cyclodextrin core and cationic arms can used to co-load hydrophobic drugs like docetaxel for obtaining stable complexes in vivo. Micellization process is not required to deliver siRNA and docetaxel by new cyclodextrin derivative. Better blood compatibility and lower toxicity has shown with this star-shaped copolymers resulting in promising drug and gene delivery system in nasopharyngeal cancer therapy (Liu et al. 2014).

Cationic cyclodextrin derivatives have been applied to form inclusion complex with hydrophobic molecule adamantane to deliver siRNA in prostate cancer therapy. A well as cationic cyclodextrins increase the half-life of siRNA due to prevent binding nanoparticles to serum proteins and escape from mononuclear phagocyte system. The prepared nanocomplexes with approximately 300 nm in size protected siRNA from degradation and were non-toxic in prostate cancer cells. Further, higher siRNA cellular uptake has evidence to the ability of cyclodextrins to form inclusion complexes with adamantane in prostate cancer cells. This nanoformulation has shown as a safe and specific-targeted delivery system for siRNA in the treatment of solid tumours (Fitzgerald et al. 2016).

Another strategy is to develop amphiphilic cyclodextrin nanoparticles using folate-targeted cyclodextrin that has an ability to deliver siRNA in tumour tissues. Folate-targeted nanoparticles enter prostate cancer cell line via the prostate specific membrane antigen (PSMA), which is overexpressed in cancer cells. They have shown to increased cellular uptake and circulation time of siRNA eight-fold (Evans et al. 2016).

The PEGylation in nanoparticulate systems have seen the best choice because of prevent the opsonization. However, optimization of PEG density and incorporation of targeting ligand is critical to increase cellular uptake of specifically-targeted nanoparticles (Hak et al. 2012). Ohyama *et al.* prepared folate-PEG-appended dendrimer (generation 4)/α-cyclodextrin conjugate to evaluate the ability of siRNA carrier in vitro and in vivo. This novel system have been found as a promising carrier as a result of higher stability, higher blood circulating ability and in vivo antitumor activity (Ohyama et al. 2016).

3.3 Conclusion

While cyclodextrins and their derivatives have already been widely used as excipients in the pharmaceutical field, the use of their derivatives for composing supramolecular assemblies for biomedical applications has emerged only recently. In view of recent studies, it is obvious that novel nanosystems based on cyclodextrin derivatives have been considered to be promising carriers for drug delivery as well as other pharmaceutical applications, in view of their successful properties for tumour targeting.

Cyclodextrin based nanocarriers have many advantages following prolonged circulation in bloodstream, higher drug loading, unique surface morphology, favorable particle size for passive targeting as well as easy modification capability for active targeting. Cyclodextrin based supramolecular systems provide enhanced solubility, functionalizing flexibility, and recognition capability for both the drug payload and the therapeutic target, as partly demonstrated by intensive studies on nanomedicines assembled from CD-containing polymers. A new cyclodextrin chemistry and self-assembly properties make it possible to develop nanocarriers such as nanoparticles, micelles, liposomes, magnetic nanoparticles as well as host–guest interactions using cyclodextrins and their derivatives.

The use of cyclodextrin based nanosized drug delivery systems shown to be promising and innovative approach that provide effective cancer therapy. On the other hand, one of the limiting factors is toxicity observed on in vitro cell culture studies. The other one is pharmacokinetic outcome for cyclodextrin-based nanocarriers in the body. Moreover, administration route and mechanism of elimination need to be further elucidated. In addition to this, systemic biocompatibility evaluation and toxicological study is necessary for candidate products. Last, scale up of cyclodextrin derivatives with high purity and yield and cost are the limiting factors for developing of new drug delivery systems.

In spite of these challenges, the future of CD-based nanosystems in drug delivery is promising, in view of the notable clinical success of new pharmaceuticals based on CDs and their derivatives, and cyclodextrin polymers as well as other controlled delivery systems.

Acknowledgements Authors would like to acknowledge TUBITAK Scientific Research Project 112S538 for financial support of this study. Nazlı Erdoğar has been supported by TUBITAK in the framework of 2214-A Foreign Postgraduate Research Scholarship Programme for her studies in Aalborg University, Denmark.

References

Agardan NB, Degim Z, Yilmaz S, Altintas L, Topal T (2016) The effectiveness of Raloxifene-loaded liposomes and Cochleates in breast cancer therapy. AAPS PharmSciTech 17:968–977. https://doi.org/10.1208/s12249-015-0429-3

Agueros M, Zabaleta V, Espuelas S, Campanero MA, Irache JM (2010) Increased oral bioavailability of paclitaxel by its encapsulation through complex formation with cyclodextrins in poly(anhydride) nanoparticles. J Control Release 145:2–8. https://doi.org/10.1016/j.jconrel.2010.03.012

Alexiou C, Tietze R, Schreiber E, Jurgons R, Richter H, Trahms L, Rahn H, Odenbach S, Lyer S (2011) Cancer therapy with drug loaded magnetic nanoparticles-magnetic drug targeting. J Magn Magn Mater 323:1404–1407. https://doi.org/10.1016/j.jmmm.2010.11.059

Anton N, Benoit JP, Saulnier P (2008) Design and production of nanoparticles formulated from nano-emulsion templates-a review. J Control Release 128:185–199. https://doi.org/10.1016/j.jconrel.2008.02.007

Badruddoza AZ, Rahman MT, Ghosh S, Hossain MZ, Shi J, Hidajat K, Uddin MS (2013) β-Cyclodextrin conjugated magnetic, fluorescent silica core-shell nanoparticles for biomedical applications. Carbohydr Polym 95:449–457. https://doi.org/10.1016/j.carbpol.2013.02.046

Bertrand N, Wu J, Xu X, Kamaly N, Farokhzad OC (2014) Cancer nanotechnology: the impact of passive and active targeting in the era of modern cancer biology. Adv Drug Deliv Rev 66:2–25. https://doi.org/10.1016/j.addr.2013.11.009

Bilensoy E, Hincal AA (2009) Recent advances and future directions in amphiphilic cyclodextrin nanoparticles. Expert Opin Drug Deliv 6:1161–1173. https://doi.org/10.1517/17425240903222218

Bilensoy E, Gurkaynak O, Dogan AL, Hincal AA (2008a) Safety and efficacy of amphiphilic beta-cyclodextrin nanoparticles for paclitaxel delivery. Int J Pharm 347:163–170. https://doi.org/10.1016/j.ijpharm.2007.06.051

Bilensoy E, Gurkaynak O, Ertan M, Sen M, Hincal AA (2008b) Development of nonsurfactant cyclodextrin nanoparticles loaded with anticancer drug paclitaxel. J Pharm Sci 97:1519–1529. https://doi.org/10.1002/jps.21111

Boztas AO, Karakuzu O, Galante G, Ugur Z, Kocabas F, Altuntas CZ, Yazaydin AO (2013) Synergistic interaction of paclitaxel and curcumin with cyclodextrin polymer complexation in human cancer cells. Mol Pharm 10:2676–2683. https://doi.org/10.1021/mp400101k

Brannon-Peppas L, Blanchette JO (2004) Nanoparticle and targeted systems for cancer therapy. Adv Drug Deliv Rev 56:1649–1659. https://doi.org/10.1016/j.addr.2004.02.014

Brewster ME, Loftsson T (2007) Cyclodextrins as pharmaceutical solubilizers. Adv Drug Deliv Rev 59:645–666. https://doi.org/10.1016/j.addr.2007.05.012

Brigger I, Dubernet C, Couvreur P (2002) Nanoparticles in cancer therapy and diagnosis. Adv Drug Deliv Rev 54:631–651

Calixto GM, Bernegossi J, de Freitas LM, Fontana CR, Chorilli M (2016) Nanotechnology-based drug delivery systems for photodynamic therapy of cancer: a review. Molecules 21:342. https://doi.org/10.3390/molecules21030342

Carrier RL, Miller LA, Ahmed I (2007) The utility of cyclodextrins for enhancing oral bioavailability. J Control Release 123:78–99. https://doi.org/10.1016/j.jconrel.2007.07.018

Cavalcanti IMF, Mendonca EAM, Lira MCB, Honrato SB, Camara CA, Amorim RVS, Mendes J, Rabello MM, Hernandes MZ, Ayala AP, Santos-Magalhaes NS (2011) The encapsulation of beta-lapachone in 2-hydroxypropyl-beta-cyclodextrin inclusion complex into liposomes: a physicochemical evaluation and molecular modeling approach. Eur J Pharm Sci 44:332–340. https://doi.org/10.1016/j.ejps.2011.08.011

Chen Y, Chen C, Xiao Y, Zhang X, Chen Y (2015) Liposomes encapsulating 10-Hydroxycamptothecin-cyclodextrin complexes and their in vitro anti-tumor activities. J Nanosci Nanotechnol 15:3786–3795

Chidambaram M, Manavalan R, Kathiresan K (2011) Nanotherapeutics to overcome conventional cancer chemotherapy limitations. J Pharm Pharm Sci 14:67–77

Conte C, Scala A, Siracusano G, Leone N, Patane S, Ungaro F, Miro A, Sciortino MT, Quaglia F, Mazzaglia A (2014) Nanoassembly of an amphiphilic cyclodextrin and Zn(II)-phthalocyanine with the potential for photodynamic therapy of cancer. RSC Adv 4:43903–43911. https://doi.org/10.1039/c4ra07847k

Davis ME (2009) Design and development of IT-101, a cyclodextrin-containing polymer conjugate of camptothecin. Adv Drug Deliv Rev 61:1189–1192. https://doi.org/10.1016/j.addr.2009.05.005

Davis ME, Brewster ME (2004) Cyclodextrin-based pharmaceutics: past, present and future. Nat Rev Drug Discov 3:1023–1035. https://doi.org/10.1038/nrd1576

Deng C, Jiang Y, Cheng R, Meng F, Zhong Z (2012) Biodegradable polymeric micelles for targeted and controlled anticancer drug delivery: promises, progress and prospects. NanoToday 7:467–480

Dharap SS, Wang Y, Chandna P, Khandare JJ, Qiu B, Gunaseelan S, Sinko PJ, Stein S, Farmanfarmaian A, Minko T (2005) Tumor-specific targeting of an anticancer drug delivery system by LHRH peptide. Proc Natl Acad Sci U S A 102:12962–12967. https://doi.org/10.1073/pnas.0504274102

di Cagno M, Terndrup Nielsen T, Lambertsen Larsen K, Kuntsche J, Bauer-Brandl A (2014) β-Cyclodextrin-dextran polymers for the solubilization of poorly soluble drugs. Int J Pharm 468:258–263. https://doi.org/10.1016/j.ijpharm.2014.04.029

Duchene D, Ponchel G, Wouessidjewe D (1999) Cyclodextrins in targeting. Application to nanoparticles. Adv Drug Deliv Rev 36:29–40

Elsabahy M, Nazarali A, Foldvari M (2011) Non-viral nucleic acid delivery: key challenges and future directions. Curr Drug Deliv 8:235–244

Erdogar N, Esendagli G, Nielsen TT, Sen M, Oner L, Bilensoy E (2016) Design and optimization of novel paclitaxel-loaded folate-conjugated amphiphilic cyclodextrin nanoparticles. Int J Pharm 509:375–390. https://doi.org/10.1016/j.ijpharm.2016.05.040

Erdoğar N, Esendağlı G, Nielsen TT, Esendaglı-Yılmaz G, Yöyen-Ermiş D, Erdoğdu B, Sargon MF, Eroglu H, Bilensoy E (2017) Therapeutic efficacy of folate receptor-targeted amphiphilic cyclodextrin nanoparticles as a novel vehicle for paclitaxel delivery in breast cancer. J Drug Target. https://doi.org/10.1080/1061186X.2017.1339194

Evans JC, Malhotra M, Guo J, O'Shea JP, Hanrahan K, O'Neill A, Landry WD, Griffin BT, Darcy R, Watson RW, O'Driscoll CM (2016) Folate-targeted amphiphilic cyclodextrin.siRNA nanoparticles for prostate cancer therapy exhibit PSMA mediated uptake, therapeutic gene silencing in vitro and prolonged circulation in vivo. Nanomedicine 12:2341–2351. https://doi.org/10.1016/j.nano.2016.06.014

Farokhzad OC, Langer R (2006) Nanomedicine: developing smarter therapeutic and diagnostic modalities. Adv Drug Deliv Rev 58:1456–1459. https://doi.org/10.1016/j.addr.2006.09.011

Farokhzad OC, Langer R (2009) Impact of nanotechnology on drug delivery. ACS Nano 3:16–20. https://doi.org/10.1021/nn900002m

Fitzgerald KA, Malhotra M, Gooding M, Sallas F, Evans JC, Darcy R, O'Driscoll CM (2016) A novel, anisamide-targeted cyclodextrin nanoformulation for siRNA delivery to prostate cancer cells expressing the sigma-1 receptor. Int J Pharm 499:131–145. https://doi.org/10.1016/j.ijpharm.2015.12.055

Frank DW, Gray JE, Weaver RN (1976) Cyclodextrin nephrosis in the rat. Am J Pathol 83:367–382

Garland MJ, Cassidy CM, Woolfson D, Donnelly RF (2009) Designing photosensitizers for photodynamic therapy: strategies, challenges and promising developments. Future Med Chem 1:667–691. https://doi.org/10.4155/fmc.09.55

Gaur S, Chen L, Yen T, Wang Y, Zhou B, Davis M, Yen Y (2012) Preclinical study of the cyclodextrin-polymer conjugate of camptothecin CRLX101 for the treatment of gastric cancer. Nanomedicine 8:721–730. https://doi.org/10.1016/j.nano.2011.09.007

Gidwani B, Vyas A (2014) Synthesis, characterization and application of epichlorohydrin-beta-cyclodextrin polymer. Colloids Surf B Biointerfaces 114:130–137. https://doi.org/10.1016/j.colsurfb.2013.09.035

Gidwani B, Vyas A (2015) A comprehensive review on cyclodextrin-based carriers for delivery of chemotherapeutic cytotoxic anticancer drugs. Biomed Res Int 2015:198268. https://doi.org/10.1155/2015/198268

Gillet A, Grammenos A, Compere P, Evrard B, Piel G (2009) Development of a new topical system: drug-in-cyclodextrin-in-deformable liposome. Int J Pharm 380:174–180. https://doi.org/10.1016/j.ijpharm.2009.06.027

Golden PL, Huwyler J, Pardridge WM (1998) Treatment of large solid tumors in mice with daunomycin-loaded sterically stabilized liposomes. Drug Deliv 5:207–212. https://doi.org/10.3109/10717549809052036

Guo J, Bourre L, Soden DM, O'Sullivan GC, O'Driscoll C (2011) Can non-viral technologies knockdown the barriers to siRNA delivery and achieve the next generation of cancer therapeutics? Biotechnol Adv 29:402–417. https://doi.org/10.1016/j.biotechadv.2011.03.003

Hak S, Helgesen E, Hektoen HH, Huuse EM, Jarzyna PA, Mulder WJ, Haraldseth O, Davies Cde L (2012) The effect of nanoparticle polyethylene glycol surface density on ligand-directed tumor targeting studied in vivo by dual modality imaging. ACS Nano 6:5648–5658. https://doi.org/10.1021/nn301630n

He Q, Wu W, Xiu KM, Zhang Q, Xu FJ, Li JS (2013) Controlled drug release system based on cyclodextrin-conjugated poly(lactic acid)-b-poly(ethylene glycol) micelles. Int J Pharm 443:110–119. https://doi.org/10.1016/j.ijpharm.2012.12.042

Hirayama F, Uekama K (1999) Cyclodextrin-based controlled drug release system. Adv Drug Deliv Rev 36:125–141

Huarte J, Espuelas S, Lai Y, He B, Tang J, Irache JM (2016) Oral delivery of camptothecin using cyclodextrin/poly(anhydride) nanoparticles. Int J Pharm 506:116–128. https://doi.org/10.1016/j.ijpharm.2016.04.045

Hu-Lieskovan S, Heidel JD, Bartlett DW, Davis ME, Triche TJ (2005) Sequence-specific knockdown of EWS-FLI1 by targeted, nonviral delivery of small interfering RNA inhibits tumor growth in a murine model of metastatic Ewing's sarcoma. Cancer Res 65:8984–8992. https://doi.org/10.1158/0008-5472.CAN-05-0565

Ito A, Shinkai M, Honda H, Kobayashi T (2005) Medical application of functionalized magnetic nanoparticles. J Biosci Bioeng 100:1–11. https://doi.org/10.1263/jbb.100.1

Kim H, Kim S, Park C, Lee H, Park HJ, Kim C (2010) Glutathione-induced intracellular release of guests from mesoporous silica nanocontainers with cyclodextrin gatekeepers. Adv Mater 22:4280–4283. https://doi.org/10.1002/adma.201001417

Kobayashi M, Wood PA, Hrushesky WJ (2002) Circadian chemotherapy for gynecological and genitourinary cancers. Chronobiol Int 19:237–251

Kwon GS, Okano T (1996) Polymeric micelles as new drug carriers. Adv Drug Deliv Rev 21:107–116. https://doi.org/10.1016/S0169-409x(96)00401-2

Lakkakula JR, Macedo Krause RW (2014) A vision for cyclodextrin nanoparticles in drug delivery systems and pharmaceutical applications. Nanomedicine (Lond) 9:877–894. https://doi.org/10.2217/nnm.14.41

Laza-Knoerr AL, Gref R, Couvreur P (2010) Cyclodextrins for drug delivery. J Drug Target 18:645–656. https://doi.org/10.3109/10611861003622552

Lemos-Senna E, Wouessidjewe D, Lesieur S, Puisieux F, Couarraze G, Duchene D (1998) Evaluation of the hydrophobic drug loading characteristics in nanoprecipitated amphiphilic cyclodextrin nanospheres. Pharm Dev Technol 3:85–94. https://doi.org/10.3109/10837459809028482

Lim HJ, Cho EC, Shim J, Kim DH, An EJ, Kim J (2008) Polymer-associated liposomes as a novel delivery system for cyclodextrin-bound drugs. J Colloid Interface Sci 320:460–468. https://doi.org/10.1016/j.jcis.2008.01.025

Lin IC, Fang JH, Lin CT, Sung SY, Su YL, Hu SH (2016) Enhanced targeted delivery of cyclodextrin-based supermolecules by core-shell nanocapsules for magnetothermal chemotherapy. Macromol Biosci 16:1273–1286. https://doi.org/10.1002/mabi.201600131

Liu T, Xue W, Ke B, Xie MQ, Ma D (2014) Star-shaped cyclodextrin-poly(l-lysine) derivative co-delivering docetaxel and MMP-9 siRNA plasmid in cancer therapy. Biomaterials 35:3865–3872. https://doi.org/10.1016/j.biomaterials.2014.01.040

Loftsson T, Brewster ME (2011) Pharmaceutical applications of cyclodextrins: effects on drug permeation through biological membranes. J Pharm Pharmacol 63:1119–1135. https://doi.org/10.1111/j.2042-7158.2011.01279.x

Loftsson T, Duchene D (2007) Cyclodextrins and their pharmaceutical applications. Int J Pharm 329:1–11. https://doi.org/10.1016/j.ijpharm.2006.10.044

Loftssona T, Jarvinen T (1999) Cyclodextrins in ophthalmic drug delivery. Adv Drug Deliv Rev 36:59–79

Lorence A, Nessler CL (2004) Camptothecin, over four decades of surprising findings. Phytochemistry 65:2735–2749. https://doi.org/10.1016/j.phytochem.2004.09.001

Lourenço LM, Pereira PM, Maciel E, Valega M, Domingues FM, Domingues MR, Neves MG, Cavaleiro JA, Fernandes R, Tome JP (2014) Amphiphilic phthalocyanine-cyclodextrin conjugates for cancer photodynamic therapy. Chem Commun (Camb) 50:8363–8366. https://doi.org/10.1039/c4cc02226b

Ma M, DeQuan L (1999) New organic nanoporous polymers and their inclusion complexes. Chem Mater 11:872–874

Maeda H, Bharate GY, Daruwalla J (2009) Polymeric drugs for efficient tumor-targeted drug delivery based on EPR-effect. Eur J Pharm Biopharm 71:409–419. https://doi.org/10.1016/j.ejpb.2008.11.010

Maestrelli F, Gonzalez-Rodriguez ML, Rabasco AM, Mura P (2006) Effect of preparation technique on the properties of liposomes encapsulating ketoprofen-cyclodextrin complexes aimed for transdermal delivery. Int J Pharm 312:53–60. https://doi.org/10.1016/j.ijpharm.2005.12.047

Malingre MM, Beijnen JH, Schellens JH (2001) Oral delivery of taxanes. Investig New Drugs 19:155–162

Matsuda H, Arima H (1999) Cyclodextrins in transdermal and rectal delivery. Adv Drug Deliv Rev 36:81–99

Mazzaglia A, Angelini N, Darcy R, Donohue R, Lombardo D, Micali N, Sciortino MT, Villari V, Scolaro LM (2003) Novel heterotopic colloids of anionic porphyrins entangled in cationic amphiphilic cyclodextrins: spectroscopic investigation and intracellular delivery. Chemistry 9:5762–5769. https://doi.org/10.1002/chem.200304861

Mccormack B, Gregoriadis G (1994) Drugs-in-cyclodextrins-in-liposomes - a novel concept in drug-delivery. Int J Pharm 112:249–258. https://doi.org/10.1016/0378-5173(94)90361-1

McCormack B, Gregoriadis G (1996) Comparative studies of the fate of free and liposome-entrapped hydroxypropyl-beta-cyclodextrin/drug complexes after intravenous injection into rats: implications in drug delivery. Biochim Biophys Acta 1291:237–244

Merisko-Liversidge E, Liversidge GG, Cooper ER (2003) Nanosizing: a formulation approach for poorly-water-soluble compounds. Eur J Pharm Sci 18:113–120

Merkus FW, Verhoef JC, Marttin E, Romeijn SG, van der Kuy PH, Hermens WA, Schipper NG (1999) Cyclodextrins in nasal drug delivery. Adv Drug Deliv Rev 36:41–57

Minko T (2004) Drug targeting to the colon with lectins and neoglycoconjugates. Adv Drug Deliv Rev 56:491–509. https://doi.org/10.1016/j.addr.2003.10.017

Minko T, Dharap SS, Pakunlu RI, Wang Y (2004) Molecular targeting of drug delivery systems to cancer. Curr Drug Targets 5:389–406

Mitsunaga M, Ogawa M, Kosaka N, Rosenblum LT, Choyke PL, Kobayashi H (2011) Cancer cell-selective *in vivo* near infrared photoimmunotherapy targeting specific membrane molecules. Nat Med 17:1685–1691. https://doi.org/10.1038/nm.2554

Mora-Huertas CE, Fessi H, Elaissari A (2010) Polymer-based nanocapsules for drug delivery. Int J Pharm 385:113–142. https://doi.org/10.1016/j.ijpharm.2009.10.018

Ohyama A, Higashi T, Motoyama K, Arima H (2016) In vitro and in vivo tumor-targeting siRNA delivery using folate-PEG-appended dendrimer (G4)/α-cyclodextrin conjugates. Bioconjug Chem 27:521–532. https://doi.org/10.1021/acs.bioconjchem.5b00545

Peltier S, Oger JM, Lagarce F, Couet W, Benoit JP (2006) Enhanced oral paclitaxel bioavailability after administration of paclitaxel-loaded lipid nanocapsules. Pharm Res 23:1243–1250. https://doi.org/10.1007/s11095-006-0022-2

Pizzimenti S, Daga M, Ciamporcero E, Toaldo C, Pettazzoni P, Osella-Abate S, Novelli M, Minelli R, Bisazza A, Gamba P, Testa G, Ullio C, Ferruti P, Ranucci E, Bernengo MG, Ferretti C, Dianzani C, Biasi F, Barrera G, Cavalli R (2015) Improved anti-tumoral therapeutic efficacy of 4-hydroxynonenal incorporated in novel lipid nanocapsules in 2D and 3D models. J Biomed Nanotechnol 11:2169–2185. https://doi.org/10.1166/jbn.2015.2131

Ross RW, Small EJ (2002) Osteoporosis in men treated with androgen deprivation therapy for prostate cancer. J Urol 167:1952–1956

Schluep T, Hwang J, Cheng J, Heidel JD, Bartlett DW, Hollister B, Davis ME (2006) Preclinical efficacy of the camptothecin-polymer conjugate IT-101 in multiple cancer models. Clin Cancer Res 12:1606–1614. https://doi.org/10.1158/1078-0432.CCR-05-1566

Sehouli J, Stengel D, Elling D, Ortmann O, Blohmer J, Riess H, Lichtenegger W, Ovarian Cancer Study Group of the Nordostdeutsche Gesellschaft für Gynäkologische Onkologie (NOGGO) (2002) First-line chemotherapy with weekly paclitaxel and carboplatin for advanced ovarian cancer: a phase I study. Gynecol Oncol 85:321–326. https://doi.org/10.1006/gyno.2002.6623

Simões SM, Veiga F, Ribeiro AC, Figueiras AR, Taboada P, Concheiro A, Alvarez-Lorenzo C (2014) Supramolecular gels of poly-α-cyclodextrin and PEO-based copolymers for controlled drug release. Eur J Pharm Biopharm 87:579–588. https://doi.org/10.1016/j.ejpb.2014.04.006

Simões SM, Rey-Rico A, Concheiro A, Alvarez-Lorenzo C (2015) Supramolecular cyclodextrin-based drug nanocarriers. Chem Commun (Camb) 51:6275–6289. https://doi.org/10.1039/c4cc10388b

Sortino S, Mazzaglia A, Scolaro LM, Merlo FM, Valveri V, Sciortino MT (2006) Nanoparticles of cationic amphiphilic cyclodextrins entangling anionic porphyrins as carrier-sensitizer system in photodynamic cancer therapy. Biomaterials 27:4256–4265. https://doi.org/10.1016/j.biomaterials.2006.03.035

Sun L, Zhou DS, Zhang P, Li QH, Liu P (2015) Gemcitabine and gamma-cyclodextrin/docetaxel inclusion complex-loaded liposome for highly effective combinational therapy of osteosarcoma. Int J Pharm 478:308–317. https://doi.org/10.1016/j.ijpharm.2014.11.052

Szente L, Szejtli J (1999) Highly soluble cyclodextrin derivatives: chemistry, properties, and trends in development. Adv Drug Deliv Rev 36:17–28

Tarasi R, Khoobi M, Niknejad H, Ramazani A, Ma'mani L, Bahadorikhalili S, Shafiee A (2016) Beta-cyclodextrin functionalized poly (5-amidoisophthalicacid) grafted Fe_3O_4 magnetic nanoparticles: a novel biocompatible nanocomposite for targeted docetaxel delivery. J Magn Magn Mater 417:451–459. https://doi.org/10.1016/j.jmmm.2016.05.080

Uekama K, Hirayama F, Irie T (1998) Cyclodextrin drug carrier systems. Chem Rev 98:2045–2076

Ünal H, d'Angelo I, Pagano E, Borrelli F, Izzo A, Ungaro F, Quaglia F, Bilensoy E (2015a) Core-shell hybrid nanocapsules for oral delivery of camptothecin: formulation development, in vitro and in vivo evaluation. J Nanopart Res 17:42. https://doi.org/10.1007/s11051–014–2838-8

Ünal H, Ozturk N, Bilensoy E (2015b) Formulation development, stability and anticancer efficacy of core-shell cyclodextrin nanocapsules for oral chemotherapy with camptothecin. Beilstein J Org Chem 11:204–212. https://doi.org/10.3762/bjoc.11.22

van der Meel R, Vehmeijer LJ, Kok RJ, Storm G, van Gaal EV (2013) Ligand-targeted particulate nanomedicines undergoing clinical evaluation: current status. Adv Drug Deliv Rev 65:1284–1298. https://doi.org/10.1016/j.addr.2013.08.012

Wang Y, Li X, Wang L, Xu Y, Cheng X, Wei P (2011) Formulation and pharmacokinetic evaluation of a paclitaxel nanosuspension for intravenous delivery. Int J Nanomedicine 6:1497–1507. https://doi.org/10.2147/IJN.S21097

Weiss GJ, Chao J, Neidhart JD, Ramanathan RK, Bassett D, Neidhart JA, Choi CH, Chow W, Chung V, Forman SJ, Garmey E, Hwang J, Kalinoski DL, Koczywas M, Longmate J, Melton RJ, Morgan R, Oliver J, Peterkin JJ, Ryan JL, Schluep T, Synold TW, Twardowski P, Davis ME, Yen Y (2013) First-in-human phase 1/2a trial of CRLX101, a cyclodextrin-containing polymer-camptothecin nanopharmaceutical in patients with advanced solid tumor malignancies. Investig New Drugs 31:986–1000. https://doi.org/10.1007/s10637-012-9921-8

Wicki A, Witzigmann D, Balasubramanian V, Huwyler J (2015) Nanomedicine in cancer therapy: challenges, opportunities, and clinical applications. J Control Release 200:138–157. https://doi.org/10.1016/j.jconrel.2014.12.030

Xu X, Ho W, Zhang X, Bertrand N, Farokhzad O (2015) Cancer nanomedicine: from targeted delivery to combination therapy. Trends Mol Med 21:223–232. https://doi.org/10.1016/j.molmed.2015.01.001

Yao HJ, Ju RJ, Wang XX, Zhang Y, Li RJ, Yu Y, Zhang L, Lu WL (2011) The antitumor efficacy of functional paclitaxel nanomicelles in treating resistant breast cancers by oral delivery. Biomaterials 32:3285–3302. https://doi.org/10.1016/j.biomaterials.2011.01.038

Ye YJ, Wang Y, Lou KY, Chen YZ, Chen R, Gao F (2015) The preparation, characterization, and pharmacokinetic studies of chitosan nanoparticles loaded with paclitaxel/dimethyl-beta-cyclodextrin inclusion complexes. Int J Nanomedicine 10:4309–4319. https://doi.org/10.2147/IJN.S83508

Yu S, Zhang Y, Wang X, Zhen X, Zhang Z, Wu W, Jiang X (2013) Synthesis of paclitaxel-conjugated β-cyclodextrin polyrotaxane and its antitumor activity. Angew Chem Int Ed Engl 52:7272–7277. https://doi.org/10.1002/anie.201301397

Zeng J, Huang H, Liu S, Xu H, Huang J, Yu J (2013) Hollow nanosphere fabricated from β-cyclodextrin-grafted α,β-poly(aspartic acid) as the carrier of camptothecin. Colloids Surf B Biointerfaces 105:120–127. https://doi.org/10.1016/j.colsurfb.2012.12.024

Zhang Z, Ding JX, Chen XF, Xiao CS, He CL, Zhuang XL, Chen L, Chen XS (2013) Intracellular pH-sensitive supramolecular amphiphiles based on host-guest recognition between benzimidazole and beta-cyclodextrin as potential drug delivery vehicles. Polym Chem 4:3265–3271. https://doi.org/10.1039/c3py00141e

Zhang L, Lu JF, Jin YM, Qiu LY (2014) Folate-conjugated beta-cyclodextrin-based polymeric micelles with enhanced doxorubicin antitumor efficacy. Colloids Surf B Biointerfaces 122:260–269. https://doi.org/10.1016/j.colsurfb.2014.07.005

Zhao D, Zhao X, Zu Y, Li J, Zhang Y, Jiang R, Zhang Z (2010) Preparation, characterization, and *in vitro* targeted delivery of folate-decorated paclitaxel-loaded bovine serum albumin nanoparticles. Int J Nanomedicine 5:669–677

Chapter 4
Cyclodextrins for Essential Oils Applications

Miriana Kfoury, Lizette Auezova, Hélène Greige-Gerges, and Sophie Fourmentin

Contents

M. Kfoury (✉)
Unité de Chimie Environnementale et Interactions sur le Vivant (UCEIV, EA 4492), Dunkerque, France

Bioactive Molecules Research Laboratory, Faculty of Sciences, Lebanese University, Beirut, Lebanon

L. Auezova · H. Greige-Gerges
Bioactive Molecules Research Laboratory, Faculty of Sciences, Lebanese University, Beirut, Lebanon

S. Fourmentin
Unité de Chimie Environnementale et Interactions sur le Vivant (UCEIV, EA 4492), Dunkerque, France

S. Fourmentin et al. (eds.), *Cyclodextrin Applications in Medicine, Food, Environment and Liquid Crystals*, Environmental Chemistry for a Sustainable World, https://doi.org/10.1007/978-3-319-76162-6_4

Abstract There is a growing interest for the aromatic and biological properties of essential oils, as alternatives to synthetic chemicals. Nonetheless, essential oils and their components are poorly soluble in water, and are highly sensitive to degradation and evaporation. Encapsulation in cyclodextrins can reduce these drawbacks and improve the properties of essential oils. Cyclodextrins are non-toxic cyclic oligosaccharides obtained by enzymatic degradation of starch. Cyclodextrins inclusion complexes find applications in the food, pharmaceutical and cosmetic industries.

This chapter reviews encapsulation of essential oils in cyclodextrins. The strength of binding between cyclodextrins and essential oils components covers a wide range of formation constants with values ranging from 13 to 166,338 M^{-1}. The encapsulation in cyclodextrins increases the aqueous solubility of essential oils up to 16-fold and reduces oil photodegradation rates up to 44-fold, while ensuring gradual release of oils. This chapter also discusses the effect of encapsulation on biological activities such as antimicrobial and antioxidant properties of essential oils. Biological effects depend on the nature and concentrations of essential oils and cyclodextrin, the tested microorganism, and other factors. Emerging cyclodextrin-based approaches for textiles and nanofibers are also discussed.

Abbreviations

CD	Cyclodextrin
CRYSMEB	Low methylated-β-cyclodextrin
EO	Essential oil
HP-β-CD	Hydroxypropyl-β-cyclodextrin
HPLC	High performance liquid chromatography
IC50	Half maximal inhibitory concentration
ITC	Isothermal titration calorimetry
LD50	Median lethal dose
MBC	Minimal bactericidal concentration
MHE	Multiple headspace extraction
MIC	Minimum inhibitory concentration
NMR	Nuclear magnetic resonance
RAMEB	Randomly methylated-β-cyclodextrin
SH-GC	Static headspace-gas chromatography
TG	Thermogravimetry

4.1 Introduction

Essential oils are mixtures of volatile organic compounds obtained from aromatic plants (Fig. 4.1). In addition to their fragrant properties, essential oils possess numerous biological properties, including antimicrobial, anticancer, anti-obesity, anti-inflammatory activities, etc. (da Silveira et al. 2014; Jayasena and Jo 2014; Costa et al. 2016; Rashed et al. 2017). About 160 essential oils are considered as Generally Recognized as Safe (GRAS) by the Food and Drug Administration (FDA)[1] (Prakash et al. 2015). Due to their natural origin, essential oils have gained great interest in the food, cosmetic and pharmaceutical industries (Astray et al. 2009; Raut and Karuppayil 2014).

The global essential oils market size exceeded USD $ 6.0 billion in 2015[2] and their chemical instability requires additional storage and quality control concern

Fig. 4.1 Examples of some vegetables and aromatic plants used for the production of essential oils

[1] FDA: Code Fed. Regul. (CFR). Title 21 Food Drugs. Chapter I – Food Drug Adm. Dep. Heal. Hum. Serv. Subchapter B – Food Hum. Consum. (Continued), Part 182–Subst. Gen. Recognized as Safe (GRAS). 2016.

[2] Essential Oil Market Analysis By Product (Orange, Corn Mint, Eucalyptus, Citronella, Peppermint, Lemon, Clove Leaf, Lime, Spearmint), By Application (Medical, Food & Beverage, Spa & Relaxation, Cleaning & Home) And Segment Forecasts To 2024.

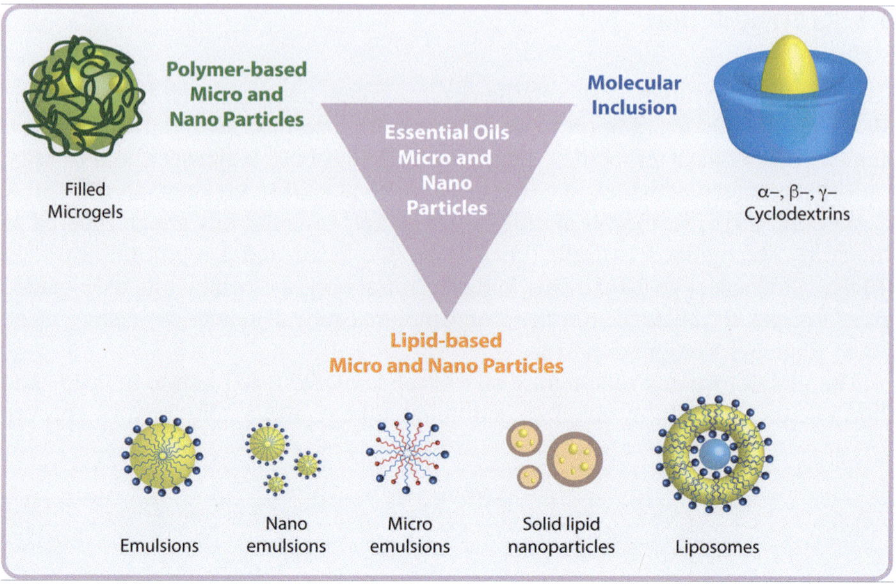

Fig. 4.2 Different systems used for the encapsulation of essential oils (adapted from Dima and Dima 2015)

(Turek and Stintzing 2013). Essential oils are also practically insoluble in aqueous systems, highly volatile and their use is often limited due to flavouring issues (Burt 2004). These drawbacks limit their applications. Encapsulation of essential oils in different systems can solve these problems. Such systems involve emulsions, beads, bioactive films, capsules, liposomes, nanocarriers and inclusion complexes (Sherry et al. 2013; Crini 2014; Pinho et al. 2014; Dima and Dima 2015; Sagiri et al. 2016) (Fig. 4.2).

Researchers are conscious of the potential of inclusion encapsulation using cyclodextrins to overcome these limitations in various sectors (Marques 2010; Sherry et al. 2013; Carvalho et al. 2016). Cyclodextrins are cyclic oligosaccharides that do not pose significant safety concern (Gould and Scott 2005; Brewster and Loftsson 2007; Stella and He 2008; Crini 2014). Moreover, they are relatively cheap biodegradable materials, and encapsulation could be performed both in solution and in solid state (Szente and Szejtli 2004; Fenyvesi et al. 2005; Loftsson and Duchêne 2007). Several procedures have been developed to prepare inclusion complexes, and various experimental methods have been described to explore and characterize the binding efficiency of cyclodextrins with essential oils (Del Valle 2004; Mura 2014).

Cyclodextrins are mainly used to increase the solubility of water-insoluble guests (Brewster and Loftsson 2007). The hydrophobic cavity of cyclodextrins also provides a microenvironment that protects the guest from volatilization and against harmful environmental factors (Del Valle 2004; Szejtli 2004; Marques 2010).

Besides, encapsulation in cyclodextrins ensures a controlled and delayed release of aromatic substances (Wang and Chen 2005; Hougeir and Kircik 2012).

Cyclodextrins are inert and do not interfere with the biological properties of essential oils (Del Valle 2004; Bilia et al. 2014). Moreover, they provide opportunity for delivering these water insoluble compounds with reproducible absorption and enhancing bioavailability (Nieddu et al. 2014; Pinho et al. 2014; Anishetty et al. 2015; Suvarna et al. 2017).

This chapter discusses the encapsulation of essential oils in cyclodextrins. The preparation and characterization of cyclodextrin/essential oil inclusion complexes are described. The binding stability between cyclodextrins and essential oils are first reviewed. Precisely, the determination of formation constants (K_f) and the factors affecting the strength and the stability of the inclusion complexes are discussed. Then, the main benefits of essential oils encapsulation in cyclodextrins (solubility, stability and release) and the literature results are summarized and examined. Moreover, the consequences of encapsulation on the biological activities and functionalities of essential oils are reviewed in order to evaluate the effectiveness of the use of inclusion complexes as bioactive agents. Thus, this work provides a wealth of information to researchers, interested in the application of encapsulated essential oils in food, agriculture, cosmetic and pharmaceutical industries.

4.2 Definition and Properties of Essential Oils

Essential oils have been used for thousands of years as incense, perfumes and cosmetics and for their culinary and medical applications (Karapinar and Aktuğ 1987; Lee et al. 2015; Sharifi-Rad et al. 2017). They give spices and herbs their specific scent and flavour and provide flowers and fruits their perfume. Essential oils can be found in all parts of the plant, including the seeds, bark, root, leaves, flowers, wood, balsam and resin (Burt 2004). The aromatic content in the plants is weak and thus highly prized. In some cases, it takes tons of plants to produce grams of essential oil.

More than 3000 types of essential oils are currently known, of which 300 are of commercial interest. Essential oils contain a mixture of related components; the monoterpenes ($C_{10}H_{16}$; two isoprene units), sesquiterpenes ($C_{15}H_{24}$; three isoprene units) and their derivatives being the major constituents. The monoterpene and sesquiterpene derivatives include cyclic and acyclic compounds from different classes, such as alcohols, esters, phenols, ketones, lactones, aldehydes and oxides (Berthelot et al. 2012; Sharifi-Rad et al. 2017). Other substantial components of essential oils are phenylpropanoids derived from the carbon skeleton of phenylalanine (Gray et al. 2012). Figure 4.3 reports some examples of the chemical structure of typical essential oil components.

In addition to their aromatic properties, essential oils and their constituents have been shown to possess various biological activities including anti-inflammatory, anticancer, antimicrobial, antidiabetic, anti-aging and insect repellent (Nazzaro et al. 2013; da Silveira et al. 2014; Raut and Karuppayil 2014; Murbach Teles

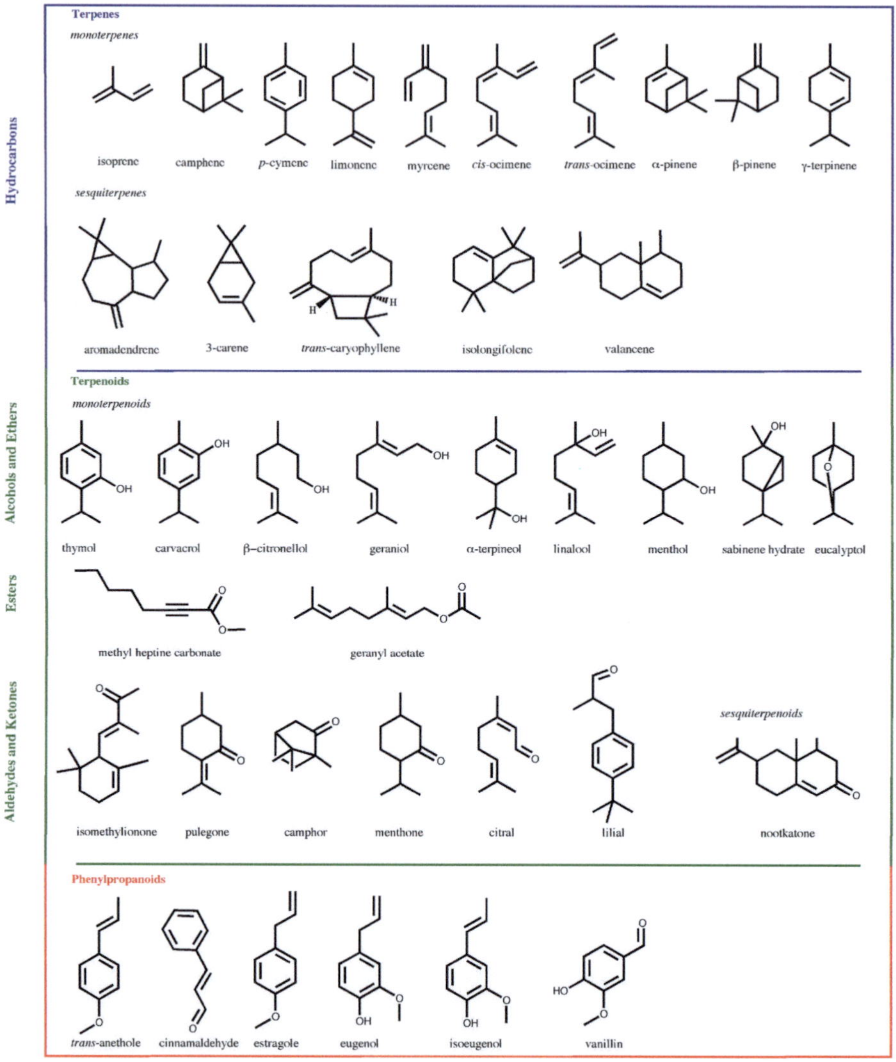

Fig. 4.3 Chemical structure of some typical essential oil components. They can be subdivided into three groups of related chemical constituents; terpenes, terpenoids and phenylpropanoids

Andrade et al. 2014; Dwivedy et al. 2016; Kfoury et al. 2016a, b; Sharifi-Rad et al. 2017). Several mechanisms have been proposed to explain their actions (Dorman and Deans 2000; Burt 2004; Li et al. 2011; Bajpai et al. 2012; Marchese et al. 2016). Essential oils have been experiencing a renaissance owing to the progression of alternative medicine practices and to the growing consumer interest in natural products as alternatives for artificial additives and pharmacological drugs (Chaves et al. 2008; Fisher and Phillips 2009). The US Food and Drug Administration (FDA) has considered 160 essential oils as ‘Generally Recognized As Safe’ (GRAS) for the use in food, drugs and cosmetics (Tisserand and Young 2014).

Nonetheless, the major issue is the low aqueous solubility of essential oils and their individual components. The logP (Octanol–water partition coefficients) values for the essential oil components fell generally in the range 1.81 to 4.48 (Griffin et al. 1999). They are also chemically unstable and susceptible to loss by volatilization, oxidative deterioration when exposed to oxygen, moisture, light, and heat or interaction with other matrix ingredients in food, cosmetic and pharmaceutical formulations (Turek and Stintzing 2013; Perricone et al. 2015; Pavela and Benelli 2016). These modifications have major drawbacks on the shelf life or the organoleptic and biological properties of the essential oil containing product.

4.3 Encapsulation of Essential Oils in Cyclodextrins

4.3.1 Characterization of Cyclodextrin/Essential Oil Inclusion Complexes in Solution

The initial step in the characterization of an inclusion complex is the determination of the stoichiometry and formation constant (Landy et al. 2000). Formation constant is also referred to as binding or association constant; its value indicates the strength of cyclodextrin-guest molecule interactions. The most common stoichiometry of cyclodextrin:guest inclusion complexes is 1:1 but higher stoichiometries can be encountered as well (Landy et al. 2007).

A wide range of chromatographic, spectroscopic and calorimetric methods are used for this purpose (Kfoury et al. 2016c). However, the determination of stoichiometry and formation constant values is not a simple approach. In all cases, obtained experimental data should be treated cautiously to ensure final reliable results. It is thoughtful to use non-linear rather than linear regression analysis. A 1:1 stoichiometry for the inclusion complex is assumed when using linear analysis to determine the formation constant value. However, inclusion complexes could have higher stoichiometry, and at equilibrium the solution could consist of a mixture of inclusion complexes with different stoichiometries (Landy et al. 2000, 2007). This could be only revealed when non-linear analysis is applied. Formation constant value is specific for each individual essential oil component and could not be determined for the entire essential oil mixture. Numerous studies evaluated formation constant for cyclodextrin/essential oil component inclusion complexes using various methods. Authors have applied high performance liquid chromatography (HPLC), static headspace-gas chromatography (SH-GC), UV-Visible spectroscopy, fluorescence spectroscopy, NMR spectroscopy and isothermal titration calorimetry (ITC) (Tanemura et al. 1998; Demian 2000; Liu and Guo 2002; Decock et al. 2006; Ciobanu et al. 2012, 2013; Kfoury et al. 2016d, e). Table 4.1 summarizes the formation constant values for cyclodextrin/essential oil component inclusion complexes.

Table 4.1 Formation constant (K_f) values of cyclodextrin/essential oil component inclusion complexes. The formation constant value indicates the strength of interactions between the cyclodextrin and the guest molecule

EO component	Log P[a]	α-CD	β-CD	γ-CD	CRYSMEB	RAMEB	HP-β-CD	Captisol®
trans-Anethole	3.096	1163[b] 927[c] 710[f]	630[b] 1040[d] 542[c] 497[f]	96[b]	740[b] 1039[c] 877[f]	1553[b] 1815[c] 1110[f]	1042[b] 712[e] 845[c] 981[f]	1210[g]
Aromadendrene	4.850	–	344 [g]	925 [g]	585 [g]	474 [g]	362 [g]	1274[g]
Benzyl alcohol	1.275	52[h]	64[h]	–	56[h]	53[h]	63[h]	–
Camphene	3.329	598[b]	4825[b]	360[b]	6625[b]	6057[b]	3033[b]	4616[i]
Camphor	2.160	184[b]	2058[b]	1048[b]	1901[b]	1194[b]	1280[b]	1091[g]
Carvacrol	3.810	454[j]	2620[j]	999[j]	2421[j]	3564[j]	2154[j]	–
3-Carene	3.450	1805[g]	3561[g]	94 [g]	5469[g]	3350 [g]	2337[g]	3887 [g]
β -Caryophyllene	5.170	–	28674[k]	4004[k]	11488[k]	14274[k]	4960[k]	11115[i]
Cinnamaldehyde	2.484	236[h]	450[h] 400[l]	–	595[h]	1696[h]	969[h]928[m]	–
β-Citronellol	3.152	223[h]	3141[h]	–	3290[h]	4048[h]	2578[h]	–
p-Cymene	3.898	140[b]	2505[b]	88[b]	2549[b]	3543[b]	2213[b]	2868[i]
Eucalyptol	2.716	13[b]	615[b]	742[b]	688[b]	673[b]	334[b]	881[i]
Eugenol	2.100	350[c] 94[h]	462[c] 264[h] 140[d]	–	454[h]	568[h]	436[c] 462[h]	–
Estragole	2.818	335[c] 478[n]	987[c] 939[n]	108[l]	1584[c] 1661[n]	1916[c] 1761[n]	1508[c] 1581[n]	1479[i]
Geraniol	3.202	90[h]	528[h]	–	977[h]	1100[h]	712[h]	–
Isoeugenol	2.379	178[c] 85[h]	364[c] 255[h]	–	263[h]	514[h]	418[c] 441[h]	–
Isolongifolene	4.700	–	–	5824 [g]	6306[g]	5673[g]	2232[g]	5884[g]
Isomethylionone	4.160	71[h]	9869[h]	–	15632[h]	13176[h]	9789[h]	–
Lilial	4.389	4387[h]	56567[h]	–	147617[h]	166338[h]	112205[h]	–
Limonene	3.615	1289[b]	3162[b] 2230[d]	116[b]	3668[b]	4386[b]	2787[b] 4730[o]	4069[i]
Linalool	3.213	32[b]	366[b]	138[b]	816[b]	833[b]	596[b] 940[o] 958[e] 720[p]	–
Menthol	3.335	82[b] 10[q]	1731[b] 2240[d]	105[b]	2396[b]	1928[b]	1079[b]	–
Menthone	3.149	35[b]	656[b] 546[d]	83[b]	989[b]	748[b]	664[b]	745[g]
Myrcene	3.994	212[b]	1431[b]	138[b]	959[b]	1286[b]	575[b] 1240[o]	916[i]
Methyl heptine carbonate	3.220	2905[h]	226[h]	–	539[h]	485[h]	325[h]	–

(continued)

Table 4.1 (continued)

EO component	Log P[a]	α-CD	β-CD	γ-CD	CRYSMEB	RAMEB	HP-β-CD	Captisol®
Nootkatone	3.67		5801[r]				4838[r]	–
cis-Ocimene	3.970	42[k]	432[k]	20[k]	622[k]	593[k]	538[k]	–
trans-Ocimene	3.970	46[k]	538[k]	26[k]	789[k]	640[k]	627[k]	–
α-Pinene	3.542	1778[b]	2588[b]	214[b]	2999[b]	2395[b]	1637[b] 5780° 1842[s]	1633[i]
β-Pinene	3.329	1018[b]	4587[b]	633[b]	5141[b]	4450[b]	3151[b] 7360° 1671[s]	5053[i]
Pulegone	2.516	30[b]	331[b]	82[b]	1025[b]	796[b]	676[b] 798[s]	635 [g]
Sabinene hydrate	2.320	108[k]	2108[k]	708[k]	1308[k]	1882[k]	772[k]	–
γ-Terpinene	3.360	37[k]	1309[k]	40[k]	1950[k]	2066[k]	1488[k]	2456[i]
α-terpineol	2.600	126[k]	1143[k]	89[k]	1223[k]	1287[k]	761[k]	–
Thymol	3.342	107[j]	1467[j]	233[j]	2386[j]	3337[j]	806[s] 1488[j]	–
Valencene	5.010	–	6421[g]	3020 [g]	8322[g]	4627[g]	2565 [g]	6897[g]
Vanillin	1.067	–	90[t] 100[u]	–	–	–	–	–

EO essential oil; *CD* cyclodextrin; *RAMEB* randomly methylated-β-cyclodextrin; *CRYSMEB* a low methylated-β-cyclodextrin; *HP-β-CD* hydroxypropyl-β-cyclodextrin; *Captisol®* sulfobutylether-β-cyclodextrin

[a]Log P: http://www.molinspiration.com/cgi-bin/properties; [b]Ciobanu et al. 2013; [c]Kfoury et al. 2014a; [d]Donze and Coleman 1993; [e]Demian 2000; [f]Kfoury et al. 2014b; [g]unpublished data; [h]Decock et al. 2008; [i]Kfoury et al. 2017; [j]Kfoury et al. 2016d; [k]Kfoury et al. 2015a; [l]Jiang et al. 2010; [m]Chen et al. 2010; [n]Kfoury et al. 2015b; [o]Tanemura et al. 1998; [p]Numanoglu et al. 2007; [q]Astray et al. 2010; [r]Kfoury et al. 2016e; [s]Kfoury et al. 2014c; [t]Ferrazza et al. 2014; [u]Zeng et al. 2012

Titration experiments, using a constant concentration of essential oil component and increasing amounts of cyclodextrin, are generally employed. Nevertheless, some alternative non-conventional approaches suitable for such low soluble compounds like essential oil components are now being developed, in particular competitive methods applied to UV-Visible and isothermal titration calorimetry (ITC) (Landy et al. 2000; Bertaut and Landy 2014). The main advantage of the competitive protocols is that they overcome solubility limitations. Moreover, they allow the determination of formation constant values, with a high accuracy, for a) inclusion complexes of essential oil component that lack any chromophore and cannot be studied by direct UV-Visible, and b) for athermic inclusion complexes that cannot be directly detected by isothermal titration calorimetry (ITC) (Bertaut and Landy 2014). Furthermore, a "rapid method" based on static headspace-gas chromatography (SH-GC) technique was developed and successfully validated to study the interactions between cyclodextrin and individual essential oil components present simultaneously in complex mixtures like essential oils (Fourmentin et al. 2013;

Kfoury et al. 2015c). This method implies that neither further purification is required nor the use of standard compounds. This approach can be also applied to other chromatographic techniques equipped with high separation efficiency columns.

Recently, a NMR method combined with an algorithmic treatment that relies on global analysis, was explored to determine formation constants (Kfoury et al. 2016d, e). This analysis uses simultaneously the chemical shifts (δ) and diffusion coefficients (D) variations for several guest protons to calculate the formation constant value.

The formation constant values for a great number of cyclodextrin/essential oil component inclusion complexes were successfully determined (Table 4.1). The obtained results cover a wide range of formation constant values from 13 to 166,338 M^{-1}. The binding efficiency could be classified as very weak, weak, moderate, strong and very strong when formation constant values are in the following ranges: 0–500, 500–1000, 1000–5000, 5000–20,000 and greater than 20,000 M^{-1}, respectively (Carrier et al. 2007).

Several studies in the literature showed a high positive correlation between the hydrophobic character of essential oil components (logP) and formation constant values, with the correlation coefficients being 0.986 for β-cyclodextrin (Decock et al. 2008), 0.855 and 0.882, for α-cyclodextrin and β-cyclodextrin, respectively (Astray et al. 2010) and 0.923 for hydroxypropyl-β-cyclodextrin (HP-β-CD) (Kfoury et al. 2014c). However, it is cautious not to establish rules based on a limited number of guests because the correlation becomes weaker when taking in consideration all values from the literature (Table 4.1) (Kfoury et al. 2016c). This indicates that this correlation could not be universally applicable for estimating or predicting the stability of a cyclodextrin/essential oil component inclusion complex.

Also, other factors, such as the space filling of the cyclodextrin cavity, which is directly related to the structure and geometrical conformation of the guest (Decock et al. 2008; Fourmentin et al. 2013), contribute substantially to the complex stability. Indeed, the essential oil component has to fit entirely or at least partially into the cyclodextrin cavity. The weak residual empty space shows sufficient contact between the cyclodextrin cavity and the essential oil component indicating a tight encapsulation (Eftink et al. 1989). Figure 4.4 illustrates the better fitting, with a very weak residual empty space, of sesquiterpenes (β-caryophyllene and valencene) for the β-CD cavity as compared to monoterpenes (carene and γ-terpinene).

Another example is the case of β-caryophyllene with the different native cyclodextrins; the β-cyclodextrin cavity can effectively accommodate β-caryophyllene as indicated by the high formation constant value (28,674 M^{-1}). The cavity of α-cyclodextrin is too small to encapsulate this bulky sesquiterpene; thus, no K_f value could be determined for α-cyclodextrin/β-caryophyllene. The wider cavity of γ-cyclodextrin is less engaged in the interaction with this guest than β-cyclodextrin leading to a lower value (4004 M^{-1}).

The importance of the steric effect is also confirmed when isomers are not likewise recognized by cyclodextrins; for example, α- and β-pinene, thymol and carvacrol, *trans*-anethole and estragole, eugenol and isoeugenol, *cis*-ocimene and *trans*-ocimene (Table 4.1) and others (Ruktanonchai et al. 2011; Miron et al. 2012; Zeng et al. 2012; Bethanis et al. 2013; Ciobanu et al. 2013; Nowakowski and Ejchart 2014).

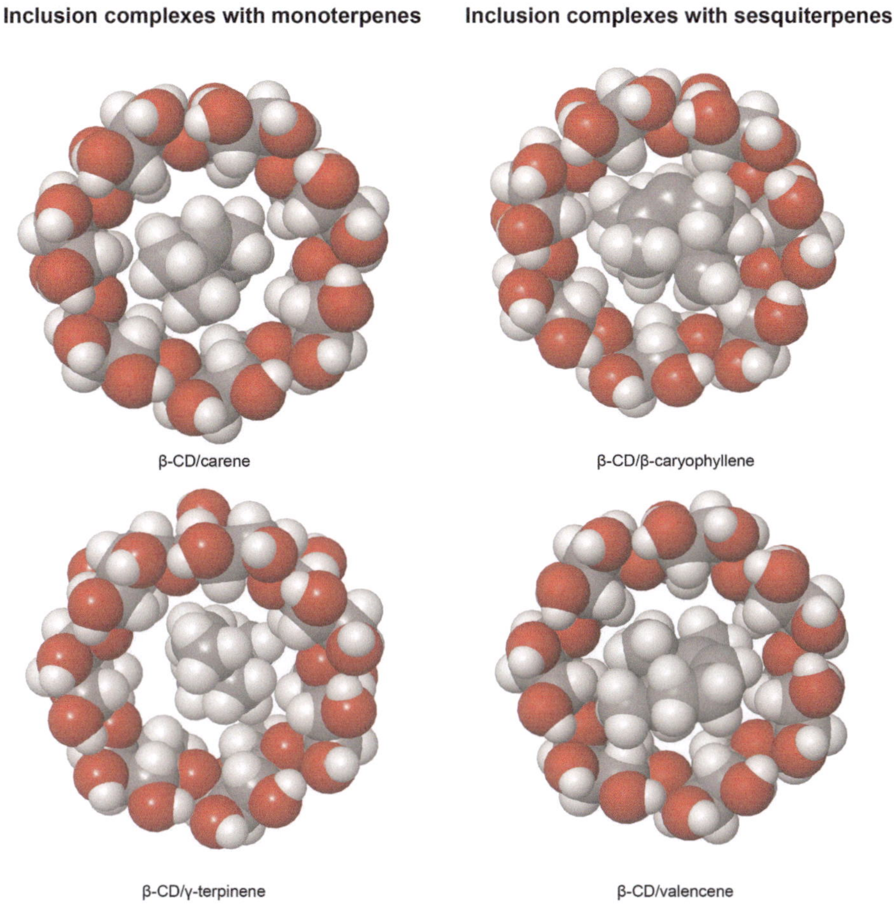

Fig. 4.4 Illustration of the space filling of the β-cyclodextrin cavity by the sesquiterpenes (β-caryophyllene and valencene) and the monoterpenes (γ-terpinene and carene). The weak residual empty space, observed between the cyclodextrin cavity and the sesquiterpenes as compared to the monoterpenes suggests that the sesquiterpenes, provides sufficient contact with the wall of the cavity and a more tight encapsulation

4.3.2 Preparation and Characterization of Solid Cyclodextrin/Essential Oil Inclusion Complexes

While the oily state of the essential oils is appropriate for some applications, there are, however, situations where the solid form would be preferred (Anishetty et al. 2015). The formulation of cyclodextrin/essential oil inclusion complexes can be also achieved in the solid form (Miller et al. 2007; Jambhekar and Breen 2016). This ensures an easier and safer handling by converting oils into solid dosage forms.

The solid cyclodextrin inclusion complexes also help solving the major limitation of the use of essential oils, their chemical instability in the presence of air, light, moisture and heating.

Solid cyclodextrin/essential oil inclusion complexes could be obtained using various techniques such as co-precipitation, freeze-drying, spray drying, co-grinding, co-evaporation, sealed-heating, complexation by using supercritical carbon dioxide and microwave assisted encapsulation (Del Valle 2004; Kfoury et al. 2016c). Solid inclusion complexes could be prepared both at the laboratory and industrial scales. The preparation parameters including the temperature, the mixing time and speed, the nature of cyclodextrin, the use of co-solvent or other additives and the drying process should be optimized. These parameters may affect the properties of the obtained complexes particularly their crystallinity and the recovery and encapsulation yields. Thus, the choice of the method depends on encapsulation yield, rapidity, simplicity, cost and the desired characteristics of the final product (Hernández-Sánchez et al. 2017).

The physicochemical properties of an inclusion complex are generally evaluated using thermal analysis, X-ray diffraction, Fourier transform infrared spectroscopy, Raman spectroscopy, scanning electron microscopy and transmission electron microscopy (Mura 2014; Kfoury et al. 2016c). Moreover, the quantification of essential oil in solid inclusion complex and the encapsulation yield could be performed. Results are generally expressed as loading capacity and encapsulation efficiency (EE%). The loading capacity is a measure of the encapsulated amount of essential oil per gram of the solid inclusion complex and the encapsulation efficiency (EE%) is the encapsulated amount of essential oil expressed as a percentage of the quantity initially used to prepare the solid inclusion complex. An appropriate evaluation of these parameters is the most important issue for the application of cyclodextrin inclusion complexes as solid dosage forms. Results from the literature regarding the loading capacity and encapsulation efficiency (EE%) of cyclodextrin/essential oil inclusion complexes are listed in Table 4.2.

UV-Visible spectroscopy is the main technique used to determine the loading capacity and encapsulation efficiency (EE%). However, this method is limited to compounds that possess a chromophore. Moreover, it could not be applied to complex mixtures such as essential oils. Chromatographic (HPLC and GC) and thermal (TG/DTG) analyses are also useful to evaluate the essential oil content in cyclodextrin inclusion complexes. However, these methods are time-consuming and expensive. In addition, thermal analysis is based on the weight loss phenomenon that could also arise upon water and solvent losses (Hădărugă et al. 2012). Recently, a new simple and solvent-free method, the multiple headspace extraction (MHE) coupled to gas chromatography, was applied for this purpose (Kfoury et al. 2016f). This method was first described and developed by Kolb and Ettre (2006). The multiple headspace extraction (MHE) consists in a series of extractions of the headspace of a diluted inclusion complex solution, then the obtained profile is compared to that of an external standard and the essential oil content is determined. This method is suitable for volatile compounds like essential oils.

Table 4.2 Loading capacity and encapsulation efficiency (EE%) values of cyclodextrin/essential oil solid inclusion complexes

EO or EO component	Cyclodextrin	Preparation ratio (EO:CD)	Preparation	Results	Technique	References
Allium sativum	β-CD	12:88(W:W)	Precipitation	20[a]	GC-fid	Ayala-Zavala et al. 2008
Achillea millefolium L. s. l.	HP-β-CD	1:10(W:W)	FD	45[a]	UV-Vis	Rakmai et al. 2017a
Benzyl isothiocyanate	β-CD	1:16(W:W)	FD	86[a]	UV-Vis	Li et al. 2015
trans-Anethole	α-CD	1:1(M:M)	FD	33[b]	UV-Vis	Kfoury et al. 2016a
trans-Anethole	β-CD	1:1(M:M)	FD	19[b]	UV-Vis	Kfoury et al. 2016a
trans-Anethole	HP-β-CD	1:1(M:M)	FD	22[b]	UV-Vis	Kfoury et al. 2016a
trans-Anethole	RAMEB	1:1(M:M)	FD	33[b]	UV-Vis	Kfoury et al. 2016a
trans-Anethole	CRYSMEB	1:1(M:M)	FD	29[b]	UV-Vis	Kfoury et al. 2016a
trans-Anethole	β-CD	1:1(M:M)	FD	20[b]	MHE	Kfoury et al. 2016f
trans-Anethole	β-CD	1:1(M:M)	Precipitation-FD	93[b]	MHE	Kfoury et al. 2016f
Black pepper	HP-β-CD	1:10(W:W)	FD	50[a]	UV-Vis	Rakmai et al. 2017b
Carvacrol	β-CD	1:1(M:M)	FD	91[a]	UV-Vis	Santos et al. 2015
Carvacrol	β-CD	1:1(M:M)	Kneading	84[a]	UV-Vis	Santos et al. 2015
Carvacrol	HP-β-CD	1:1(M:M)	FD	83[b]	UV-Vis	Kamimura et al. 2014
Carvacrol	HP-β-CD	1:1(M:M)	Kneading	77[b]	UV-Vis	Kamimura et al. 2014
Carvacrol	β-CD	1:1(M:M)	Paste	34[a]	HPLC	Menezes et al. 2016
Carvacrol	β-CD	1:1(M:M)	Slurry	72[a]	HPLC	Menezes et al. 2016
Cinnamomum zeylanicum	β-CD	16:84(W:W)	Precipitation	13[a]	GC-fid	Ayala-Zavala et al. 2008
Cinnamon	β-CD	1:1(M:M)	FD	42[a]	UV-vis	Hill et al. 2013
Citrus sinensis	β-CD	1:1(M:M)	Kneading	74[b]	TG/DTG	Galvão et al. 2015
Citrus sinensis	β-CD	1:1(M:M)	Precipitation	73[b]	TG/DTG	Galvão et al. 2015
Clove	β-CD	1:1(M:M)	FD	78[a]	UV-vis	Hill et al. 2013
Clove	HP-β-CD	1:1(M:M)	Kneading	97[a]		Cetin Babaoglu et al. 2017
Estragole	β-CD	1:1(M:M)	Precipitation	93[b]	UV-Vis	Yang et al. 2017
Estragole	α-CD	1:1(M:M)	FD	23[b]	UV-Vis	Kfoury et al. 2015b

(continued)

Table 4.2 (continued)

EO or EO component	Cyclodextrin	Preparation ratio (EO:CD)	Preparation	Results	Technique	References
Estragole	β-CD	1:1(M:M)	FD	19[b]	UV-Vis	Kfoury et al. 2015b
Estragole	HP-β-CD	1:1(M:M)	FD	37[b]	UV-Vis	Kfoury et al. 2015b
Estragole	RAMEB	1:1(M:M)	FD	41[b]	UV-Vis	Kfoury et al. 2015b
Estragole	CRYSMEB	1:1(M:M)	FD	42[b]	UV-Vis	Kfoury et al. 2015b
Eucalyptol	HP-β-CD	1:1(M:M)	FD	88[a]	HPLC	Kfoury et al. 2014c
Eugenol	α-CD	1:1(M:M)	FD	120[b]	UV-Vis	Kfoury et al. 2016a
Eugenol	β-CD	1:1(M:M)	FD	120[b]	UV-Vis	Kfoury et al. 2016a
Eugenol	HP-β-CD	1:1(M:M)	FD	98[b]	UV-Vis	Kfoury et al. 2016a
Eugenol	RAMEB	1:1(M:M)	FD	110[b]	UV-Vis	Kfoury et al. 2016a
Eugenol	CRYSMEB	1:1(M:M)	FD	120[b]	UV-Vis	Kfoury et al. 2016a
Geraniol	HP-β-CD	1:1(M:M)	FD	85[a]	HPLC	Kfoury et al. 2014c
Hyptis martiusii	β-CD	1:1(M:M)	Paste	60[b]	TG/DTG	Andrade et al. 2017
Hyptis martiusii	β-CD	1:1(M:M)	Slurry	69[b]	TG/DTG	Andrade et al. 2017
Isoeugenol	α-CD	1:1(M:M)	FD	81[b]	UV-Vis	Kfoury et al. 2016a
Isoeugenol	β-CD	1:1(M:M)	FD	61[b]	UV-Vis	Kfoury et al. 2016a
Isoeugenol	HP-β-CD	1:1(M:M)	FD	110[b]	UV-Vis	Kfoury et al. 2016a
Isoeugenol	RAMEB	1:1(M:M)	FD	100[b]	UV-Vis	Kfoury et al. 2016a
Isoeugenol	CRYSMEB	1:1(M:M)	FD	100[b]	UV-Vis	Kfoury et al. 2016a
Limonene	HP-β-CD	1:1(M:M)	FD	15[a]	HPLC	Kfoury et al. 2014c
Linalool	HP-β-CD	1:1(M:M)	FD	90[a]	HPLC	Kfoury et al. 2014c
Lippia gracilis	HP-β-CD	1:1(M:M)	Slurry	63[a]	TG	Marreto et al. 2008
Lippia gracilis	HP-β-CD	1:1(M:M)	Paste	100[a]	TG	Marreto et al. 2008
Lippia graveolens	β-CD	14.3:100(W:W)	Spray drying	54[a]	Distillation	Arana-Sánchez et al. 2010
Mentha x villosa	β-CD	1:9(W:W)	Precipitation	140[b]	GC-MS	Martins et al. 2007
α-Pinene	HP-β-CD	1:1(M:M)	FD	25[a]	HPLC	Kfoury et al. 2014c

β-Pinene	HP-β-CD	1:1(M:M)	FD	21[a]	HPLC	Kfoury et al. 2014c
Pulegone	HP-β-CD	1:1(M:M)	FD	82[a]	HPLC	Kfoury et al. 2014c
Salvia sclarea L.	β-CD	1:1(M:M)	Precipitation	140[b]	UV-Vis	Tian et al. 2008
Terpinen-4-ol	β-CD	1:1(M:M)	FD	117[b]	UV-Vis	Yang et al. 2015
Thyme	β-CD	1:1(M:M)	FD	96[b]	UV-Vis	Tao et al. 2014
Thyme	β-CD	1:1(M:M)	Kneading	84[b]	UV-Vis	Tao et al. 2014
Thymol	β-CD	1:1(M:M)	FD	100[b]	UV-Vis	Tao et al. 2014
Thymol	β-CD	1:1(M:M)	Kneading	90[b]	UV-Vis	Tao et al. 2014
Thymol	HP-β-CD	1:1(M:M)	FD	91[a]	HPLC	Kfoury et al. 2014c
Menthol	HP-β-CD	1:5(W:W)	FD	84[b]	TG	Zhu et al. 2016
Vanillin	β-CD	1:1(M:M)	Precipitation	33[a]	GC-fid	Zeng et al. 2012

EO essential oil; *CD* cyclodextrin; *RAMEB* randomly methylated-β-cyclodextrin; *CRYSMEB* a low methylated-β-cyclodextrin; *HP-β-CD* hydroxypropyl-β-cyclodextrin; *FD* Freeze-Drying; *TG* thermogravimetry; *DTG* derivative thermogravimetry; *UV-Vis* UV-Visible spectroscopy
[a]Encapsulation Efficiency (EE%); [b]loading capacity: mg essential oil or essential oil component/g inclusion complex

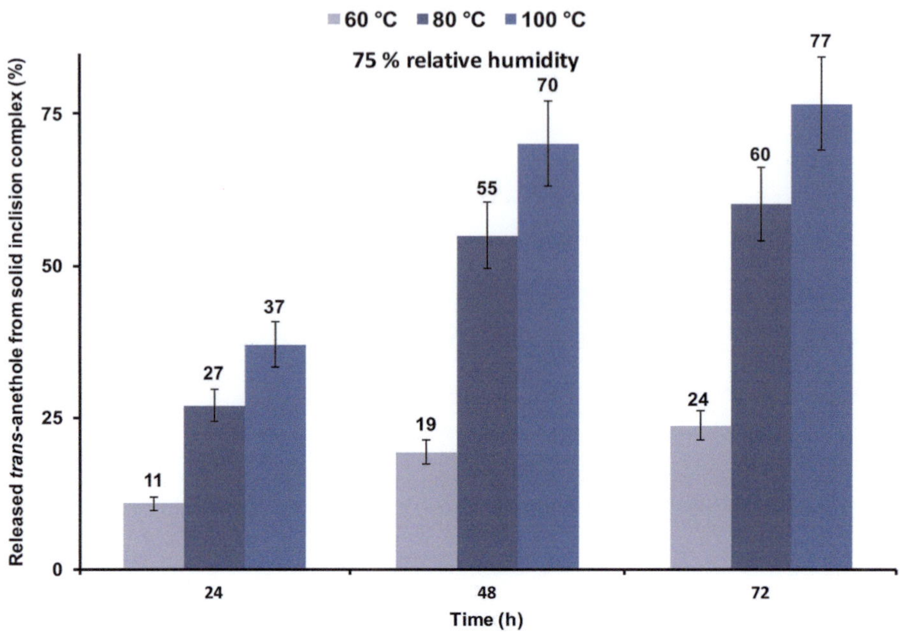

Fig. 4.5 Release of *trans*-anethole (%) from cyclodextrin inclusion complex exposed to 75% relative humidity at three different temperatures. Note that the released fraction of *trans*-anethole increased when increasing the temperature and the exposure time

The theoretical values for the loading capacity values of essential oils in β-cyclodextrin inclusion complexes are presumed to be between 8 and 12% (Pagington 1987). The values listed in Table 4.2 fit well with this range.

Solid inclusion complexes present several advantages: they could improve the handling of oily essential oils and could be used as standardized dosage form. However, the major task is to use inclusion complexes to design bio-based active food and drug packaging systems (Kapetanakou and Skandamis 2016; Ribeiro-Santos et al. 2017). Essential oils could be used in active packaging either as aromatic additives or bioactive agents. When essential oils are applied as cyclodextrin inclusion complexes they could benefit from a gradual release into the product (Martina et al. 2013). This release may be influenced by specific factors mainly the temperature and the relative humidity (Reineccius et al. 2002; Neoh et al. 2006; Ayala-Zavala et al. 2008; Ho et al. 2011; Yamamoto et al. 2012; Yang et al. 2015). Therefore, several studies attempted to evaluate the release performance of essential oils from solid inclusion complexes under specific temperature and humidity conditions. Figure 4.5 illustrates the release of *trans*-anethole from β-cyclodextrin inclusion complex as a function of time and temperature in a humid atmosphere (75%).

Table 4.3 presents a review of the main results on the release of essential oils from solid cyclodextrin inclusion complexes under specific temperature and humidity conditions.

Table 4.3 Released fractions of essential oils from solid cyclodextrin inclusion complexes under specific temperature and relative humidity (RH%) conditions

EO or EO component	Cyclodextrin	Temperature	RH%	Exposure time	Released fraction	References
Allium sativum L.	β-CD	37 °C	–	60 h	76%	Wang et al. 2011
Allium sativum L.	β-CD	50 °C	–	60 h	100%	Wang et al. 2011
trans-Anethole	β-CD	60 °C	75%	72 h	24%	Kfoury et al. 2016f
trans-Anethole	β-CD	80 °C	75%	72 h	60%	Kfoury et al. 2016f
trans-Anethole	β-CD	100 °C	75%	72 h	77%	Kfoury et al. 2016f
Cinnamaldehyde	β-CD	25 °C	97%	70 days	30%	Ponce Cevallos et al. 2010
Estragole	β-CD	40 °C	75%	576 h	36%	Yang et al. 2017
Estragole	β-CD	60 °C	75%	576 h	52%	Yang et al. 2017
Estragole	β-CD	80 °C	75%	144 h	85%	Yang et al. 2017
Estragole	β-CD	100 °C	75%	24 h	100%	Yang et al. 2017
Terpinen-4-ol	β-CD	40 °C	70%	720 h	30%	Yang et al. 2015
Terpinen-4-ol	β-CD	80 °C	70%	144 h	100%	Yang et al. 2015
Terpinen-4-ol	β-CD	100 °C	70%	24 h	100%	Yang et al. 2015
Thymol	β-CD	20 °C	100%	3 weeks	76%	Del Toro-Sanchez et al. 2010
Thymol	β-CD	25 °C	97%	70 days	27%	Ponce Cevallos et al. 2010

EO essential oil; *CD* cyclodextrin; *RH%* relative humidity%

The literature data presented in Table 4.3 proved that inclusion complexes gradually release the encapsulated essential oils. This ensures the shelf life extension of essential oils and their delivery in a prolonged and delayed manner (Arfat et al. 2015; Lee et al. 2015; Morelli et al. 2015). Both elevated humidity and temperature favour the release of essential oils from inclusion complexes over time. The release of essential oils from cyclodextrin inclusion complexes in the presence of high humidity levels is due to the interaction of water molecules with the polar groups of the cyclodextrin. This leads to a disturbed equilibrium and the displacement of the essential oil component from the cyclodextrin cavity (Neoh et al. 2006; Ho et al. 2011; Mascheroni et al. 2013). Interestingly, the release of the antimicrobial agents, essential oils, from cyclodextrin inclusion complexes triggered by the high humidity normally present in food, is essential to counteract the microbial growth favoured by these conditions (Ayala-Zavala et al. 2008). Concerning the effect of the temperature on the release of essential oils from inclusion complexes, it could be explained by the increase in the volatility of essential oils. The dissimilar results obtained by different authors for the same cyclodextrin/essential oil inclusion complex e.g. β-cyclodextrin/thymol (Del Toro-Sanchez et al. 2010; Ponce Cevallos et al. 2010), could be explained by the difference in the cyclodextrin:guest ratios used for its preparation. This consequently affects the essential oil content and release from the inclusion complex. Another crucial factor that influences the release kinetics is the preparation method (Kfoury et al. 2016f).

Some applications of these solid inclusion complexes could be found in literature. For example, the incorporation of β-cyclodextrin/*trans*-cinnamaldehyde inclusion complex into a based chitosan edible coating improved the shelf life of fresh-cut melon (Moreira et al. 2014) and papaya (Brasil et al. 2012). They can also ensure an optimal flavour and nutritional quality of fruits and vegetables until consumption.

4.4 Effect of Encapsulation on the Physicochemical Properties of Essential Oils

4.4.1 Solubility Enhancement

Cyclodextrins are generally preferred to organic solvents to solubilize essential oils, because they are safer. Moreover, unlike cyclodextrins, the solubilizing efficiency of organic solvents is often lost by precipitation when introduced into an aqueous medium (Gould and Scott 2005).

Phase solubility studies are generally carried out to describe and quantify the effect of cyclodextrin concentration on the guest's solubility (Higuchi and Connors 1965). In general, cyclodextrins solubilize guests as a linear function of their concentration based on the formation of inclusion complexes (Brewster and Loftsson 2007). Phase solubility studies can be performed successfully with a single essential oil component using a variety of techniques mainly the UV-Visible spectroscopy as well as fluorescence spectroscopy and chromatography (HPLC and GC) to determine the amount of solubilized guest (Waleczek et al. 2003; Karathanos et al. 2007; Mourtzinos et al. 2008; Mazzobre et al. 2011; Liang et al. 2012; Zeng et al. 2012; Hill et al. 2013; Ansari et al. 2014; Tao et al. 2014; Kfoury et al. 2014a; Santos et al. 2015). These techniques allow precise measurements leading to an accurate quantification of the essential oil component's solubility. However, essential oils are mixtures of a great number of components. Therefore, carrying out phase solubility studies with essential oils is a challenge. Nonetheless, phase solubility studies have been commonly carried out to investigate the solubilizing efficiency of cyclodextrins toward different essential oils such as chamomile essential oil (Waleczek et al. 2003), clove essential oil (Hernandez-Sanchez et al. 2012; Hill et al. 2013),

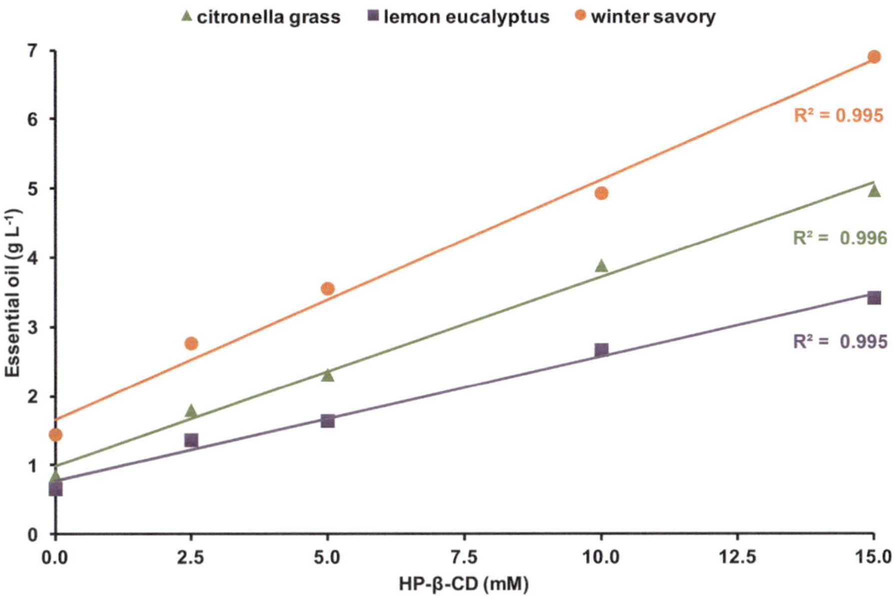

Fig. 4.6 Phase solubility diagrams of *citronella grass, lemon eucalyptus* and *winter savory* essential oils with hydroxypropyl-β-cyclodextrin (HP-β-CD). The profiles show a linear increase in the essential oils solubility as a function of cyclodextrin concentration (Kfoury et al. 2016g; reprinted with permission)

cinnamon bark extract (Hill et al. 2013), clary essential oil (Tian et al. 2008), and garlic essential oil (Bai et al. 2010). The authors have used analytical methods based on the assessment of the response of a single essential oil component and concluded on the solubility of the entire essential oil. However, this is not cautious and does not reflect the behaviour of the entire essential oil. A multitude of equilibriums could take place when adding cyclodextrin to essential oil because each essential oil component presents a distinct S_0, particular stoichiometry and formation constant value (Kfoury et al. 2015c). Recently, a new Total Organic Carbon (TOC) method was developed to perform phase solubility studies for essential oils (Kfoury et al. 2016g). Total Organic Carbon is a non-specific, simple, economic and rapid method. It allows determination of the concentration of organic carbon in aqueous samples, mainly in monitoring water quality (Bourgeois et al. 2001). Total Organic Carbon provides a precise determination of the solubility of the essential oil because it measures the total organic carbon content leading to the calculation of the massic concentration of encapsulated essential oil. Figure 4.6 illustrates the phase solubility diagrams for three different essential oils, *citronella grass, lemon eucalyptus* and *winter savory*, with hydroxypropyl-β-cyclodextrin (Kfoury et al. 2016g).

Table 4.4 is a summary of the results of the phase solubility studies of essential oils performed with hydroxypropyl-β-cyclodextrin (HP-β-CD) using Total Organic Carbon (TOC).

Table 4.4 Results of the phase solubility studies of essential oils with hydroxypropyl-β-cyclodextrin (HP-β-CD). Studies are carried out using Total Organic Carbon (TOC) measurements

Essential oil	Botanical name	Equation	R^2	S_0[a]	S_t/S_0[b]
Basil[c]	*Ocimum basilicum Var. basilicum*	Y = 0.22X + 1.09	0.989	0.82	4.09
Bay rum tree[c]	*Pimenta racemosa*	Y = 0.19X + 1.77	0.959	1.00	3.36
Caraway[d]	*Carum carvi*	Y = 0.10X + 0.91	0.994	0.91	2.05
Citronella grass[c]	*Cymbopogon nardus*	Y = 0.27X + 0.98	0.996	0.85	4.59
Dill[d]	*Anethum graveolens*	Y = 0.09X + 0.70	0.992	0.70	2.28
Ho Wood[c]	*Cinnamomum camphora CT linalol*	Y = 0.24X + 2.37	0.975	1.91	2.61
Lemon eucalyptus[c]	*Eucalyptus citriodora*	Y = 0.18X + 0.77	0.995	0.65	4.07
Lime[e]	*Citrus aurantifolia*	Y = 0.18X + 0.10	0.989	0.10	14.51
Mandarin orange[c]	*Citrus reticulata Blanco*	Y = 0.32X + 0.33	0.999	0.21	16.67
Mandarin[e]	*Citrus reticulata Blanco*	Y = 0.40X + 0.18	0.996	0.18	16.86
Marjoram[c]	*Origanum majorana CT thujanol*	Y = 0.17X + 2.30	0.973	1.97	1.98
Niaouli[e]	*Melaleuca quinquenervia*	Y = 0.33X + 4.13	0.994	3.48	2.19
Oregano[c]	*Origanum compactum*	Y = 0.25X + 2.30	0.985	0.98	4.01
Rosemary[c]	*Rosmarinus officinalis Cineoliferum*	Y = 0.26X + 1.77	0.989	2.38	3.01
Tarragon[c]	*Artemisia dracunculus*	Y = 0.27X + 0.77	0.988	0.43	8.45
Tea Tree[e]	*Melaleuca alternifolia*	Y = 0.54X + 4.06	0.992	2.97	3.04
Winter savory[c]	*Satureja montana*	Y = 0.35X + 1.66	0.995	1.44	3.43

[a]experimental S_0 of the essential oil (g/l); [b]St and S_0 are the essential oil solubility in 10 mM hydroxypropyl-β-cyclodextrin (HP-β-CD) solution and in water, respectively.
[c]Kfoury et al. 2016g; [d]Unpublished data; [e]Kfoury et al. 2017.

These results proved that cyclodextrins are efficient to improve the solubility of essential oils. The increase in the essential oil's solubility with cyclodextrin concentration was linear ($R^2 > 0.95$). The intrinsic solubility (S_0) varied among the essential oils because of their different compositions. These results also lead to the conclusion that the solubilizing potential of cyclodextrins (S_t/S_0) is inversely proportional to the essential oil's intrinsic solubility (S_0) (Kfoury et al. 2014a; 2016g).

Phase-solubility diagrams are mainly used to determine the cyclodextrin concentration needed to solubilize the guest in solution. Moreover, they could be expanded to solid dosage forms by determining the complexation efficiency (CE) based on the phase solubility diagram (Loftsson et al. 2005). The feasibility of formulating a given guest as solid cyclodextrin inclusion complex depends on two factors, the dosage and the complexation efficiency (CE). If the complexation efficiency (CE) is 0.1, it means that 1 out of every 11 cyclodextrins molecules forms a complex with the drug and if the complexation efficiency (CE) is 0.01 then only 1 out of every 100 cyclodextrins forms a complex (Loftsson et al. 2005). The formulation bulk is then estimated based on the CE. They are inversely proportional. Altogether, these parameters reflect the possibility of using cyclodextrins in the formulation of solid dosage forms (Loftsson et al. 2005).

4.4.2 Retention

The loss of essential oils and their volatile components by evaporation is still a key drawback in their use in the different formulations (Marques 2010; Ciobanu et al. 2012). Indeed, the chemical composition of an essential oil ensures its specific scent and activity. Thus, the loss of some volatile components may have a strong impact on the flavour and the activity of an essential oil. Hence, essential oils should be retained in formulations. This ensures the stability of the organoleptic and sensorial properties of the product and thus its acceptability by the consumers. Retention of essential oils also provides convenience in maintaining an active dose of essential oils or their components when they are incorporated in active formulations. A recent study showed that, among several additives, cyclodextrins showed the strongest effect on aroma retention by inclusion complexation (Baránková and Dohnal 2016). Several release studies revealed that cyclodextrins reduce the volatility, control and delay the release of essential oils and their components (Decock et al. 2008; Ciobanu et al. 2013; Kfoury et al. 2015c, 2016f). The encapsulation in β-cyclodextrin improved the retention of citral (26-fold enhancement) and menthol (86-fold enhancement) in fruit leathers and hard candies, respectively (Reineccius et al. 2004). Carvone and limonene, the major components of caraway essential oil, were efficiently retained during heat treatment (Partanen et al. 2002). Moreover, the retention of essential oils could be beneficial for active essential oils that have, strong odour, pungent taste, high flavour impact and low flavour threshold such as clove essential oil (Hernández-Sánchez et al. 2017), garlic essential oil (Wang et al. 2011), oregano and thyme essential oils (Burt 2004). The encapsulation of essential oils allows a dynamic equilibrium between the free and the encapsulated forms in the formulation. Consequently, the odour and flavour of the encapsulated part will be masked until it is progressively released (Qi and Hedges 1995). The incorporation of cyclodextrins in an energy drink, for example, reduced the bitterness of ginseng up to 50% (Tamamoto et al. 2010). Cyclodextrins were also used to optimize the debittering process of soybean antioxidant hydrolysates (Hou et al. 2013). β-Cyclodextrin extended the release period of lemon aroma in fat free yogurt (Kant et al. 2004) and menthol in chewing gums (Yoshii et al. 2007). Cyclodextrin/essential oil inclusion complexes have also found application in taste masking technology (Pein et al. 2014). For example, formulations with β-cyclodextrin/anethole inclusion complex efficiently contributed to mask the bitterness of zinc acetate dihydrate and, also, retained pleasant flavour longer (Eby 1992, US 5095035 A). Another study showed that the incorporation of β-cyclodextrin into orodispersible tablets of nicergoline containing 7–9% of camphor, improved the taste acceptability of the drug (Elbary et al. 2013). In conclusion, encapsulation in cyclodextrins could be efficient to maintain and retain essential oils in the formulations and allow their gradual release (Martina et al. 2013; Mascheroni et al. 2013).

4.4.3 Stability Enhancement

Essential oils are highly volatile compounds susceptible to oxygen, light, heat and humidity (Turek and Stintzing 2013) and prone to hydrolysis, oxidation, heat degradation, evaporation and reaction with the matrix ingredients (Hyldgaard et al. 2012). Degradation of essential oils could lead to the deterioration of their sensory properties as well as their biologic performance. Therefore, developing formulations to preserve essential oils is considered an essential task for their application in the various fields. Despite many years of research, manufactures still struggle with improving the shelf life of essential oil containing formulations.

The protection of essential oils could be achieved by encapsulation. Cyclodextrins offer essential oils a safer and longer life within a protected environment, the apolar cavity while maintaining their properties (De Vos et al. 2010; Liang et al. 2012; Hill et al. 2013; Pinho et al. 2014; Costa et al. 2015). Encapsulation also prevents off note development (off-flavours) and the production of toxic isomers by reducing contact between essential oil components and oxygen, ions or other formulation ingredients and avoiding direct exposure to light (Szejtli and Szente 2005). For example, encapsulation in hydroxypropyl-β-cyclodextrin (HP-β-CD) improved significantly the stability of basil and tarragon essential oils against UV irradiations (Fig. 4.7) (Kfoury et al. 2015b).

It has been demonstrated that the addition of β-cyclodextrin preserves the colour intensity and reduces the browning of pear juice without any significant decrease in the aromatic quality (Lopez-Nicolas et al. 2009; Andreu-Sevilla et al. 2011). This could be explained by the fact that cyclodextrins protect the aromatic compounds from their environment (Lopez-Nicolas et al. 2014). Cyclodextrins prevent the cyclization of citral (Szente and Szejtli 2004) and the isomerization of *trans*-anethole (Kfoury et al. 2014a), estragole (Kfoury et al. 2015b), eugenol and isoeugenol (Kfoury et al. 2016a) under UV-irradiation. Cyclodextrins also stabilized geraniol and thymol (Mourtzinos et al. 2008), curcumin (Paramera et al. 2011) and nootkatone (Kfoury et al. 2016e) against photodegradation. It has been also proved that the stability of the encapsulated guests was enhanced at high temperature and in humidity conditions. The encapsulation of Patchouli alcohol in β-cyclodextrin resulted in a slower degradation rate under 40 °C and 70% relative humidity conditions (Xu et al. 2017). Moreover, the encapsulation in β-cyclodextrin enhanced the thermal and oxidative stability of *Salmo salar L.* essential oil as evaluated by the means of thermal and Karl Fischer titration (KFT) methods (Hădărugă et al. 2016).

Thus, encapsulation in cyclodextrins might be also one of the promising tools to protect essential oils against the harmful environment.

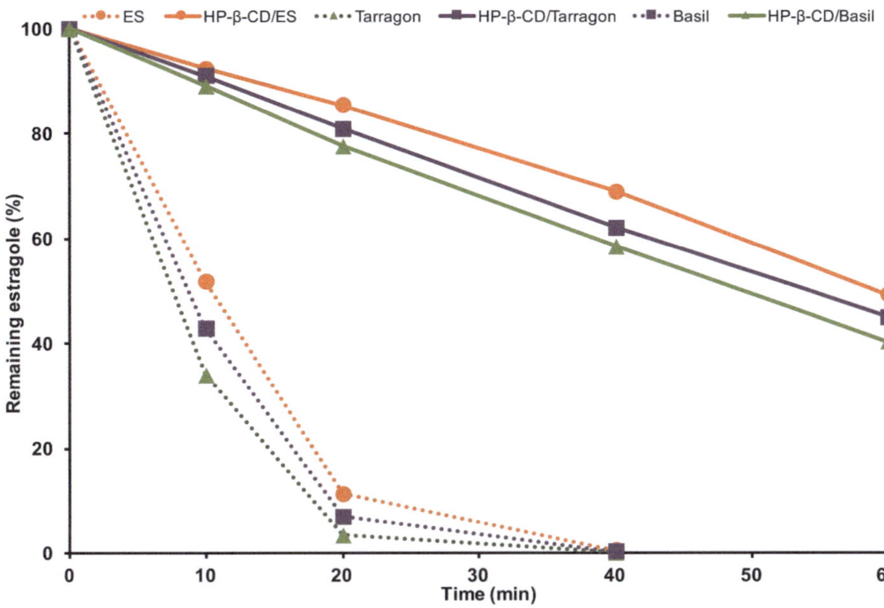

Fig. 4.7 Photodegradation profiles of pure estragole and estragole present in tarragon and basil essential oils in water and in hydroxypropyl-β-cyclodextrin (HP-β-CD) solution under UVC irradiation. The protective effect of cyclodextrin is clear and the degradation of estragole was slowed down by encapsulation. Free estragole was completely disappeared after 40 min of irradiation whereas more than 40% was still remaining in the inclusion complexes solutions after 1 h of irradiation (Kfoury et al. 2015b; reprinted with permission)

4.5 Effect of Encapsulation on the Biological Properties of Essential Oils

The exploration and understanding of the benefits of essential oils in nutrition and medicine are still in progress. According to the Biopharmaceutics Classification System, volatile compounds (such as essential oils and essential oil components) belong to class 2. Drugs that belong to this class have low solubility and high permeability; therefore, they mainly undergo passive transport through the biological membranes (Amidon et al. 1995).

Inclusion complexes are essential to design essential oils containing formulations because cyclodextrins do not interfere with their activity (Lucas-Abellán et al. 2008; Costa et al. 2015). Also, cyclodextrins do not permeate hydrophobic membranes *via* passive diffusion since they are large hydrophilic oligosaccharides with numerous hydrogen bond donors and acceptors (Lipinski 2000, 2004; Loftsson and Brewster 2010; Kurkov and Loftsson 2013). They act as penetration enhancers. They increase the solubility and the concentration of the active agent at the biological membrane. This, consequently, ameliorates its diffusion (Kurkov and

Loftsson 2013). One of the primary sites of action of essential oils is the biological membrane (Bakkali et al. 2008; Raut and Karuppayil 2014). Cyclodextrin inclusion complexes enhance the access of active compounds to this region (Kurkov and Loftsson 2013). Accordingly, a lower concentration of active agent could be used to achieve the effective concentration at the target sites. Several studies attempted to evaluate the effect of encapsulation on the antimicrobial and antioxidant properties of essential oils. Table 4.5 summarizes the main published results regarding the antimicrobial properties of essential oils or their individual components upon encapsulation in cyclodextrins.

The formulations containing cyclodextrin encapsulated essential oils showed a broad-spectrum of antimicrobial activity against a wide range of pathogenic microorganisms; this is evidently due to the different essential oils chemical compositions (Arana-Sánchez et al. 2010). Many studies have been carried out on main essential oil components as main contributors to the properties of essential oils such as eugenol in clove oil, thymol in thyme oil, carvacrol in oregano oil, anethole in anis oil, allyl disulfide in garlic oil, etc. (Huang et al. 2010; Hill et al. 2013; Llana-Ruiz-Cabello et al. 2015; Cetin Babaoglu et al. 2017).

The results are summarized in Table 4.5. They can be divided into three categories. Firstly, the encapsulation improved the antimicrobial activity of essential oils by decreasing the active concentration (MIC, MBC, IC50 and LD50) or increasing the inhibition of growth. This improvement of the antimicrobial properties could be due to the increased water solubility of the active compounds upon encapsulation, which in turn enhanced the surface area of contact with the pathogen (Bilia et al. 2014; Fenyvesi et al. 2014). Essential oils could also bind to the hydrophobic constituents such as lipids and proteins that reduce their activity (Burt et al. 2005; Tao et al. 2014). But, cyclodextrins could reduce these interactions and preserve the inhibitory properties of essential oils (Budryn et al. 2015). Also, the molecular size of the inclusion complex may increase the cellular absorption mechanisms and thus increasing the antimicrobial activity (Bilia et al. 2014; Rakmai et al. 2017a). Secondly, the encapsulation does not affect the antimicrobial activity. Finally, the encapsulation maintained some antimicrobial properties of the essential oils. This could be explained by the fact that cyclodextrins themselves are able to increase the growth of some pathogens since they could be used as a carbohydrate source for the microorganisms (Ayala-Zavala et al. 2008: Kfoury et al. 2016a).

Aiming a potential application of essential oils as antioxidant agents, it is generally intended to reduce their concentration as much as possible while achieving a notable activity, especially when they have a strong pungent taste and smell. The effect of encapsulation on the antioxidant properties of essential oils and their components was evaluated in several studies. A synopsis of the main results is presented in Table 4.6.

In general, the encapsulation of essential oils in cyclodextrins either maintained or increased the ability of essential oils to reduce free radicals. However, few studies showed a decrease in the potential of essential oils to scavenge the free radicals. It seems that cyclodextrins could block the functional groups of essential oils, which

Table 4.5 Literature data related to the antimicrobial activity of free and encapsulated essential oils

EO or EO component	Cyclodextrin	Microorganism	Free	Encapsulated	References
Achillea millefolium L. s. l.	HP-β-CD	*Escherichia coli*	500[a]	62.5[a]	Rakmai et al. 2017a
Achillea millefolium L. s. l.	HP-β-CD	*Staphylococcus aureus*	250[a]	62.5[a]	Rakmai et al. 2017a
Carvacrol	α-CD	*Escherichia coli*	0.5[a]	0.125[a]	Liang et al. 2012
Carvacrol	β-CD	*Escherichia coli*	0.5[a]	0.25[a]	Liang et al. 2012
Carvacrol	HP-β-CD	*Escherichia coli*	0.5[a]	0.125[a]	Liang et al. 2012
Carvacrol	α-CD	*Staphylococcus aureus*	0.5[a]	0.5[a]	Liang et al. 2012
Carvacrol	β-CD	*Staphylococcus aureus*	0.5[a]	0.5[a]	Liang et al. 2012
Carvacrol	HP-β-CD	*Staphylococcus aureus*	0.5[a]	0.125[a]	Liang et al. 2012
Carvacrol	α-CD	*Bacillus subtilis*	0.5[a]	0.5[a]	Liang et al. 2012
Carvacrol	β-CD	*Bacillus subtilis*	0.5[a]	0.5[a]	Liang et al. 2012
Carvacrol	HP-β-CD	*Bacillus subtilis*	0.5[a]	0.125[a]	Liang et al. 2012
Carvacrol	α-CD	*Saccharomyces cerevisiae*	0.5[a]	0.25[a]	Liang et al. 2012
Carvacrol	β-CD	*Saccharomyces cerevisiae*	0.5[a]	0.25[a]	Liang et al. 2012
Carvacrol	HP-β-CD	*Saccharomyces cerevisiae*	0.5[a]	0.25[a]	Liang et al. 2012
Carvacrol	β-CD	*Escherichia coli K12*	1000[a]	300[a]	Santos et al. 2015
Carvacrol	β-CD	*Saccharomyces cerevisiae*	0.5[a]	0.25[a]	Liang et al. 2012
Carvacrol	HP-β-CD	*Saccharomyces cerevisiae*	0.5[a]	0.25[a]	Liang et al. 2012
Carvacrol	β-CD	*Escherichia coli K12*	1000[a]	300[a]	Santos et al. 2015
Carvacrol	β-CD	*Salmonella typhimurium*	1100[a]	300[a]	Santos et al. 2015
Carvacrol	HP- β-CD	*Salmonella typhimurium LT2*	1150[a]	300[a]	Kamimura et al. 2014
Carvacrol	HP- β-CD	*Escherichia coli K12*	1000[a]	300[a]	Kamimura et al. 2014
Cinnamon	β-CD	*Salmonella typhimurium LT2*	400[a]	166[a]	Hill et al. 2013

(continued)

Table 4.5 (continued)

EO or EO component	Cyclodextrin	Microorganism	Free	Encapsulated	References
Citrus sinensis (L.) Osbeck	β-CD	*Aedes aegypti* Larvae	21.5[b]	23[b]	Galvão et al. 2015
Clove	β-CD	*Salmonella typhimurium LT2*	>1000[a]	281[a]	Hill et al. 2013
Coriandrum sativum L	β-CD	*Bacillus subtilis*	4.2[c]	1.2[c]	Dima et al. 2014
Coriandrum sativum L	β-CD	*Bacillus cereus*	4.1[c]	1.1[c]	Dima et al. 2014
Coriandrum sativum L	β-CD	*Rhodorula glutinis*	4.0[c]	1.8[c]	Dima et al. 2014
Coriandrum sativum L	β-CD	*Candida utilis*	4.3[c]	1.2[c]	Dima et al. 2014
Coriandrum sativum L	β-CD	*Saccharomyces cerevisiae*	4.7[c]	0.9[c]	Dima et al. 2014
Coriandrum sativum L	β-CD	*Aspergillus Niger*	7.5[c]	3.1[c]	Dima et al. 2014
Coriandrum sativum L	β-CD	*Penicillium glaucum*	6.5[c]	3.0[c]	Dima et al. 2014
Coriandrum sativum L	β-CD	*Geotrichum candidum*	5.3[c]	1.1[c]	Dima et al. 2014
Estragole	HP-β-CD	*Fusarium oxysporum*	540[d]	880[d]	Kfoury et al. 2016a
Estragole	HP-β-CD	*Botrytis cinerea*	527[d]	765[d]	Kfoury et al. 2016a
Eugenol	HP-β-CD	*Fusarium oxysporum*	151[d]	676[d]	Kfoury et al. 2016a
Eugenol	HP-β-CD	*Botrytis cinerea*	190[d]	326[d]	Kfoury et al. 2016a
Eugenol	α-CD	*Escherichia coli*	5[a]	5[a]	Liang et al. 2012
Eugenol	β-CD	*Escherichia coli*	5[a]	5[a]	Liang et al. 2012
Eugenol	HP-β-CD	*Escherichia coli*	5[a]	2.5[a]	Liang et al. 2012
Eugenol	α-CD	*Staphylococcus aureus*	2.5[a]	1.25[a]	Liang et al. 2012
Eugenol	β-CD	*Staphylococcus aureus*	2.5[a]	1.25[a]	Liang et al. 2012
Eugenol	HP-β-CD	*Staphylococcus aureus*	2.5[a]	1.25[a]	Liang et al. 2012
Eugenol	α-CD	*Bacillus subtilis*	1.25[a]	0.625[a]	Liang et al. 2012
Eugenol	β-CD	*Bacillus subtilis*	1.25[a]	0.625[a]	Liang et al. 2012

(continued)

Table 4.5 (continued)

EO or EO component	Cyclodextrin	Microorganism	Free	Encapsulated	References
Eugenol	HP-β-CD	*Bacillus subtilis*	1.25[a]	0.625[a]	Liang et al. 2012
Eugenol	α-CD	*Saccharomyces cerevisiae*	10[a]	10[a]	Liang et al. 2012
Eugenol	β-CD	*Saccharomyces cerevisiae*	10[a]	10[a]	Liang et al. 2012
Eugenol	HP-β-CD	*Saccharomyces cerevisiae*	10[a]	5[a]	Liang et al. 2012
Eugenol	β-CD	*Salmonella typhimurium LT2*	1000[a]	693[a]	Hill et al. 2013
Isoeugenol	HP-β-CD	*Fusarium oxysporum*	133[d]	482[d]	Kfoury et al. 2016a
Isoeugenol	HP-β-CD	*Botrytis cinerea*	82[d]	733[d]	Kfoury et al. 2016a
Linalool	α-CD	*Escherichia coli*	5[a]	5[a]	Liang et al. 2012
Linalool	β-CD	*Escherichia coli*	5[a]	5[a]	Liang et al. 2012
Linalool	HP-β-CD	*Escherichia coli*	5[a]	2.5[a]	Liang et al. 2012
Linalool	α-CD	*Staphylococcus aureus*	2.5[a]	1.25[a]	Liang et al. 2012
Linalool	β-CD	*Staphylococcus aureus*	2.5[a]	1.25[a]	Liang et al. 2012
Linalool	HP-β-CD	*Staphylococcus aureus*	2.5[a]	1.25[a]	Liang et al. 2012
Linalool	α-CD	*Bacillus subtilis*	2.5[a]	1.25[a]	Liang et al. 2012
Linalool	β-CD	*Bacillus subtilis*	2.5[a]	1.25[a]	Liang et al. 2012
Linalool	HP-β-CD	*Bacillus subtilis*	2.5[a]	2.5[a]	Liang et al. 2012
Linalool	α-CD	*Saccharomyces cerevisiae*	5[a]	5[a]	Liang et al. 2012
Linalool	β-CD	*Saccharomyces cerevisiae*	5[a]	5[a]	Liang et al. 2012
Linalool	HP-β-CD	*Saccharomyces cerevisiae*	5[a]	1.25[a]	Liang et al. 2012
Origanum vulgare	β-CD	*Microsporum gypseum*	62.04[e]	67.09[e]	Papajani et al. 2015
Origanum vulgare	β-CD	*Microsporum canis*	53.72[e]	59.06[e]	Papajani et al. 2015
Origanum vulgare	β-CD	*Arthroderma cajetani*	41.94[e]	51.89[e]	Papajani et al. 2015

(continued)

Table 4.5 (continued)

EO or EO component	Cyclodextrin	Microorganism	Free	Encapsulated	References
Origanum vulgare	β-CD	*Trichophyton mentagrophytes*	75.78[e]	89.81[e]	Papajani et al. 2015
Origanum vulgare	β-CD	*Epidermophyton floccosum*	94.59[e]	99.56[e]	Papajani et al. 2015
Origanum vulgare	β-CD	*Trichophyton rubrum*	67.44[e]	74.44[e]	Papajani et al. 2015
Origanum vulgare	β-CD	*Botrytis cinerea*	36.29[e]	45.22[e]	Papajani et al. 2015
Origanum vulgare	β-CD	*Pyricuhria oryzae*	61.11[e]	74.09[e]	Papajani et al. 2015
Rosmarinus officinalis	β-CD	*Microsporum gypseum*	14.58[e]	13.91[e]	Papajani et al. 2015
Rosmarinus officinalis	β-CD	*Microsporum canis*	0[e]	1.8[e]	Papajani et al. 2015
Rosmarinus officinalis	β-CD	*Tricholosporum violaceum*	20.45[e]	21.65[e]	Papajani et al. 2015
Rosmarinus officinalis	β-CD	*Trichophyton mentagrophytes*	10.71[e]	11.20[e]	Papajani et al. 2015
Rosmarinus officinalis	β-CD	*Trichophyton rubrum*	38.21[e]	43.14[e]	Papajani et al. 2015
Rosmarinus officinalis	β-CD	*Trichophyton tonsurans*	4.26[e]	6.23[e]	Papajani et al. 2015
Rosmarinus officinalis	β-CD	*Pyricuhria oryzae*	17.53[e]	18.47[e]	Papajani et al. 2015
Terpinen-4-ol	β-CD	*Staphylococcus aureus*	6-15[c]	>15[c]	Yang et al. 2015
Terpinen-4-ol	β-CD	*Pseudomonas aeruginosa*	6-15[c]	>15[c]	Yang et al. 2015
Terpinen-4-ol	β-CD	*Escherichia coli*	6-15[c]	>15[c]	Yang et al. 2015
Thyme	β-CD	*Escherichia coli K12*	640[a]	470-850[a]	Tao et al. 2014
Thymol	β-CD	*Escherichia coli K12*	730[a]	370-900[a]	Tao et al. 2014
trans-Anethole	HP-β-CD	*Fusarium oxysporum*	306[d]	341[d]	Kfoury et al. 2016a
trans-Anethole	HP-β-CD	*Botrytis cinerea*	358[d]	620[d]	Kfoury et al. 2016a
trans-Cinnamaldehyde	β-CD	*Salmonella typhimurium LT2*	400[a]	479[a]	Hill et al. 2013
Estragole	β-CD	*Staphylococcus aureus*	1900[a]	14000[a]	Yang et al. 2017
Estragole	β-CD	*Bacillus subtilis*	2100[a]	15000[a]	Yang et al. 2017

(continued)

Table 4.5 (continued)

EO or EO component	Cyclodextrin	Microorganism	Free	Encapsulated	References
Estragole	β-CD	*Escherichia coli*	2500[a]	19000[a]	Yang et al. 2017
Lippia graveolens	β-CD	*Escherichia coli*	0.2[f]	0.05[f]	Arana-Sánchez et al. 2010
Lippia graveolens	β-CD	*Pseudomonas aeruginosa*	0.2[f]	0.1[f]	Arana-Sánchez et al. 2010
Lippia graveolens	β-CD	*Staphylococcus aureus*	0.1[f]	0.05[f]	Arana-Sánchez et al. 2010

EO essential oil; *CD* cyclodextrin; *HP-β-CD* hydroxypropyl-β-cyclodextrin
[a]minimum inhibitory concentration MIC (μg/ml)
[b]median lethal dose LD50 (ppm)
[c]inhibition zone (mm)
[d]half maximal inhibitory concentration IC50 (μg/ml)
[e]% inhibition of growth
[f]minimal bactericidal concentration MBC (μg/ml)

become less available to reduce the radical species (Kamimura et al. 2014; Rakmai et al. 2017a, b). On the contrary, the preservation and improvement of the antioxidant activity upon encapsulation may be due to the formation of inclusion complexes (Lucas-Abellán et al. 2008; Lu et al. 2009). While inclusion complexes may guarantee excellent protection of essential oils against degradation or evaporation, they do not interfere with the functional groups of the active compounds (Kfoury et al. 2014a; Rakmai et al. 2017a, b). An increase in the antioxidant activity might be observed when intermolecular hydrogen bonds are formed between cyclodextrin and guest resulting in a stable radical formation upon reaction with free radicals (Lucas-Abellán et al. 2008; Zhao et al. 2010).

Also, cyclodextrins could act as secondary antioxidants that enhance the anti-radical efficiency of guests (Nunez-Delicado et al. 1997; Lopez-Nicolas et al. 2007).

Overall, these results provide encouraging evidence of the potential effectiveness of cyclodextrin encapsulated essential oils to be used as stable natural antimicrobial and antioxidant agents. With the increase of consumer concerns regarding limited availability of natural products and environment protection, the use of cyclodextrin/essential oil inclusion complexes will gain further interest in detriment to synthetic materials. Importantly, both cyclodextrins and essential oils are natural and obtained from renewable resources.

Table 4.6 Literature data on the antioxidant activity of free and encapsulated essential oils

EO or EO component	Cyclodextrin	Radical	Free	Encapsulated	References
trans-Anethole	α-CD	DPPH•	6[a]	8[a]	Kfoury et al. 2014a
trans-Anethole	β-CD	DPPH•	6[a]	9[a]	Kfoury et al. 2014a
Carvacrol	HP-β-CD	ABTS+•	7491[b]	6797[b]	Kamimura et al. 2014
Carvacrol	HP-β-CD	ABTS+•	7491[b]	7042[b]	Santos et al. 2015
Carvacrol	β-CD	DPPH•	0.03[c]	0.02[c]	Miguel et al. 2010
Carvacrol	β-CD	ABTS+•	7057[b]	11542[b]	Kfoury et al. 2016d
Carvacrol	HP-β-CD	ABTS+•	7057[b]	12086[b]	Kfoury et al. 2016d
Cinnamomum camphora CT linalol	β-CD	ABTS+•	83[b]	244[b]	Kfoury et al. 2015c
Citrus reticulata Blanco	β-CD	ABTS+•	112[b]	286[b]	Kfoury et al. 2015c
Clove	HP-β-CD	DPPH•	25.2[a]	28.7[a]	Cetin Babaoglu et al. 2017
Coriandrum sativum L	β-CD	DPPH•	87.89[a]	79.77[a]	Dima et al. 2014
Cymbopogon nardus	β-CD	ABTS+•	146[b]	352[b]	Kfoury et al. 2015c
Estragole	α-CD	DPPH•	2[a]	10[a]	Kfoury et al. 2014a
Estragole	β-CD	DPPH•	2[a]	3[a]	Kfoury et al. 2014a
Eucalyptus citriodora	β-CD	ABTS+•	99[b]	195[b]	Kfoury et al. 2015c
Eugenol	α-CD	DPPH•	54[a]	50[a]	Kfoury et al. 2014a
Eugenol	β-CD	DPPH•	54[a]	54[a]	Kfoury et al. 2014a
Isoeugenol	α-CD	DPPH•	58[a]	57[a]	Kfoury et al. 2014a
Isoeugenol	β-CD	DPPH•	58[a]	57[a]	Kfoury et al. 2014a
Lippia graveolens	β-CD	DPPH•	88.11[a]	45.8[a]	Arana-Sánchez et al. 2010
Origanum compactum	β-CD	ABTS+•	4120[b]	4760[b]	Kfoury et al. 2015c
Origanum majorana CT thujanol	β-CD	ABTS+•	84[b]	211[b]	Kfoury et al. 2015c
Pimenta racemosa	β-CD	ABTS+•	7608[b]	8292[b]	Kfoury et al. 2015c
Piper nigrum L.	HP-β-CD	DPPH•	54[a]	50[a]	Rakmai et al. 2017b
Rosmarinus officinalis cineoliferum	β-CD	ABTS+•	62[b]	230[b]	Kfoury et al. 2015c
Satureja montana	β-CD	ABTS+•	3063[b]	4674[b]	Kfoury et al. 2015c
Thymol	β-CD	DPPH•	0.03[c]	0.02[c]	Miguel et al. 2010
Thymol	β-CD	ABTS+•	14927[b]	20764[b]	Kfoury et al. 2016d
Thymol	HP-β-CD	ABTS+•	14927[b]	21201[b]	Kfoury et al. 2016d

EO essential oil; *CD* cyclodextrin; *HP-β-CD* hydroxypropyl-β-cyclodextrin; *ABTS+•* 2,2′-azino-bis(3- ethylbenzothiazoline-6-sulphonic acid) radical cation; *DPPH•* 1,1-diphenyl-2-picryl-hydrazyl radical

[a]Antioxidant Activity AA (%)

[b]Trolox equivalent antioxidant capacity TEAC (μmol/g)

[c]half maximal inhibitory concentration IC50 (mg)

4.6 Emerging Cyclodextrin-Based Technologies for Essential Oils Encapsulation

Cyclodextrins are also considered as promising carriers in textiles, cosmetotextiles and medical textiles (Voncina and Vivod 2013; Radu et al. 2016). Cyclodextrins, generally grafted to textiles, provide them hosting cavities that can encapsulate a wide variety of functional guest molecules (Szejtli 2003; Voncina and Vivod 2013). The incorporation of slimming, hydrating, or perfuming agents as inclusion complexes allows their controlled release and ensures, consequently, their progressive effect on the skin (Rakshit 2011). Textile material containing β-cyclodextrin/cedar oil, as example, showed a prolonged insect repellent activity compared to textile materials containing free cedar oil[3]. Also the odour release intensity of perfume from PET textile was considerably postponed when the material was treated by cyclodextrins (Voncina and Vivod 2013).

More recently, the electrospinning has attracted considerable attention to produce nanofibers and nanowebs having high surface-to-volume ratio and highly porous structure. The electrospun nanofibers have large applications in food packaging, wound dressing and biomedical use etc. Cyclodextrins are interesting in this field. The self-assembly and aggregation characteristics of these host molecules in concentrated solutions allow the production of nanofibers whitout any polymer matrix (Aytac et al. 2016b). Thus, the nanofibers and nanowebs could be effectively functionalized with cyclodextrin/essential oil inclusion complexes. Figure 4.8 shows a schematic representation of the electrospinning of nanofibers from zein and cyclodextrin/thymol inclusion complex (Aytac et al. 2017).

The electrospinning technology is very simple; a continuous filament is electrospun from concentrated inclusion complex solutions (with or without polymer) under a very high electrical field resulting in nanofibers (Aytac et al. 2016a, b). Electrospun nanofibers were successfully obtained with inclusion complexes of geraniol (Aytac et al. 2016a), vanillin (Celebioglu et al. 2016), limonene (Aytac et al. 2016b).

These nanofibers have shown fast-dissolving properties, high thermal stability, improved water solubility and enhanced antimicrobial and antioxidant properties. An interesting feature of these electrospun nanofibers is their free-standing nature useful for a simple and longer storage of essential oils. Electrospinnig nanofibers from cyclodextrin/essential oil inclusion complexes would be extremely interesting for food application because of the edible nature of native cyclodextrins and for further potential medical, packaging, textile and agriculture use. Electrospun polystyrene textile fibers containing cyclodextrin/menthol inclusion complex retained

[3]Vraz Kresevic S, Voncina B, Gersak J. Insect resistant and eco-friendly textile materials. In: Dragcevic Zvonko (ed.): 4th International Textile, Clothing & Design Conference: Magic world of textiles : book of proceedings, ITC&DC, 05-08 October, 2008, Dubrovnik, Croatia. Zagreb: Faculty of Textile Technology, University of Zagreb; 2008.

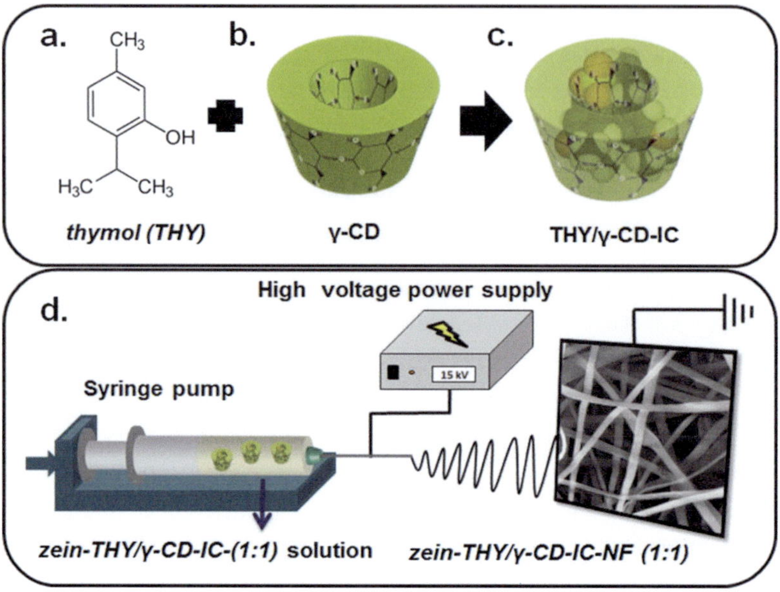

Fig. 4.8 The chemical structure of (**a**) thymol (THY); (**b**) schematic representation of γ-cyclodextrin (γ-CD), (**c**) γ-cyclodextrin/thymol inclusion complex (THY/ γ -CD-IC) formation, and (**d**) electrospinning of nanofibers from zein-THY/c-CD- IC (1:1) solution (Aytac et al. 2017; (reprinted with permission)

menthol and enhanced its temperature stability (Uyar et al. 2009). Also elecrospun zein-thymol/cyclodextrin inclusion complex showed a gradual release of thymol and had strong antibacterial activity (Aytac et al. 2017).

4.7 Conclusion

The encapsulation in cyclodextrins is a promising approach for the incorporation of essential oils in food, pharmaceutical, cosmetic, textile, and other products. Cyclodextrins are able to overcome the major limitation of essential oils use, the low aqueous solubility, and to increase their chemical stability in the presence of light, oxygen, humidity and heat. Importantly, cyclodextrins can successfully encapsulate essential oils both in solution and in solid state. Moreover, formation of inclusion complexes ensures an easier handling of oily essential oils and solid dosage formulation. Cyclodextrins, also, allow a delayed and controlled release of essential oils under specific conditions of humidity and temperature. In addition, encapsulation in cyclodextrins maintains or even enhances the biological properties and functionalities of essential oils. The use of essential oils coming from natural sources and having relevant biological properties will encourage the transition to a more clean and sustainable environment.

Acknowledgments Authors thank Marc Fourmentin from the university of the Littoral Opal Coast (ULCO), France for the illustrations.

References

Amidon GL, Lennernäs H, Shah VP, Crison JR (1995) A theoretical basis for a biopharmaceutic drug classification: the correlation of in vitro drug product dissolution and in vivo bioavailability. Pharm Res 12:413–420. https://doi.org/10.1023/A:1016212804288

Andrade TA, Freitas TS, Araújo FO, Menezes PP, Anne G, Dória A, Rabelo AS, Quintans-Júnior LJ, Santos MRV, Bezerra DP, Serafini MR, Menezes IRS, Nunes PS, Araújo AAS, Costa MS, Campina FF, Santos ATL, Silva ARP, Coutinho HDM (2017) Physico-chemical characterization and antibacterial activity of inclusion complexes of *Hyptis martiusii Benth* essential oil in β-cyclodextrin. Biomed Pharmacother 89:201–207. https://doi.org/10.1016/j.biopha.2017.01.158

Andreu-Sevilla AJ, López-Nicolás JM, Carbonell-Barrachina ÁA, García-Carmona F (2011) Comparative effect of thea of α-, β-, or γ-cyclodextrin on main sensory and physico-chemical parameters. J Food Sci 76(5):347–353. https://doi.org/10.1111/j.1750-3841.2011.02190.x

Anishetty R, Puri S, Garg V, Yadav AK, Mittal A (2015) Approaches for overcoming poor oral bioavailability of drugs: nanoarchitectures in focus. Int J Pharma Bio Sci 6(4):306–334

Ansari MJ, Kohli K, Ali J, Anwer MK, Jamil S, Ahmed MM (2014) Physicochemical characterizations and dissolution behavior of curcumin and α-cyclodextrin molecular inclusion complexes. Der Pharm Lett 6(6):1–7

Arana-Sánchez A, Estarrón-Espinosa M, Obledo-Vázquez EN, Padilla-Camberos E, Silva-Vázquez R, Lugo-Cervantes E (2010) Antimicrobial and antioxidant activities of Mexican oregano essential oils (*Lippia graveolens* H. B. K.) with different composition when microencapsulated in β-cyclodextrin. Lett Appl Microbiol 50(6):585–590. https://doi.org/10.1111/j.1472-765X.2010.02837.x

Arfat YA, Benjakul S, Vongkamjan K, Sumpavapol P, Yarnpakdee S (2015) Shelf-life extension of refrigerated sea bass slices wrapped with fish protein isolate/fish skin gelatin-ZnO nanocomposite film incorporated with basil leaf essential oil. J Food Sci Technol 52:6182–6193. https://doi.org/10.1007/s13197-014-1706-y

Astray G, Gonzalez-Barreiro C, Mejuto JC, Rial-Otero R, Simal-Gándara J (2009) A review on the use of cyclodextrins in foods. Food Hydrocoll 23(7):1631–1640. https://doi.org/10.1016/j.foodhyd.2009.01.001

Astray G, Mejuto JC, Morales J, Rial-Otero R, Simal-Gandara J (2010) Factors controlling flavors binding constants to cyclodextrins and their applications in foods. Food Res Int 43(4):1212–1218. https://doi.org/10.1016/j.foodres.2010.02.017

Ayala-Zavala JF, Soto-Valdez H, Gonzalez-Leon A, Alvarez-Parrilla E, Martin-Belloso O, Gonzalez-Aguilar GA (2008) Microencapsulation of cinnamon leaf (*Cinnamomum zeylanicum*) and garlic (*Allium sativum*) oils in β-cyclodextrin. J Incl Phenom Macrocycl Chem 60:359–368. https://doi.org/10.1007/s10847-007-9385-1

Aytac Z, Yildiz ZI, Kayaci-Senirmak F, San Keskin NO, Tekinayde T, Uyar T (2016a) Electrospinning of polymer-free cyclodextrin/geraniol-inclusion complex nanofibers: enhanced shelf-life of geraniol with antibacterial and antioxidant properties. RSC Adv 6(52):46089–46099. https://doi.org/10.1039/c6ra07088d

Aytac Z, Yildiz ZI, Kayaci-Senirma F, San KNO, Kusku SI, Durgun E, Tekinay T, Uyat T (2016b) Fast-dissolving, prolonged release, and antibacterial cyclodextrin/limonene-inclusion complex nanofibrous webs via polymer-free electrospinning. J Agric Food Chem 64(39):7325–7334. https://doi.org/10.1021/acs.jafc.6b02632

Aytac Z, Ipek S, Durgun E, Tekinay T, Uyar T (2017) Antibacterial electrospun zein nanofibrous web encapsulating thymol/cyclodextrin-inclusion complex for food packaging. Food Chem 233:117–124. https://doi.org/10.1016/j.foodchem.2017.04.095
Bai Y, Yu B, Xu X, Jin Z, Tian Y, Lu L (2010) Comparison of encapsulation properties of major garlic oil components by hydroxypropyl β-cyclodextrin. Eur Food Res Technol 231(4):519–524. https://doi.org/10.1007/s00217-010-1307-6
Bajpai VK, Baek K, Chul S (2012) Control of salmonella in foods by using essential oils : a review. Food Res Int 45(2):722–734. https://doi.org/10.1016/j.foodres.2011.04.052
Bakkali F, Averbeck S, Averbeck D, Idaomar M (2008) Biological effects of essential oils – a review. Food Chem Toxicol 46:446–475. https://doi.org/10.1016/j.fct.2007.09.106
Baránková E, Dohnal V (2016) Effect of additives on volatility of aroma compounds from dilute aqueous solutions. Fluid Phase Equilib 407:217–223. https://doi.org/10.1016/j.fluid.2015.05.038
Bertaut E, Landy D (2014) Improving ITC studies of cyclodextrin inclusion compounds by global analysis of conventional and non-conventional experiments. Beilstein J Org Chem 10:2630–2641. https://doi.org/10.3762/bjoc.10.275
Berthelot K, Estevez Y, Deffieux A, Peruch F (2012) Isopentenyl diphosphate isomerase: a checkpoint to isoprenoid biosynthesis. Biochimie 94(8):1621–1634. https://doi.org/10.1016/j.biochi.2012.03.021
Bethanis K, Tzamalis P, Tsorteki F, Kokkinou A, Christoforides E, Mentzafos D (2013) Structural study of the inclusion compounds of thymol, carvacrol and eugenol in β-cyclodextrin by X-ray crystallography. J Incl Phenom Macro 77(1–4):163–173. https://doi.org/10.1007/s10847-012-0230-9
Bilia AR, Guccione C, Isacchi B, Righeschi C, Firenzuoli F, Bergonzi MC (2014) Essential oils loaded in nanosystems : a developing strategy for a successful therapeutic approach. Evid Based Complement Alternat Med. Article ID 651593. https://doi.org/10.1155/2014/651593
Bourgeois W, Burgess JE, Stuetz RM (2001) On-line monitoring of wastewater quality: a review. J Chem Technol Biotechnol 76(4):337–348. https://doi.org/10.1002/jctb.393
Brasil IM, Gomes C, Puerta-gomez A, Castell-perez ME, Moreira RG (2012) Polysaccharide-based multilayered antimicrobial edible coating enhances quality of fresh-cut papaya. Food Sci Technol 47(1):39–45. https://doi.org/10.1016/j.lwt.2012.01.005
Brewster ME, Loftsson T (2007) Cyclodextrins as pharmaceutical solubilizers. Adv Drug Deliv Rev 59(7):645–666. https://doi.org/10.1016/j.addr.2007.05.012
Budryn G, Zaczyńska D, Rachwał-Rosiak D, Oracz J (2015) Changes in properties of food proteins after interaction with free and β encapsulated hydroxycinnamic acids. Eur Food Res Technol 240:1157–1166. https://doi.org/10.4315/0362-028X-68.5.919
Burt S (2004) Essential oils: their antibacterial properties and potential applications in foods — a review. Int J Food Microbiol 94:223–253. https://doi.org/10.1016/j.ijfoodmicro.2004.03.022
Burt SA, Vlielander R, Haagsman HP, Veldhuizen EJA (2005) Increase in activity of essential oil components carvacrol and thymol against *Escherichia coli* O157:H7 by addition of food stabilizers. J Food Prot 68(5):919–926. https://doi.org/10.4315/0362-028X-68.5.919
Carrier RL, Miller LA, Ahmed I (2007) The utility of cyclodextrins for enhancing oral bioavailability. J Control Release 123:78–99. https://doi.org/10.1016/j.jconrel.2007.07.018
Carvalho IT, Estevinho BN, Santos L (2016) Application of microencapsulated essential oils in cosmetic and personal healthcare products–a review. Int J Cosmet Sci 38:109–119. https://doi.org/10.1111/ics.12232
Celebioglu A, Kayaci-Senirmak F, İpek S, Durgunab E, Uyar T (2016) Polymer-free nanofibers from vanillin/cyclodextrin inclusion complexes: high thermal stability, enhanced solubility and antioxidant property. Food Funct 7:3141–3153. https://doi.org/10.1039/C6FO00569A
Cetin Babaoglu H, Bayrak A, Ozdemir N, Ozgun N (2017) Encapsulation of clove essential oil in hydroxypropyl beta-cyclodextrin for characterization, controlled release, and antioxidant activity. J Food Process Preserv. *In press*. https://doi.org/10.1111/jfpp.13202

Chaves AV, He ML, Yang WZ, Hristov AN, McAllister TA, Benchaar C (2008) Effects of essential oils on proteolytic, deaminative and methanogenic activities of mixed ruminal bacteria. Can J Anim Sci 88:117–122. https://doi.org/10.4141/CJAS07061

Chen H, Ji H, Zhou X, Wang L (2010) Green synthesis of natural benzaldehyde from cinnamon oil catalyzed by hydroxypropyl-β-cyclodextrin. Tetrahedron 66(52):9888–9893. https://doi.org/10.1016/j.tet.2010.10.063

Ciobanu A, Mallard I, Landy D, Brabie G, Nistor D, Fourmentin S (2012) Inclusion interactions of cyclodextrins and crosslinked cyclodextrin polymers with linalool and camphor in *Lavandula angustifolia* essential oil. Carbohydr Polym 87(3):1963–1970. https://doi.org/10.1016/j.carbpol.2011.10.005

Ciobanu A, Landy D, Fourmentin S (2013) Complexation efficiency of cyclodextrins for volatile flavor compounds. Food Res Int 53(1):110–114. https://doi.org/10.1016/j.foodres.2013.03.048

Costa P, Medronho B, Gonc S, Romano A (2015) Cyclodextrins enhance the antioxidant activity of essential oils from three Lamiaceae species. Ind Crop Prod 70:341–346. https://doi.org/10.1016/j.indcrop.2015.03.065

Costa G, Gidaro MC, Vulla D, Supuran CT, Alcaro S (2016) Active components of essential oils as anti-obesity potential drugs investigated by in silico techniques. J Agric Food Chem 64(26):5295–5300. https://doi.org/10.1021/acs.jafc.6b02004

Crini G (2014) Review: a history of cyclodextrins. Chem Rev 114(21):10940–10975. https://doi.org/10.1021/cr500081p

da Silveira E, Sá RDC, Andrade LN, De Oliveira RDRB, De Sousa DP (2014) A review on anti-inflammatory activity of phenylpropanoids found in essential oils. Molecules 19(2):1459–1480. https://doi.org/10.3390/molecules19021459

De Vos P, Faas MM, Spasojevic M, Sikkema J (2010) Encapsulation for preservation of functionality and targeted delivery of bioactive food components. Int Dairy J 20(4):292–302. https://doi.org/10.1016/j.idairyj.2009.11.008

Decock G, Fourmentin S, Surpateanu GG, Landy D, Decock P, Surpateanu G (2006) Experimental and theoretical study on the inclusion compounds of aroma components with β-cyclodextrins. Supramol Chem 18(6):477–482. https://doi.org/10.1080/10610270600665749

Decock G, Landy D, Surpateanu G, Fourmentin S (2008) Study of the retention of aroma components by cyclodextrins by static headspace gas chromatography. J Incl Phenom Macrocycl Chem 62(3–4):297–302. https://doi.org/10.1007/s10847-008-9471-z

Del Valle EMM (2004) Cyclodextrins and their uses: a review. Process Biochem 39(9):1033–1046. https://doi.org/10.1016/S0032-9592(03)00258-9

Demian BA (2000) Correlation of the solubility of several aromatics and terpenes in aqueous hydroxypropyl-β-cyclodextrin with steric and hydrophobicity parameters. Carbohydr Res 328:635–639. https://doi.org/10.1016/S0008-6215(00)00139-7

Dima C, Dima S (2015) Essential oils in foods: extraction, stabilization, and toxicity. Curr Opin Food 5:29–35. https://doi.org/10.1016/j.cofs.2015.07.003

Dima C, Cotarlet M, Tiberius B, Bahrim G, Alexe P, Dima S (2014) Encapsulation of coriander essential oil in beta-cyclodextrin: antioxidant and antimicrobial properties evaluation. Rom Biotechnol Lett 19(2):9128–9141. https://doi.org/10.1016/j.foodhyd.2016.11.014

Donze C, Coleman W (1993) β-CD inclusion complexes: relative selectivity of terpene and aromatic guest molecules studied by competitive inclusion experiments. J Incl Phenom Macrocycl Chem 62(3–4):297–302. https://doi.org/10.1007/BF00708758

Dorman HJD, Deans SG (2000) Antimicrobial agents from plants: antibacterial activity of plant volatile oils. J Appl Microbiol 88:308–316. https://doi.org/10.1046/j.1365-2672.2000.00969.x

Dwivedy AK, Kumar M, Upadhyay N, Prakash B, Dubey NK (2016) Plant essential oils against food borne fungi and mycotoxins. Curr Opin Food Sci 11:16–21. https://doi.org/10.1016/j.cofs.2016.08.010

Eby GAIII (1992) Taste-masked zinc acetate com- positions for oral absorption. US patent 5,095,035. March 10:1992

Eftink MR, Andy ML, Bystrom K, Perlmutter HD, Kristol DS (1989) Cyclodextrin inclusion complexes: studies of the variation in the size of slicyclic guests. J Am Chem Soc 111(17):6765–6772. https://doi.org/10.1021/ja00199a041

Elbary AA, El Nabarawi MA, Ali AMA, Hassan AH (2013) Formulation and evaluation of taste-masked orally disintegrating tablets of nicergoline based on β-cyclodextrin inclusion complexation. Int. J Drug Deliv 5(1):110–120

Fenyvesi E, Gruiz K, Verstichel S, BDe W, Leitgib L, Csabai K, Szaniszlo N (2005) Biodegradation of cyclodextrins in soil. Chemosphere 60:1001–1008. https://doi.org/10.1016/j.chemosphere.2005.01.026

Fenyvesi E, Szemán J, Csabai K, Malanga M, Szente L (2014) Methyl-beta-cyclodextrins: the role of number and types of substituents in solubilizing power. J Pharm Sci 103(5):1443–1452. https://doi.org/10.1002/jps.23917

Ferrazza R, Rossi B, Guella G (2014) DOSY-NMR and Raman investigations on the self-aggregation and cyclodextrin complexation of vanillin. J Phys Chem B 118(25):7147–7155

Fisher K, Phillips C (2009) The mechanism of action of a citrus oil blend against *Enterococcus faecium* and *Enterococcus faecalis*. J Appl Microbiol 106:1343–1349. https://doi.org/10.1111/j.1365-2672.2008.04102.x

Fourmentin S, Ciobanu A, Landy D, Wenz G (2013) Space filling of β-cyclodextrin and β-cyclodextrin derivatives by volatile hydrophobic guests. Beilstein J Org Chem 9:1185–1191. https://doi.org/10.3762/bjoc.9.133

Galvão JG, Silva VF, Ferreira SG, França FRM, Santos DA, Freitas LS, Alves PB, Araújo AAS, Cavalcanti SCH, Nunes RS (2015) β-cyclodextrin inclusion complexes containing *Citrus sinensis* (L.) Osbeck essential oil: an alternative to control *Aedes aegypti* larvae. Thermochim Acta 608:14–19. https://doi.org/10.1016/j.tca.2015.04.001

Gould S, Scott RC (2005) 2-Hydroxypropyl-β-cyclodextrin (HP- β -CD): a toxicology review. Food Chem Toxicol 43(10):1451–1459. https://doi.org/10.1016/j.fct.2005.03.007

Gray J, Caparros-Ruiz D, Grotewold E (2012) Grass phenylpropanoids: regulate before using! Plant Sci 184:112–120. https://doi.org/10.1016/j.plantsci.2011.12.008

Griffin S, Wyllie SG, Markham J (1999) Determination of octanol – water partition coefficient for terpenoids using reversed-phase high-performance liquid chromatography. J Chromatogr A 864:221–228. https://doi.org/10.1016/S0021-9673(99)01009-2

Hădărugă NG, Hădărugă DI, Isengard HD (2012) Water content of natural cyclodextrins and their essential oil complexes : a comparative study between Karl Fischer titration and thermal methods. Food Chem 132:1741–1748. https://doi.org/10.1016/j.foodchem.2011.11.003

Hădărugă DI, Ünlüsayin M, Gruia AT, Bir C, Rusu G, Nicoleta GH (2016) Thermal and oxidative stability of Atlantic salmon oil (*Salmo salar L.*) and complexation with β –cyclodextrin. Beilstein J Org Chem 12:179–191. https://doi.org/10.3762/bjoc.12.20

Hernandez-Sanchez P, Lopez-Miranda S, Lucas-Abellan C, Nunez-Delicado E (2012) Complexation of eugenol (EG), as main component of clove oil and as pure compound, with β- and HP-β-CDs. Food Nutr Sci 3:716–723. https://doi.org/10.4236/fns.2012.36097

Hernández-Sánchez P, López-Miranda S, Guardiola L, Serrano-Martínez A, Gabaldón JA, Nuñez-Delicado E (2017) Optimization of a method for preparing solid complexes of essential clove oil with β-cyclodextrins. J Sci Food Agric 97(2):420–426. https://doi.org/10.1002/jsfa.7781

Higuchi T, Connors KA (1965) Phase-solubility techniques. Adv Anal Chem Instrum 4:117–212

Hill LE, Gomes C, Taylor TM (2013) Characterization of beta-cyclodextrin inclusion complexes containing essential oils (*trans* -cinnamaldehyde , eugenol , cinnamon bark , and clove bud extracts) for antimicrobial delivery applications. Food Sci Technol 51(1):86–93. https://doi.org/10.1016/j.lwt.2012.11.011

Ho BT, Joyce DC, Bhandari BR (2011) Release kinetics of ethylene gas from ethylene–α--cyclodextrin inclusion complexes. Food Chem 129(2):259–266. https://doi.org/10.1016/j.foodchem.2011.04.035

Hou L, Wang J, Zhang D (2013) Optimization of debittering of soybean antioxidant hydrolysates with β-cyclodextrins using response surface methodology. J Food Sci Technol 50(3):521–527. https://doi.org/10.1007/s13197-011-0358-4

Hougeir FG, Kircik L (2012) A review of delivery systems in cosmetics. Dermatol Ther 25(3):234–237. https://doi.org/10.1111/j.1529-8019.2012.01501.x

Huang Y, Zhao J, Zhou L, Wang J, Gong Y, Chen X, Guo Z, Wang Q, Jiang W (2010) Antifungal activity of the essential oil of *Illicium verum* fruit and its main component trans-anethole. Molecules 15(11):7558–7569. https://doi.org/10.3390/molecules15117558

Hyldgaard M, Mygind T, Meyer RL, Debabov D (2012) Essential oils in food preservation: mode of action, synergies, and interactions with food matrix components. Front Microbiol 3:1–24. https://doi.org/10.3389/fmicb.2012.00012

Jambhekar SS, Breen P (2016) Cyclodextrins in pharmaceutical formulations II : solubilization, binding constant, and complexation efficiency. Drug Discov Today 21(2):363–368. https://doi.org/10.1016/j.drudis.2015.11.016

Jayasena DD, Jo C (2014) Potential application of essential oils as natural antioxidants in meat and meat products: a review. Food Rev Int 30(1):71–90. https://doi.org/10.1080/87559129.2013.853776

Jiang S, Li JN, Jiang ZT (2010) Inclusion reactions of β-cyclodextrin and its derivatives with cinnamaldehyde in *Cinnamomum loureirii* essential oil. Eur Food Res Technol 230(4):543–550. https://doi.org/10.1007/s00217-009-1192-z

Kamimura JA, Santos EH, Hill LE, Gomes CL (2014) Antimicrobial and antioxidant activities of carvacrol microencapsulated in hydroxypropyl-beta-cyclodextrin. Food Sci Technol 57(2):701–709. https://doi.org/10.1016/j.lwt.2014.02.014

Kant A, Linforth RST, Hort J, Taylor AJ (2004) Effect of β-cyclodextrin on aroma release and flavor perception. J Agric Food Chem 52(7):2028–2035. https://doi.org/10.1021/jf0307088

Kapetanakou AE, Skandamis PN (2016) Applications of active packaging for increasing microbial stability in foods: natural volatile antimicrobial compounds. Curr Opin Food Sci 12:1–12. https://doi.org/10.1016/j.cofs.2016.06.001

Karapinar M, Aktuğ SE (1987) Inhibition of foodborne pathogens by thymol, eugenol, menthol and anethole. Int J Food Microbiol 4(2):161–166. https://doi.org/10.1016/0168-1605(87)90023-7

Karathanos VT, Mourtzinos I, Yannakopoulou K, Andrikopoulos NK (2007) Study of the solubility, antioxidant activity and structure of inclusion complex of vanillin with beta-cyclodextrin. Food Chem 101(2):652–658. https://doi.org/10.1016/j.foodchem.2006.01.053

Kfoury M, Landy D, Auezova L, Greige-Gerges H, Fourmentin S (2014a) Effect of cyclodextrin complexation on phenylpropanoids' solubility and antioxidant activity. Beilstein J Org Chem 10:2322–2331. https://doi.org/10.3762/bjoc.10.241

Kfoury M, Auezova L, Greige-Gerges H, Ruellan S, Fourmentin S (2014b) Cyclodextrin, an efficient tool for *trans*-anethole encapsulation: chromatographic, spectroscopic, thermal and structural studies. Food Chem 164:454–461. https://doi.org/10.1016/j.foodchem.2014.05.052

Kfoury M, Auezova L, Fourmentin S, Greige-Gerges H (2014c) Investigation of monoterpenes complexation with hydroxypropyl-beta-cyclodextrin. J Incl Phenom Macrocycl Chem 80:51–60. https://doi.org/10.1007/s10847-014-0385-7

Kfoury M, Balan R, Landy D, Nistor D, Fourmentin S (2015a) Investigation of the complexation of essential oil components with cyclodextrins. Supramol Chem 27(9):1–10. https://doi.org/10.1080/10610278.2015.1051977

Kfoury M, Auezova L, Ruellan S, Greige-Gerges H, Fourmentin S (2015b) Complexation of estragole as pure compound and as main component of basil and tarragon essential oils with cyclodextrins. Carbohydr Polym 118:156–164. https://doi.org/10.1016/j.carbpol.2014.10.073

Kfoury M, Auezova L, Greige-Gerges H, Fourmentin S (2015c) Promising applications of cyclodextrins in food: improvement of essential oils retention, controlled release and antiradical activity. Carbohydr Polym 131:264–272. https://doi.org/10.1016/j.carbpol.2015.06.014

Kfoury M, Lounès-Hadj Sahraoui A, Bourdon N, Laruelle F, Fontaine J, Auezova L, Greige-Gerges H, Fourmentin S (2016a) Solubility, photostability and antifungal activity of phenylpropanoids encapsulated in cyclodextrins. Food Chem 196:518–525. https://doi.org/10.1016/j.foodchem.2015.09.078

Kfoury M, Borgie M, Verdin A, Ledoux F, Courcot D, Auezova L, Fourmentin S (2016b) Essential oil components decrease pulmonary and hepatic cells inflammation induced by air pollution particulate matter. Environ Chem Lett 14:345–351. https://doi.org/10.1007/s10311-016-0572-4

Kfoury M, Hadaruga NG, Hadaruga DI, Fourmentin S (2016c) Cyclodextrins as encapsulation material for flavors and aroma. A volume in nanotechnology in the agri-food industry.. ISBN: 978-0-12-804307-3.

Kfoury M, Landy D, Ruellan S, Auezova L, Greige-Gerges H, Fourmentin S (2016d) Determination of formation constants and structural characterization of cyclodextrin inclusion complexes with two phenolic isomers: carvacrol and thymol. Beilstein J Org Chem 12(12):29–42. https://doi.org/10.3762/bjoc.12.5

Kfoury M, Landy D, Ruellan S, Auezova L, Greige-gerges H, Fourmentin S (2016e) Nootkatone encapsulation by cyclodextrins : effect on water solubility and photostability. Food Chem 236:41–48. https://doi.org/10.1016/j.foodchem.2016.12.086

Kfoury M, Auezova L, Greige-Gerges H, Larsen KL, Fourmentin S (2016f) Release studies of *trans*-anethole from β-cyclodextrin solid inclusion complexes by multiple headspace extraction. Carbohydr Polym 151:1245–1250. https://doi.org/10.1016/j.carbpol.2016.06.079

Kfoury M, Auezova L, Greige-Gerges H, Fourmentin S (2016g) Development of a Total organic carbon method for the quantitative determination of solubility enhancement by cyclodextrins : application to essential oils. Anal Chim Acta 918:21–25. https://doi.org/10.1016/j.aca.2016.03.013

Kfoury M, Pipkin JD, Antle V, Fourmentin S (2017) Captisol®: an efficient carrier and solubilizing agent for essential oils and their components. Flavour Fragr J 32(5):340–346. https://doi.org/10.1002/ffj.3395

Kolb B, Ettre LS (2006) Static headspace-gas chromatography theory and practice, 2nd edn. Wiley, Hoboken

Kurkov SV, Loftsson T (2013) Cyclodextrins. Int J Pharm 453(1):167–180. https://doi.org/10.1016/j.ijpharm.2012.06.055

Landy D, Fourmentin S, Salome M, Surpateanu G (2000) Analytical improvement in measuring formation constants of inclusion complexes between β-cyclodextrin and phenolic compounds. J Incl Phenom 38(1–4):187–198. https://doi.org/10.1023/A:1008156110999

Landy D, Tetart F, Truant E, Blach F, Fourmentin S, Surpateanu G (2007) Development of a competitive continuous variation plot for the determination of inclusion compounds stoichiometry. J Incl Phenom Macrocycl Chem 57:409–413. https://doi.org/10.1007/s10847-006-9226-7

Lee JH, Lee J, Song KB (2015) Development of a chicken feet protein film containing essential oils. Food Hydrocoll 46:208–215. https://doi.org/10.1016/j.foodhyd.2014.12.020

Li W, Du L, Li M (2011) Alkaloids and flavonoids as α-adrenergic receptor antagonists. Curr Med Chem 18:4923–4932. https://doi.org/10.2174/092986711797535209

Li W, Liu X, Yang Q, Zhang N, Du Y, Zhu H (2015) Preparation and characterization of inclusion complex of benzyl isothiocyanate extracted from papaya seed with β-cyclodextrin. Food Chem 184:99–104. https://doi.org/10.1016/j.foodchem.2015.03.091

Liang H, Yuan Q, Vriesekoop F, Lv F (2012) Effects of cyclodextrins on the antimicrobial activity of plant-derived essential oil compounds. Food Chem 135(3):1020–1027. https://doi.org/10.1016/j.foodchem.2012.05.054

Lipinski CA (2000) Drug-like properties and the cause of poor solubility and poor permeability. J Pharmacol Toxicol Methods 44:235–249. https://doi.org/10.1016/S1056-8719(00)00107-6

Lipinski CA (2004) Lead- and drug-like compounds: the rule-of-five revolution. Drug Discov Today Technol 1:337–341. https://doi.org/10.1016/j.ddtec.2004.11.007

Liu L, Guo QX (2002) The driving forces in the inclusion complexation of cyclodextrins. J Incl Phenom Macrocycl Chem 42(1):1–14. https://doi.org/10.1023/A:1014520830813

Llana-Ruiz-Cabello M, Gutiérrez-Praena D, Puerto M, Pichardo S, Jos Á, Cameán AM (2015) In vitro pro-oxidant/antioxidant role of carvacrol, thymol and their mixture in the intestinal Caco-2 cell line. Toxicol In Vitro 29(4):647–656. https://doi.org/10.1016/j.tiv.2015.02.006

Loftsson T, Brewster ME (2010) Pharmaceutical applications of cyclodextrins: basic science and product development. J Pharm Pharmacol 62(11):1607–1621. https://doi.org/10.1111/j.2042-7158.2010.01030.x

Loftsson T, Duchêne D (2007) Cyclodextrins and their pharmaceutical applications. Int J Pharm 329:1–11. https://doi.org/10.1016/j.ijpharm.2006.10.044

Loftsson T, Hreinsdóttir D, Másson M (2005) Evaluation of cyclodextrin solubilization of drugs. Int J Pharm 302(1–2):18–28. https://doi.org/10.1016/j.ijpharm.2005.05.042

Lopez-Nicolas JM, Perez-Lopez AJ, Carbonell-Barrachina A, Garcia-Carmona F (2007) Use of natural and modified cyclodextrins as inhibiting agents of peach juice enzymatic browning. J Agric Food Chem 55(13):5312–5319. https://doi.org/10.1021/jf070499h

Lopez-Nicolas JM, Andreu-Sevilla AJ, Carbonell-Barrachina ÁA, García-Carmona F (2009) Effects of addition of α-cyclodextrin on the sensory quality, volatile compounds, and color parameters of fresh pear juice. J Agric Food Chem 57(20):9668–9675. https://doi.org/10.1021/jf901919t

Lopez-Nicolas JM, Rodríguez-Bonilla P, García-Carmona F (2014) Cyclodextrins and antioxidants. Crit Rev Food Sci 54(2):251–276. https://doi.org/10.1080/10408398.2011.582544

Lu Z, Cheng B, Hu Y, Zhang Y, Zou G (2009) Complexation of resveratrol with cyclodextrins: solubility and antioxidant activity. Food Chem 113(1):17–20. https://doi.org/10.1016/j.foodchem.2008.04.042

Lucas-Abellán C, Mercader-Ros MT, Zafrilla MP, Fortea MI, Gabaldon JA, Nunez-Delicado E (2008) ORAC-fluorescein assay to determine the oxygen radical absorbance capacity of resveratrol complexed in cyclodextrins. J Agric Food Chem 56:2254–2259. https://doi.org/10.1021/jf0731088

Marchese A, Erdogan I, Daglia M, Barbieri R, Di A, Fazel S, Gortzi O, Izadi M, Mohammad S (2016) Antibacterial and antifungal activities of thymol: a brief review of the literature. Food Chem 210:402–414. https://doi.org/10.1016/j.foodchem.2016.04.111

Marques HMC (2010) A review on cyclodextrin encapsulation of essential oils and volatiles. Flavour Fragr J 25(5):313–326. https://doi.org/10.1002/ffj.2019

Marreto RN, Almeida EECV, Alves PB, Niculau ES, Nunes RS, Matos CRS, Araújo AAS (2008) Thermal analysis and gas chromatography coupled mass spectrometry analyses of hydroxypropyl-β-cyclodextrin inclusion complex containing *Lippia gracilis* essential oil. Thermochim Acta 475:53–58. https://doi.org/10.1016/j.tca.2008.06.015

Martina K, Binello A, Lawson D, Jicsinszky L, Cravotto G (2013) Recent applications of cyclodextrins as food additives and in food processing. Curr Nutr Food Sci 9(3):167–179. https://doi.org/10.2174/1573401311309030001

Martins ADP, Craveiro AA, MacHado MIL, Raffin FN, Moura TF, Novák C, Ehen Z (2007) Preparation and characterisation of *Mentha x villosa Hudson* oil–β-cyclodextrin complex. J Therm Anal Calorim 88(2):363–371. https://doi.org/10.1007/s10973-005-7407-z

Mascheroni E, Fuenmayor CA, Cosio MS, Di Silvestro G, Piergiovanni L, Mannino S, Schiraldi A (2013) Encapsulation of volatiles in nanofibrous polysaccharide membranes for humidity-triggered release. Carbohydr Polym 98(1):17–25. https://doi.org/10.1016/j.carbpol.2013.04.068

Mazzobre MF, dos Santos CI, Buera M (2011) Solubility and stability of β- cyclodextrin-terpineol inclusion complex as affected by water. Food Biophys 6(2):274–280. https://doi.org/10.1007/s11483-011-9208-1

Menezes PDP, Serafini MR, de Carvalho YMBG, Soares Santana DV, Lima BS, Quintans-Júnior LJ, Marreto RN, de Aquino TM, Sabino AR, Scotti L, Scotti MT, Grangeiro-Júnior S, de Souza Araújo AA (2016) Kinetic and physical-chemical study of the inclusion complex of β- cyclodextrin containing carvacrol. J Mol Struct 1125:323–330. https://doi.org/10.1016/j.molstruc.2016.06.062

Miguel MG, Dandlen SA, Figueiredo AC, Pedro LG, Barroso JG, Marques MH (2010) Comparative evaluation of the antioxidant activities of thymol and carvacrol and the corresponding β-cyclodextrin complexes. Acta Hortic 853:363–368. https://doi.org/10.17660/ActaHortic.2010.853.44

Miller LA, Carrier RL, Ahmed I (2007) Practical considerations in development of solid dosage forms that contain cyclodextrin. J Pharm Sci 96(7):1691–1707. https://doi.org/10.1002/jps.20831

Miron D, Battisti F, Schwengber Ten Caten C, Mayorga P, Scherman Schapoval EE (2012) Spectrophotometric simultaneous determination of citral isomers in cyclodextrin complexes with partial least squares supported approach. Curr Pharm Anal 8(4):401–408. https://doi.org/10.2174/157341212803341735

Moreira SM, de Carvalho W, Alexandrino AC, de Paula HCB, Rodrigues MCP, de Figueiredo RW, Maia GA, de Figueiredo EMAT, Brasil IM (2014) Int J Food Sci Technol 49(10):2192–2203. https://doi.org/10.1111/ijfs.12535

Morelli CL, Mahrous M, Belgacem MN, Branciforti MC, Bretas RES, Bras J (2015) Natural copaiba oil as antibacterial agent for bio-based active packaging. Ind Crop Prod 70:134–141. https://doi.org/10.1016/j.indcrop.2015.03.036

Mourtzinos I, Kalogeropoulos N, Papadakis SE, Konstantinou K, Karathanos VT (2008) Encapsulation of nutraceutical monoterpenes in β-cyclodextrin and modified starch. J Food Sci 73(1):89–94. https://doi.org/10.1111/j.1750-3841.2007.00609.x

Mura P (2014) Analytical techniques for characterization of cyclodextrin complexes in aqueous solution: a review. J Pharm Biomed Anal 10:238–250. https://doi.org/10.1016/j.jpba.2014.02.022

Murbach Teles Andrade BF, Conti BJ, Santiago KB, Fernandes Junior A, Sforcin JM (2014) *Cymbopogon martinii* essential oil and geraniol at noncytotoxic concentrations exerted immunomodulatory/ anti-inflammatory effects in human monocytes. J Pharm Pharmacol 66(10):1491–1496. https://doi.org/10.1111/jphp.12278

Nazzaro F, Fratianni F, Martino LD (2013) Effect of essential oils on pathogenic bacteria. Pharmaceuticals 6(12):1451–1474. https://doi.org/10.3390/ph6121451

Neoh TL, Yoshii H, Furuta T (2006) Encapsulation and release characteristics of carbon dioxide in α-cyclodextrin. J Incl Phenom Macrocycl Chem 56:125–133. https://doi.org/10.1007/s10847-006-9073-6

Nieddu M, Rassu G, Boatto G, Bosi P, Trevisi P, Giunchedi P, Carta A, Gavini E (2014) Improvement of thymol properties by complexation with cyclodextrins: In vitro and in vivo studies. Carbohydr Polym 102:393–399. https://doi.org/10.1016/j.carbpol.2013.10.084

Nowakowski M, Ejchart A (2014) Complex formation of fenchone with α- cyclodextrin: NMR titrations. J Incl Phenom Macro 79(3–4):337–342. https://doi.org/10.1007/s10847-013-0356-4

Numanoglu U, Şen T, Tarimci N, Kartal M, Koo OMY, Önyüksel H (2007) Use of cyclodextrins as a cosmetic delivery system for fragrance materials: linalool and benzyl acetate. AAPS PharmSciTech 8(4):85. https://doi.org/10.1208/pt0804085

Nunez-Delicado E, Sanchez-Ferrer A, Garcia-Carmona F (1997) Cyclodextrins as secondary antioxidants: synergism with ascorbic acid. J Agric Food Chem 45(8):2830–2835. https://doi.org/10.1021/jf9609800

Pagington JS (1987) β-Cyclodextrin: the success of molecular inclusion. Chem Br 23(5):455–458

Papajani V, Haloci E, Goci E, Shkreli R, Manfredini S (2015) Evaluation of antifungal activity of *Origanum vulgare* and *Rosmarinus officinalis* essential oil before and after inclusion in β-cyclodextrine. Int J Pharm Pharm Sci 7(5):5–8

Paramera EI, Konteles SJ, Karathanos VT (2011) Stability and release properties of curcumin encapsulated in *Saccharomyces cerevisiae*, β- cyclodextrin and modified starch. Food Chem 125(3):913–922. https://doi.org/10.1016/j.foodchem.2010.09.071

Partanen R, Ahro M, Hakala M, Kallio H, Forssell P (2002) Microencapsulation of caraway extract in β-cyclodextrin and modified starches. Eur Food Res Technol 214(3):242–247. https://doi.org/10.1007/s00217-001-0446-1

Pavela R, Benelli G (2016) Essential oils as ecofriendly biopesticides ? Challenges and constraints. Trends Plant Sci 21(12):1000–1007. https://doi.org/10.1016/j.tplants.2016.10.005

Pein M, Preis M, Eckert C, Kiene FE (2014) Taste-masking assessment of solid oral dosage forms- a critical review. Int J Pharm 465(1–2):239–254. https://doi.org/10.1016/j.ijpharm.2014.01.036

Perricone M, Arace E, Corbo MR, Sinigaglia M, Bevilacqua A (2015) Bioactivity of essential oils: a review on their interaction with food components. Front Microbiol 6:76. https://doi.org/10.3389/fmicb.2015.00076

Pinho E, Grootveld M, Soares G, Henriques M (2014) Cyclodextrins as encapsulation agents for plant bioactive compounds. Carbohydr Polym 101:121–135. https://doi.org/10.1016/j.carbpol.2013.08.078

Ponce Cevallos PA, Buera MP, Elizalde BE (2010) Encapsulation of cinnamon and thyme essential oils components (cinnamaldehyde and thymol) in β-cyclodextrin: effect of interactions with water on complex stability. J Food Eng 99(1):70–75. https://doi.org/10.1016/j.jfoodeng.2010.01.039

Prakash B, Kedia A, Mishra PK, Dubey NK (2015) Plant essential oils as food preservatives to control moulds, mycotoxin contamination and oxidative deterioration of Agri-food commodities – potentials and challenges. Food Control 47:381–391. https://doi.org/10.1016/j.foodcont.2014.07.023

Qi ZH, Hedges AR (1995) Use of cyclodextrins for flavours. In: Ho CT, Tan CT, Tong CH (eds) Flavour technology: physical chemistry, modification and process, ACS symposium series 610. American Chemical Society, Washington, DC, pp 231–243

Radu CD, Parteni O, Ochiuz L (2016) Applications of cyclodextrins in medical textiles — review. J Control Release 224:146–157. https://doi.org/10.1016/j.jconrel.2015.12.046

Rakmai J, Cheirsilp B, Torrado-agrasar A, Simal Gandara J, Mejuto JC, Torrado-agrasar A (2017a) Encapsulation of yarrow essential oil in hydroxypropyl-beta-cyclodextrin: physiochemical characterization and evaluation of bio-efficacies. CyTA J Food 0:1–9. https://doi.org/10.1080/19476337.2017.1286523

Rakmai J, Cheirsilp B, Carlos J, Torrado-agrasar A (2017b) Physico-chemical characterization and evaluation of bio efficacies of black pepper essential oil encapsulated in hydroxypropyl-beta-cyclodextrin. Food Hydrocoll 65:157–164. https://doi.org/10.1016/j.foodhyd.2016.11.014

Rakshit R (2011) Skin care textiles - a review. Man-Made Text India 39(3):81–85

Rashed AA, Mohd Nawi MN, Sulaiman K (2017) Assessment of essential oil as a potential anti-obesity agent: a narrative review assessment of essential oil as a potential anti- obesity agent: a narrative review. J Essent Oil Res 29(1):1–10. https://doi.org/10.1080/10412905.2016.1213668

Raut JS, Karuppayil SM (2014) A status review on the medicinal properties of essential oils. Ind Crop Prod 62:250–264. https://doi.org/10.1016/j.indcrop.2014.05.055

Reineccius TA, Reineccius GA, Peppard TL (2002) Encapsulation of flavors using cyclodextrins: comparison of flavor retention in alpha, beta, and gamma types. J Food Sci 67(9):3271–3279. https://doi.org/10.1111/j.1365-2621.2002.tb09577.x

Reineccius TA, Reineccius GA, Peppard TL (2004) Utilization of β-cyclodextrin for improved flavor retention in thermally processed foods. J Food Sci 69(1):58–62. https://doi.org/10.1111/j.1365-2621.2004.tb17856.x

Ribeiro-Santos R, Andrade M, Sanches-Silva A (2017) Application of encapsulated essential oils as antimicrobial agents in food packaging. Curr Opin Food Sci 14:78–84. https://doi.org/10.1016/j.cofs.2017.01.012

Ruktanonchai UR, Srinuanchai W, Saesoo S, Sramala I, Puttipipatkhachorn S, Soottitantawat A (2011) Encapsulation of citral isomers in extracted lemongrass oil with cyclodextrins: molecular modeling and physicochemical characterizations. Biosci Biotechnol Biochem 75(12):2340–2345. https://doi.org/10.1271/bbb.110523

Sagiri SS, Anis A, Pal K (2016) Review on encapsulation of vegetable oils: strategies, preparation methods, and applications. Polym Plast Technol Eng 55(3):291–311. https://doi.org/10.1080/03602559.2015.1050521

Santos EH, Kamimura JA, Hill LE, Gomes CL (2015) Characterization of carvacrol beta-cyclodextrin inclusion complexes as delivery systems for antibacterial and antioxidant applications. Food Sci Technol 60(1):583–592. https://doi.org/10.1016/j.lwt.2014.08.046

Sharifi-rad J, Sureda A, Tenore GC, Daglia M, Sharifi-rad M, Valussi M, Tundis R, Sharifi-Rad Loizzo MR, Ademiluyi AO, Sharifi-Rad R, Ayatollahi SA, Iriti M (2017) Biological activities of essential oils: from plant chemoecology to traditional healing systems. Molecules 22(1):70. https://doi.org/10.3390/molecules22010070

Sherry M, Charcosset C, Fessi H, Greige-Gerges H (2013) Essential oils encapsulated in liposomes: a review. J Liposome Res 23(4):268–275. https://doi.org/10.3109/08982104.2013.819888

Stella VJ, He Q (2008) Cyclodextrins. Toxicol Pathol 36(1):30–42. https://doi.org/10.1177/0192623307310945

Suvarna V, Gujar P, Murahari M (2017) Complexation of phytochemicals with cyclodextrin derivatives – an insight. Biomed Pharmacother 88:1122–1144. https://doi.org/10.1016/j.biopha.2017.01.157

Szejtli J (2003) Cyclodextrins in the textile industry. Starch 55(5):191–196. https://doi.org/10.1002/star.200390050

Szejtli J (2004) Past, present and future of cyclodextrin research. Pure Appl Chem 76:1825–1844

Szejtli J, Szente L (2005) Elimination of bitter, disgusting tastes of drugs and foods by cyclodextrins. Eur J Pharm Biopharm 61(3):115–125. https://doi.org/10.1016/j.ejpb.2005.05.006

Szente L, Szejtli J (2004) Cyclodextrins as food ingredients. Trends Food Sci Technol 15(3–4):137–142. https://doi.org/10.1016/j.tifs.2003.09.019

Tamamoto LC, Schmidt SJ, Lee SY (2010) Sensory properties of ginseng solutions modified by masking agents. J Food Sci 75(7):341–347. https://doi.org/10.1111/j.1750-3841.2010.01749.x

Tanemura I, Saito Y, Ueda H, Sato T (1998) Solubility method using static head-space gas chromatography for determination of the stability constants of fragrance materials with 2-hydroxypropyl-β-cyclodextrin. Chem Pharm Bull 46(3):540–541

Tao F, Hill LE, Peng Y, Gomes CL (2014) Synthesis and characterization of β-cyclodextrin inclusion complexes of thymol and thyme oil for antimicrobial delivery applications. Food Sci Technol 59(1):247–255. https://doi.org/10.1016/j.lwt.2014.05.037

Tian XN, Jiang ZT, Li R (2008) Inclusion interactions and molecular microcapsule of Salvia sclarea L. essential oil with β-cyclodextrin derivatives. Eur Food Res Technol 227(4):1001–1007. https://doi.org/10.1007/s00217-007-0813-7

Tisserand R, Young R (2014) Essential oil safety. A guide for health care professionals, 2nd edn. Churchill Livingstone, London, pp 23–38

Del Toro-Sanchez CL, Ayala-Zavala JF, Machi L, Santacruz H, Villegas-Ochoa MA, Alvarez-Parrilla E, Gonzalez-Aguilar GA (2010) Controlled release of antifungal volatiles of thyme essential oil from β -cyclodextrin capsules. J Incl Phenom Macrocycl Chem 67:431–441. https://doi.org/10.1007/s10847-009-9726-3

Turek C, Stintzing FC (2013) Stability of essential oils: a review. Compr Rev Food Sci Food Saf 12(1):40–53. https://doi.org/10.1111/1541-4337.12006

Uyar T, Hacaloglu J, Besenbacher F (2009) Reactive & Functional Polymers Electrospun polystyrene fibers containing high temperature stable volatile fragrance / flavor facilitated by cyclodextrin inclusion complexes. React Funct Polym 69(3):145–150. https://doi.org/10.1016/j.reactfunctpolym.2008.12.012

Voncina B, Vivod V (2013) Cyclodextrins in textile finishing. In: Günay M (ed) Textile dyeing. InTech, Tijeka, Croatia., Chapter 3, p 53. https://doi.org/10.5772/53777

Waleczek KJ, Marques HMC, Hempel B, Schmidt PC (2003) Phase solubility studies of pure (−) - α-bisabolol and camomile essential oil with β –cyclodextrin. Eur J Pharm Biopharm 55:247–251. https://doi.org/10.1016/S0939-6411(02)00166-2

Wang CX, Chen SL (2005) Fragrance-release property of β-cyclodextrin inclusion compounds and their application in aromatherapy. J Ind Text 34(3):157–166

Wang J, Cao Y, Sun B, Wang C (2011) Physicochemical and release characterisation of garlic oil-β-cyclodextrin inclusion complexes. Food Chem 127(4):1680–1685. https://doi.org/10.1016/j.foodchem.2011.02.036

Xu F, Yang Q, Wu L, Qi R, Wu Y, Li Y, Tang L, Gua D, Liu B (2017) Investigation of inclusion complex of patchouli alcohol with β–cyclodextrin. PLoS One 35:1–10. https://doi.org/10.1371/journal.pone.0169578

Yamamoto C, Neoh TL, Honbou H, Yoshii H, Furuta T (2012) Kinetic analysis and evaluation of controlled release of D-limonene encapsulated in spray-dried cyclodextrin powder under linearly ramped humidity. Dry Technol 30(11–12):1283–1291. https://doi.org/10.1080/07373937.2012.681089

Yang Z, Xiao Z, Ji H (2015) Solid inclusion complex of terpinen-4-ol/ β-cyclodextrin: kinetic release, mechanism and its antibacterial activity. Flavour Frag J 30(2):179–187. https://doi.org/10.1002/ffj.3229

Yang Z, Huang L, Ji H (2017) Host-guest complexes of estragole with β-cyclodextrin: an experimental and theoretical investigation. Flavour Frag J 32(2):102–111. https://doi.org/10.1002/ffj.3358

Yoshii H, Sakane A, Kawamura D, Neoh T, Kajiwara H, Furuta T (2007) Release kinetics of (−)-menthol from chewing gum. J Incl Phenom Macrocycl Chem 57(1–4):591–596. https://doi.org/10.1007/s10847-006-9279-7

Zeng Z, Fang Y, Jin H (2012) Side chain influencing the interaction between β-cyclodextrin and vanillin. Flavour Fragr J 27:378–385. https://doi.org/10.1002/ffj.3115

Zhao M, Wang H, Yang B, Tao H (2010) Identification of cyclodextrin inclusion complex of chlorogenic acid and its antimicrobial activity. Food Chem 120(4):1138–1142. https://doi.org/10.1016/j.foodchem.2009.11.044

Zhu G, Xiao Z, Zhu G, Niu Y (2016) Encapsulation of l-menthol in hydroxypropyl-β-cyclodextrin and release characteristics of the inclusion complex. Pol J Chem Technol 18(3):110–116. https://doi.org/10.1515/pjct-2016-0056

Chapter 5
Chemosensors for Water Contaminants Based on Chromophore-Appended Cyclodextrins

Hiroshi Ikeda

Contents

Abstract The development of new chemical sensing methods for the recognition of molecules or ions in water is an important theme in many fields including environmental, biological, and clinical applications and simple methods for detecting organic molecules in water is desired. Chemosensors have been attracting considerable attention among the variety of methods for detecting chemical species. Many kinds of chromophore-appended cyclodextrins have been reported for chemosensors. Here, we review the molecule-sensing abilities of chromophore-appended cyclodextrins as chemosensors. First, turn-off fluorescent chemosensors using chromophore-appended cyclodextrins are discussed. The fluorescence intensities of turn-off fluorescent cyclodextrin chemosensors always decrease upon analyte addition,

H. Ikeda (✉)
School of Life Science and Technology, Tokyo Institute of Technology, Yokohama, Japan
e-mail: hikeda@bio.titech.ac.jp

S. Fourmentin et al. (eds.), *Cyclodextrin Applications in Medicine, Food, Environment and Liquid Crystals*, Environmental Chemistry for a Sustainable World, https://doi.org/10.1007/978-3-319-76162-6_5

and the analyte-induced variation of the fluorescence intensities of the turn-off fluorescent cyclodextrin chemosensors depend on a analyte. The sensitivities of the turn-off cyclodextrin chemosensors roughly only depend on the binding affinity for a analyte. Next, dye-appended cyclodextrins are discussed as color-change chemosensors, which are effective for detection by the naked eye. Then, turn-on fluorescent chemosensors using chromophore-appended cyclodextrins are discussed. The fluorescence intensities of turn-on fluorescent cyclodextrin chemosensors increase upon addition of special analytes, which have a comparatively spherical shape. The turn-on fluorescent β-cyclodextrin chemosensor is not sensitive to bile acids, which are strongly bound by β-cyclodextrin. The shape and size of analytes have a greater influence on the fluorescence intensity than the binding affinity. Finally, chromophore-appended cyclodextrins for detection of anion or cation species are discussed.

Abbreviations

AdCOOH	Adamantanecarboxylic acid
AdOH	Adamantanol
$AdNH_2$	Adamantanamine
Bor	Borneol
CA	Cholic acid
Cam	Camphor
CD	Cyclodextrin
CDCA	Chenodeoxycholic acid
c-HexOH	Cyclohexanol
c-OctOH	Cyclooctanol
DCA	Deoxycholic acid
DNS-L-Leu-βCD	*N*-dansyl-L-leucine-appended β-cyclodextrin
DNS-D-Leu-βCD	*N*-dansyl-D-leucine-appended β-cyclodextrin
DNS-L-Phe-βCD	*N*-dansyl-L-phenylalanine-appended β-cyclodextrin
DNS-D-Phe-βCD	*N*-dansyl-D-phenylalanine-appended β-cyclodextrin
DNS-L-Val-βCD	*N*-dansyl-L-valine-appended β-cyclodextrin
DNS-D-Val-βCD	*N*-dansyl-D-valine-appended β-cyclodextrin
Fen	Fenchone
Ger	Geraniol
HDCA	Hyodeoxycholic acid
LCA	Lithocholic acid
Men	Menthol
NC0αCD	(7-nitrobenz-2-oxa-1,3-diazol-4-yl)amine-appended α-cyclodextrin
NC0βCD	(7-nitrobenz-2-oxa-1,3-diazol-4-yl)amine-appended β-cyclodextrin
NC0γCD	(7-nitrobenz-2-oxa-1,3-diazol-4-yl)amine-appended γ-cyclodextrin
NC4βCD	(7-nitrobenz-2-oxa-1,3-diazol-4-yl)aminobutylamine-appended β-cyclodextrin

NC4γCD	(7-nitrobenz-2-oxa-1,3-diazol-4-yl)aminobutylamine-appended γ-cyclodextrin
Ner	Nerol
UDCA	Ursodeoxycholic acid

5.1 Introduction

Chemical sensing plays an important role in chemical, clinical, biological, environmental, and security tests. In the field of environmental science, it is well-known that mercury, lead and cadmium are toxic to living organisms, and the detection of them in the environment has been well studied. The detection of organic molecules such as benzene, halomethane, and agricultural chemicals in water is also important; therefore, simple methods for detecting organic molecules in water is desired.

Among the variety of methods for detecting chemical species, chemosensors have been attracting considerable attention (de Silva et al. 1997; Lavigne and Anslyn 2001; Anslyn 2007; Mirsky and Yatsimirsky 2011; Prodi et al. 2011; You et al. 2015). The concept of a "chemosensor" was introduced by Czarnik (Czarnik 1993; Desvergne and Czarnik 1997). Previously wording of organic chemists to refer to new molecular indicators as "sensors" has been criticized because Webster's dictionary defined a sensor as a mechanical device sensitive to light, temperature, radiation level or the like. Czarnik defined a fluorescent chemosensor as a compound of abiotic origin that complexes to an analyte reversibly with a concomitant fluorescent signal transduction wherein it constitutes only the active transduction unit of a sensor. Changes in fluorescence intensity or color changes of chemosensors produced by the interaction of chemosensors with analytes can be detected with the naked eye. Chemosensors also have advantages including portability, operational simplicity, rapid response and cost effectiveness.

Extensive efforts have been made to construct chemosensors that selectively interact with analytes to produce detectable changes in signals. Several chromophore-appended cyclodextrins have been prepared for this purpose. Cyclodextrin chemosensors can be used to detect colorless neutral molecules in water via changes in the intensity of their fluorescence, absorption, or circular dichroism.

Cyclodextrins are torus-shaped cyclic oligosaccharides consisting of six, seven, and eight D-glucopyranose units for α-cyclodextrin, β-cyclodextrin, and γ-cyclodextrin, respectively (Fig. 5.1). Various values for the internal diameter of cyclodextrins have been reported using different estimation methods (Table 5.1) (Saenger 1980; Hall et al. 1988; Müller and Wenz 2007). Müller and Wenz calculated the minimum internal diameters of cyclodextrins by semi-empirical AM1 and PM3 calculations and showed that the stability of an inclusion complex of an α-cyclodextrin with a guest is strongly correlated with the cross-sectional diameter of the α-cyclodextrin and the guest (Müller and Wenz 2007). In aqueous solution, cyclodextrins can accommodate various organic compounds in their central cavities.

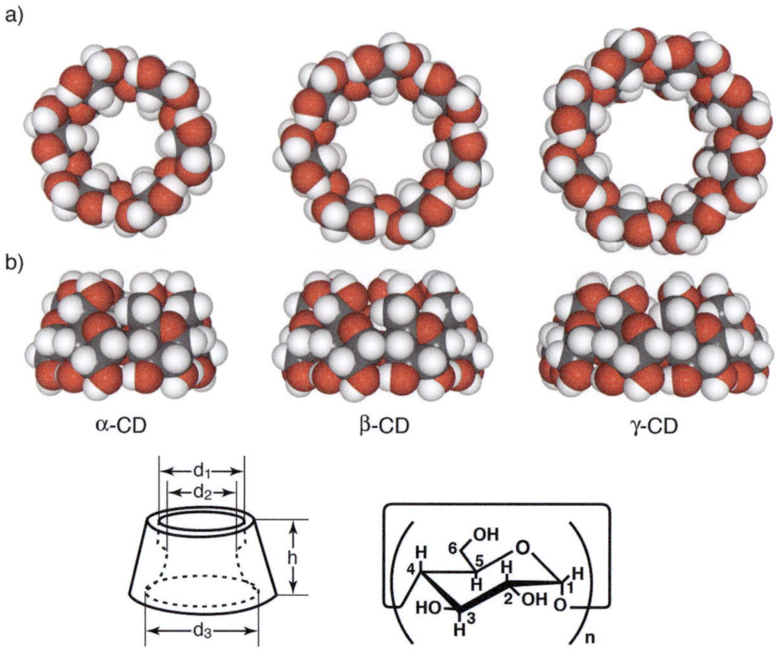

Fig. 5.1 Structures of α-cyclodextrin (α-CD), β-cyclodextrin (β-CD), and γ-cyclodextrin (γ-CD): (**a**) view from the secondary hydroxy side of cyclodextrin, (**b**) the side view of cyclodextrin

Table 5.1 Properties of cyclodextrins

	n	d_1 [Å][a]	d_2 [Å][a]	d_3 [Å][a]	h [Å][a]	d_1 [Å][b]	d_3 [Å][b]	d_2 [Å] AM1[c]	d_2 [Å] PM3[c]
α-CD	6	5.6	4.2	8.8	7.8	4.7	5.2	4.4	4.4
β-CD	7	6.8	5.6	10.8	7.8	6.0	6.4	5.8	6.5
γ-CD	8	8.0	6.8	12.0	7.8	7.5	8.3	7.4	8.1

[a]Hall et al. 1988, [b]Saenger 1980, [c]Müller and Wenz 2007

The hydrophobic, van der Waals forces, and dipole–dipole interactions between the cyclodextrin and guest are the major driving forces of inclusion phenomena, and the stability of the inclusion complex is affected by the fitness of shape and size of the guest to the cyclodextrin cavity.

Many types of chemosensors can be constructed using cyclodextrins that recognize the shape and size of analytes. In this chapter, the molecule-sensing abilities of chromophore-appended cyclodextrins will be reviewed.

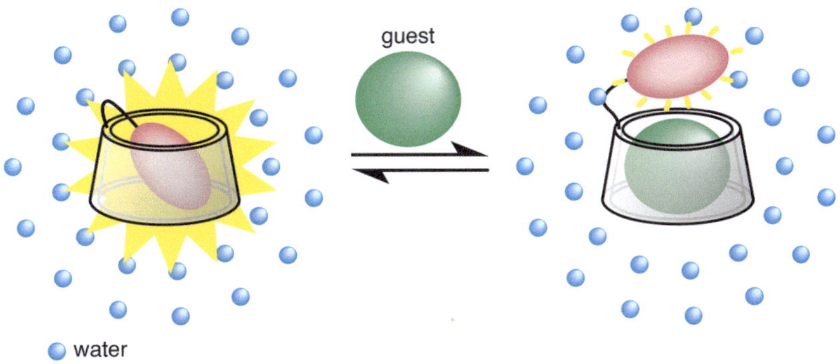

Fig. 5.2 A two-state equilibrium model for guest-induced conformational change of chromophore-appended cyclodextrins in aqueous solution; The self-inclusion state is usually the major conformation in aqueous solution (*left*). An induced-fit conformational change of the chromophore-appended cyclodextrin occurs in association with accommodation of the guest (*right*). In the case of a fluorophore, the fluorescent cyclodextrin exhibits a strong fluorescence in the self-inclusion state due to the hydrophobic environment of the cyclodextrin cavity, and exclusion of the fluorophore from the cyclodextrin cavity to the bulk water weakens its fluorescence intensity. (Adapted from Ueno et al. 1990)

5.2 Turn-Off Fluorescent Chemosensors

Ueno proposed a new system for a chemosensor to detect molecules (Ueno et al. 1990). A chromophore was connected to a cyclodextrin with a suitable spacer unit. The mechanism of this chemosensor is shown in Fig. 5.2. Although the chromophore-appended cyclodextrins can adopt some conformations in aqueous solution, the conformation equilibrium can be explained by the simplified two-state model shown in Fig. 5.2. The self-inclusion state is usually the major conformation in aqueous solution because a fluorophore unit is hydrophobic and is more stable in the central cavity of the cyclodextrin than in the bulk water. An induced-fit conformational change of the chromophore-appended cyclodextrin occurs in association with accommodation of the guest; this conformational change displaces the chromophore from inside to outside of the cyclodextrin cavity. The 'non-self-inclusion state' increases with an increase in the guest concentration. In the case of a fluorophore, the fluorescent cyclodextrin exhibits a strong fluorescence in the self-inclusion state due to the hydrophobic environment of the cyclodextrin cavity, and exclusion of the fluorophore from the cyclodextrin cavity to the bulk water weakens its fluorescence intensity. The extent of variation in fluorescence intensity depends on the affinity of the chemosensor for a guest. This chemosensor system is effective for detecting molecules.

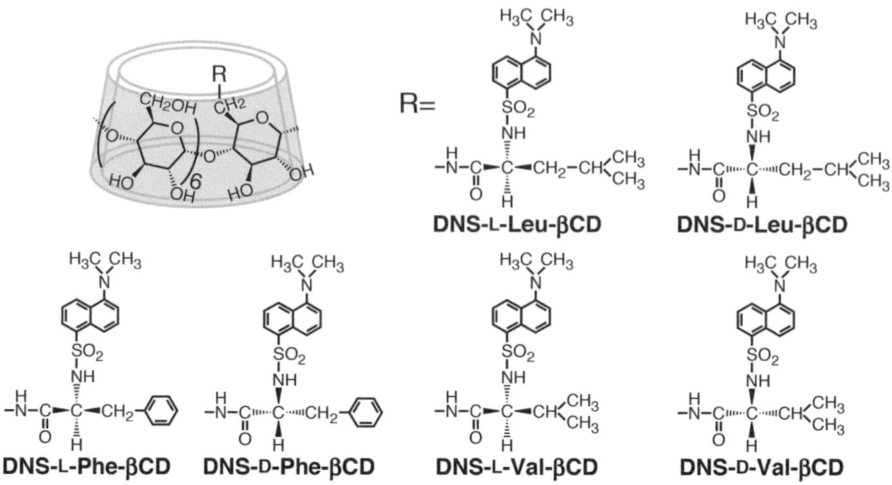

Fig. 5.3 Structures of dansyl-amino acid-cyclodextrin triad systems; *N*-dansyl-L-leucine-appended β-cyclodextrin (DNS-L-Leu-βCD), *N*-dansyl-D-leucine-appended β-cyclodextrin (DNS-D-Leu-βCD), *N*-dansyl-L-phenylalanine-appended β-cyclodextrin (DNS-L-Phe-βCD), *N*-dansyl-D-phenylalanine-appended β-cyclodextrin (DNS-D-Phe-βCD), *N*-dansyl-L-valine-appended β-cyclodextrin (DNS-L-Val-βCD), and *N*-dansyl-D-valine-appended β-cyclodextrin (DNS-D-Val-βCD)

5.2.1 Fluorophore-Amino Acid-Cyclodextrin Triad Systems for Fluorescent Chemosensors

Cyclodextrin-leucine-chromophore triad systems are examples of fluorescent cyclodextrin chemosensors (Fig. 5.3). These chemosensors contain a leucine moiety as a spacer between the cyclodextrin cavity and the dansyl moiety which acts as a fluorophore. The hydrophobic side chain of the leucine moiety is expected to increase binding affinity due to a hydrophobic cap effect. The chirality of the leucine moiety is also expected to affect the binding affinity for guests.

Ikeda et al. first investigated conformational changes of cyclodextrin chemosensors upon addition of guests via fluorescence decay and NMR techniques (Ikeda et al. 1996, 1997). While the conformational interconversion occurs too rapidly to be followed by NMR spectroscopy, analysis of the fluorescence decay of the pendant fluorophore can provide useful information with respect to conformational features because the fluorescence lifetimes of many fluorophores are of a measurable magnitude (nanoseconds) for each conformation. The dansyl moiety is sensitive to the hydrophobicity around it and has longer lifetimes when located inside the cavity as opposed to in the bulk water solution. All of the dansyl moiety fluorescence decays were analyzed by a simple double exponential function. This means that there are two kinds of observable conformational isomers which are in equilibrium. The longer and shorter lived species are the ones with the dansyl moiety inside and outside the cavity, respectively (Fig. 5.4).

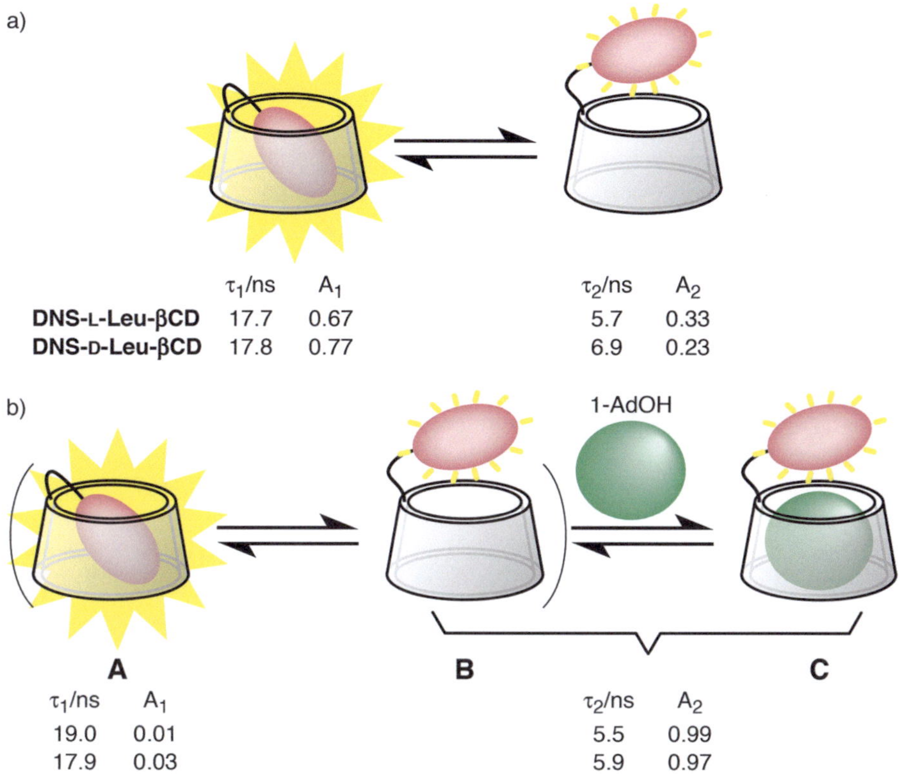

Fig. 5.4 Conformational equilibria of *N*-dansyl-L-leucine-appended β-cyclodextrin (DNS-L-Leu-βCD) and *N*-dansyl-D-leucine-appended β-cyclodextrin (DNS-D-Leu-βCD) and their guest-induced conformational changes in aqueous solution with fluorescence decay data (Adapted from Ikeda et al. 1996)

The fluorescence decay is expressed by Eq. 5.1:

$$D(\mathrm{t}) = A_1 \exp(-\mathrm{t}/\tau_1) + A_2 \exp(-\mathrm{t}/\tau_2) \tag{5.1}$$

where A_i is a pre-exponential factor contributing to the signal at zero time and τ_i is the lifetime of the *i*th component.

Alternatively, A_1/A_2 can be expressed by Eq. 5.2;

$$A_1 / A_2 = (\varepsilon_1 C_1 \Phi_1 / \tau_1) / (\varepsilon_2 C_2 \Phi_2 / \tau_2) = (\varepsilon_1 C_1 / \tau_{01}) / (\varepsilon_2 C_2 / \tau_{02}) \tag{5.2}$$

where C_i is the concentration of species i, ε_i is the molar extinction coefficient, Φ_i is the quantum efficiency, and τ_{0i} is the intrinsic fluorescence lifetime.

$$\text{If } \varepsilon_1 = \varepsilon_2 \text{ and } \tau_{01} = \tau_{02}, \tag{5.3}$$

then

$$A_1 / A_2 = C_1 / C_2 \tag{5.4}$$

A reasonable approximation of the ratio of the intrinsic lifetimes is unity because the fluorescence of the self-inclusion and non-self-inclusion states arise from the same unit (Li et al. 1975; Hashimoto and Thomas 1985; Nelson et al. 1988; Huang and Bright 1990; Dunbar and Bright 1994). Therefore, the equilibrium can be quantified from parameters obtained directly from the analysis of fluorescence decay curves. This estimation is useful for discussion of the equilibrium semi-quantitatively. However it should be noted that τ_0 of the dansyl moiety depends slightly on solvent polarity; therefore, Eq. 5.4 is only a rough approximation (Li et al. 1975).

The function ratio of two species indicates that the self-inclusion states of *N*-dansyl-L-leucine-appended β-cyclodextrin and *N*-dansyl-D-leucine-appended β-cyclodextrin are more stable than their non-self-inclusion states. The chirality of the leucine moiety also affects the stability of the self-inclusion form. The self-inclusion form of *N*-dansyl-D-leucine-appended β-cyclodextrin having the D-leucine moiety is more stable than that of *N*-dansyl-L-leucine-appended β-cyclodextrin having the L-leucine moiety.

The fractions of the shorter lifetime species of *N*-dansyl-L-leucine-appended β-cyclodextrin and *N*-dansyl-D-leucine-appended β-cyclodextrin increase upon addition of 1-adamantanol, while the fractions of the longer lifetime species of the hosts decrease (Fig. 5.4). This indicates that the induced-fit conformational change of the chromophore-appended cyclodextrin occurs in association with guest accommodation, excluding the chromophore from the inside to the outside of the cyclodextrin cavity. The fact that the fluorescence decay curves of the hosts are still double-exponential indicates that the fluorescence lifetimes of species B and C in Fig. 5.4 are similar and are not resolved by this lifetime measurement system.

Figure 5.5 shows the sensitivity parameters expressed as $\Delta I/I_0$ (where $\Delta I = I - I_0$, with I and I_0 being the fluorescence intensities in the presence and absence of the guest, respectively). Compounds *N*-dansyl-L-leucine-appended β-cyclodextrin and *N*-dansyl-D-leucine-appended β-cyclodextrin exhibit similar trends in guest dependency of the $\Delta I/I_0$ value, but the $\Delta I/I_0$ value of *N*-dansyl-L-leucine-appended β-cyclodextrin is larger than that of *N*-dansyl-D-leucine-appended β-cyclodextrin in all cases because the binding constants of *N*-dansyl-L-leucine-appended β-cyclodextrin are over twice as large as those of *N*-dansyl-D-leucine-appended β-cyclodextrin for most of the guests (Fig. 5.5).

Four steroidal compounds were detected by *N*-dansyl-L-leucine-appended β-cyclodextrin and *N*-dansyl-D-leucine-appended β-cyclodextrin with the sensitivity order cholic acid < deoxycholic acid < chenodeoxycholic acid < ursodeoxycholic acid. This order is the same as the order of their binding constants. These compounds have the same steroidal framework, and ursodeoxycholic acid and chenodeoxycholic acid are isomers with differing stereochemistry of the hydroxyl group at C-7. Deoxycholic acid is the regioisomer of ursodeoxycholic acid and chenodeoxycholic acid; it has one hydroxyl group at C-12 instead of C-7. Cholic acid has one more hydroxyl group than the other compounds.

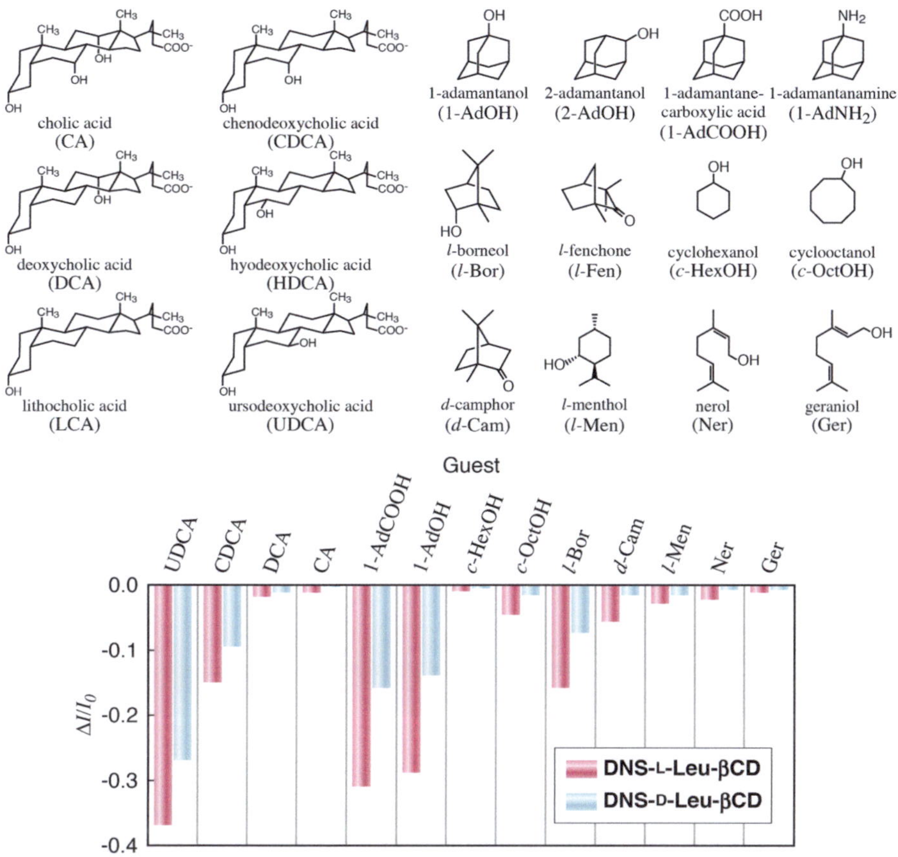

Fig. 5.5 Sensitivity parameters ($\Delta I/I_0$) of *N*-dansyl-L-leucine-appended β-cyclodextrin (DNS-L-Leu-βCD) and *N*-dansyl-D-leucine-appended β-cyclodextrin (DNS-D-Leu-βCD): [DNS-L-Leu-βCD] = [DNS-D-Leu-βCD] = 2×10^{-6} M, [guest] = 1×10^{-5} M; $\Delta I/I_0 = (I - I_0)/I_0$ (where I and I_0 are the fluorescence intensities in the presence and absence of the guest, respectively) (Adapted from Ikeda et al. 1996)

The NMR spectra of *N*-dansyl-L-leucine-appended β-cyclodextrin and *N*-dansyl-D-leucine-appended β-cyclodextrin provide information about their structures in the self-inclusion state. The dansyl moieties of *N*-dansyl-L-leucine-appended β-cyclodextrin and *N*-dansyl-D-leucine-appended β-cyclodextrin are located inside the cavity in the major conformational state. The ^{1}H NMR spectra of *N*-dansyl-L-leucine-appended β-cyclodextrin and *N*-dansyl-D-leucine-appended β-cyclodextrin were assigned using various 1D and 2D NMR techniques. Their structures were estimated from NOE data and the degree of the anisotropic ring current effect from the dansyl moiety in the ^{1}H resonances of the cyclodextrin protons (Ikeda et al. 1997). The dansyl moiety of *N*-dansyl-D-leucine-appended β-cyclodextrin is located deeper within the cavity than that of *N*-dansyl-L-leucine-appended β-cyclodextrin (Fig. 5.6). This deeper inclusion into its cavity makes *N*-dansyl-D-leucine-appended

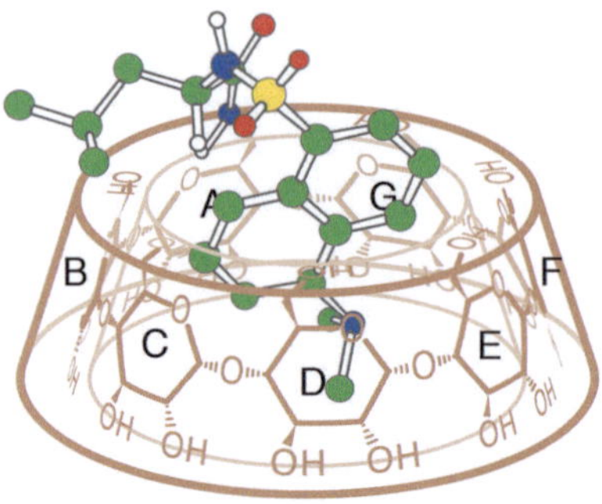

Fig. 5.6 Estimated structures of *N*-dansyl-L-leucine-appended β-cyclodextrin (DNS-L-Leu-βCD) and *N*-dansyl-D-leucine-appended β-cyclodextrin (DNS-D-Leu-βCD) (Adapted from Ikeda et al. 1996)

β-cyclodextrin the more stable self-inclusion complex, and this higher stability is also reflected in its larger fluorescence intensity and lower binding abilities.

5.2.2 *Chiral Recognition by Fluorophore-Amino Acid-Cyclodextrin Triad Systems*

There is interest in whether the influential effect of the chirality of the leucine spacer unit on the stability of the self-inclusion state is shared by other amino acid spacer units. This was investigated by preparing four kinds of fluorophore-amino acid-cyclodextrin triad systems having either phenylalanine or valine as the spacer (Fig. 5.3) (Ikeda et al. 2006a). The fluorescence intensity of *N*-dansyl-D-phenylalanine-appended β-cyclodextrin was much larger than that of *N*-dansyl-L-phenylalanine-appended β-cyclodextrin. This influence of the spacer chirality on the fluorescence intensity is similar to the *N*-dansyl-leucine-appended β-cyclodextrin. On the other hand, the fluorescence intensity of *N*-dansyl-L-valine-appended β-cyclodextrin was similar to that of *N*-dansyl-D-valine-appended β-cyclodextrin, both of which were larger than that of *N*-dansyl-L-leucine-appended β-cyclodextrin or *N*-dansyl-D-leucine-appended β-cyclodextrin. The isopropyl side chain of valine is smaller than the isobutyl side chain of leucine and scarcely inhibits the self-inclusion of the dansyl moiety into the cyclodextrin cavity, whereas the benzyl moiety of phenylalanine affects the self-inclusion depth of the dansyl moiety.

There is also interest in the ability of chemosensors having a chiral amino acid to engage in enantioselective recognition. Enantioselective chemosensors are currently of great interest to determine the enantiomeric excess of samples in drug discovery or of products in high-throughput screening of enantioselective catalysts and biocatalysts. Fig. 5.7 shows the chiral discrimination abilities of the four chemosensors for some norbornane and cyclohexane derivatives. *N*-dansyl-L-

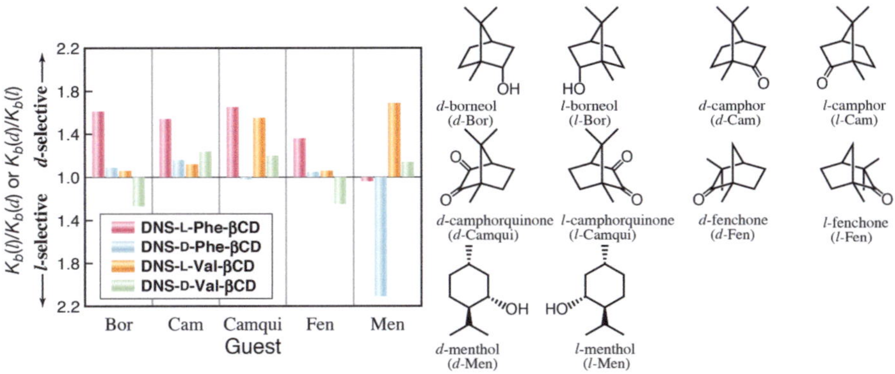

Fig. 5.7 Chiral discrimination abilities of *N*-dansyl-L-phenylalanine-appended β-cyclodextrin (DNS-L-Phe-βCD), *N*-dansyl-D-phenylalanine-appended β-cyclodextrin (DNS-D-Phe-βCD), *N*-dansyl-L-valine-appended β-cyclodextrin (DNS-L-Val-βCD), and *N*-dansyl-D-valine-appended β-cyclodextrin (DNS-D-Vale-βCD) (Adapted from Ikeda et al. 2006a)

phenylalanine-appended β-cyclodextrin exhibited good *d*-selectivity for borneol, camphor, camphorquinone, and fenchone but poor selectivity for menthol. *N*-dansyl-D-phenylalanine-appended β-cyclodextrin and *N*-dansyl-L-valine-appended β-cyclodextrin exhibited good *l*-selectivity and *d*-selectivity for menthol, respectively. Only menthol has a cyclohexane skeleton, whereas the other guests have the same norbornane framework. *N*-dansyl-L-phenylalanine-appended β-cyclodextrin exhibited high selectivity for the norbornane derivatives but does not show selectivity for the cyclohexane derivative. Each of the four chemosensors shows a different selectivity pattern for each of the norbornane derivatives having the same framework but a different functional group.

5.3 Color-Change Chemosensors

Ueno's group prepared guest responsive color-change chemosensors using dye-appended cyclodextrins (Ueno et al. 1992; Kuwabara et al. 1993, 1994, 1996, 1998, 1999; Aoyagi et al. 1997; Matsushita et al. 1997). Color-change is effective for detection by the naked eye. Methyl red-, *p*-nitrophenol-, alizarin yellow-, and phenolphthalein-appended cyclodextrins have been reported as color-change chemosensors, each acting in acidic, neutral, or alkaline solution. Fig. 5.8 shows the pH of the solution for working as a chemosensor and the color change upon addition of a guest to theses color-change chemosensors. The dye in the cyclodextrin cavity is generally protected from protonation or deprotonation. Accommodation of a guest in the cyclodextrin cavity migrates the dye from the hydrophobic cyclodextrin cavity to the bulk water. This causes protonation or deprotonation of the dye, producing the color change.

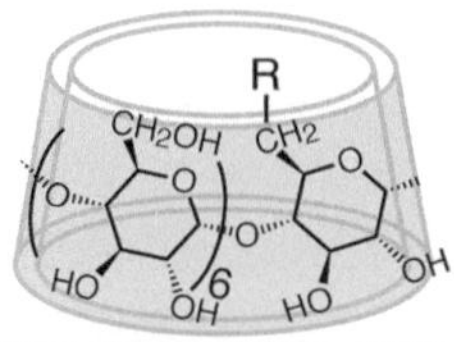

R =	without a guest	with a guest	solution pH	
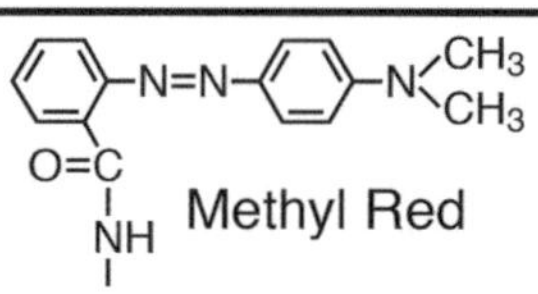 Methyl Red	yellow	red	1.6	a), c), f)
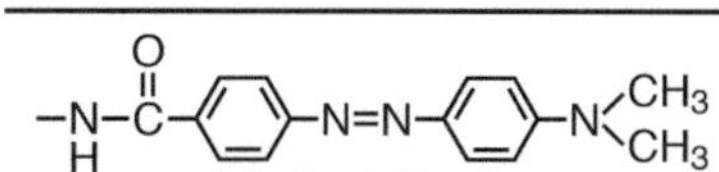 *p*-Methyl Red	orange	red	2.4	c), f)
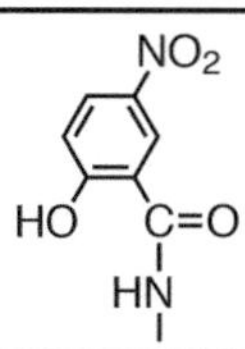 *p*-Nitrophenol	yellow	colorless	5.0	h)
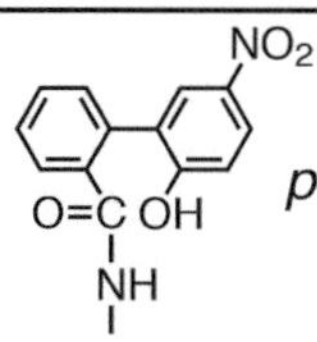 *p*-Nitrophenol	yellow	colorless	6.5	b) h)
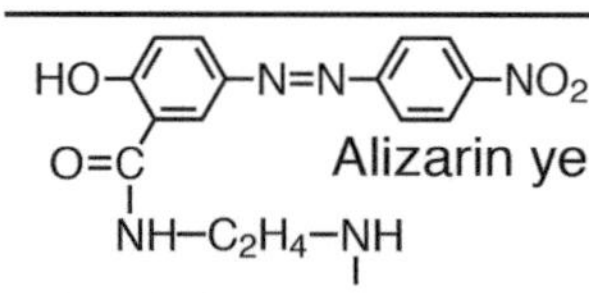 Alizarin yellow	yellow	red	8.3	g)
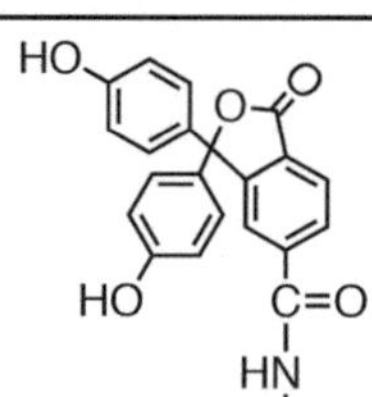 Phenolphtalein	colorless	purple	9.7	d), e)

a) Ueno 1992, b) Kuwabara 1993, c) Kuwabara1994, d) Kuwabara 1996,
e) Kuwabara 1998, f) Kuwabara1999, g) Aoyagi 1997, h) Matsushita 1997

Fig. 5.8 Structures of guest-responsive color-change chemosensors

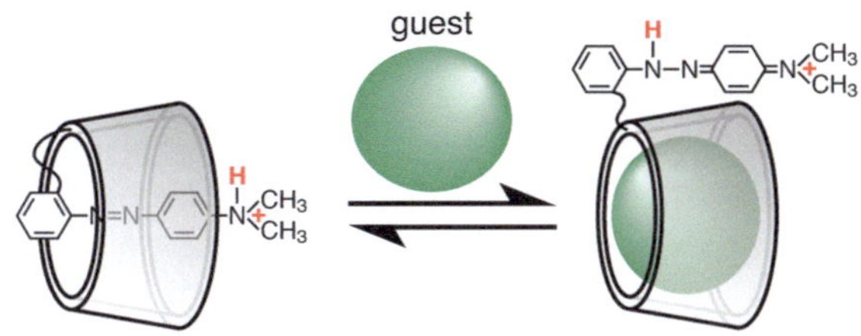

Fig. 5.9 A two-state equilibrium model for guest-induced conformational change of methyl red-appended β-cyclodextrin that causes the color change; The self-inclusion state is usually the major conformation and the ammonium form is the major form in acidic solution (left). An induced-fit conformational change of the methyl red-appended cyclodextrin occurs in association with accommodation of the guest (right). This conformational change causes the azo group to change to the azonium form, which is a tautomer of the ammonium form. This change induces the color change from yellow to red. (Adapted from Ueno et al. 1992)

For example, the azo group of methyl red-appended β-cyclodextrin is protected from protonation by the cyclodextrin cavity even in acidic solution (Fig. 5.9). Because the methyl red moiety is longer than the cyclodextrin cavity, the dimethylamino group is placed outside the cyclodextrin cavity and is easily protonated. Upon guest binding, the methyl red moiety is displaced from the cyclodextrin cavity to the acidic bulk water; the azo group changes to the azonium form, which is a tautomer of the ammonium form. This change causes the color change from yellow to red. Neutral organic molecules can be detected by these color-change chemosensors in aqueous solution. These sensing abilities for various guests are roughly parallel to the binding affinity for the guest.

5.4 Turn-On Fluorescent Chemosensors

A turn-off mechanism is greatly effective for producing a detectable change in a signal upon guest accommodation into the cyclodextrin cavity, but it has the following defects.

(1) Self-inclusion of the chromophore can inhibit accommodation of the guest. For example, the self-inclusion state of *N*-dansyl-D-leucine-appended β-cyclodextrin is twice as stable as that of *N*-dansyl-L-leucine-appended β-cyclodextrin, and the binding ability of the former for a guest is about half that of the latter.
(2) The guest selectivity of the chemosensor mainly depends on the selectivity of the cyclodextrin itself. The affinities of both β-cyclodextrin and γ-cyclodextrin for bile acid derivatives are greater than their affinities for adamantane derivatives. Therefore, the sensitivity parameter of turn-off chemosensors cannot identify whether adamantanol exists or not in a mixed solution of a bile acid and adamantanol.

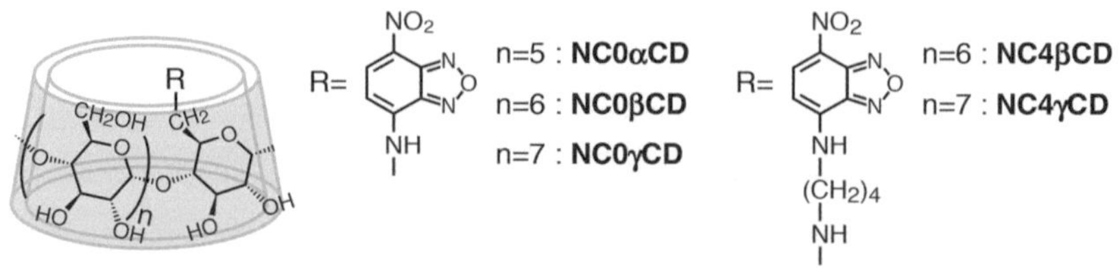

Fig. 5.10 Structures of (7-nitrobenz-2-oxa-1,3-diazol-4-yl)amine-appended cyclodextrins and (7-nitrobenz-2-oxa-1,3-diazol-4-yl)aminobutylamine-appended cyclodextrins

(3) The detection of a guest is accompanied by a decrease in the fluorescence intensity for turn-off chemosensors, although an increase in the emission intensity caused by guest response is more effective for chemical sensing systems.

Therefore, a new fluorescent chemosensor was prepared to overcome these defects. (7-Nitrobenz-2-oxa-1,3-diazol-4-yl)amine (NBDamine) was selected as a fluorophore for a new type of chemosensor (Ikeda et al. 2005). (7-Nitrobenz-2-oxa-1,3-diazol-4-yl)amine displays the interesting property of fluorescing weakly in water and strongly in organic solvents, membranes, or hydrophobic environments (Uchiyama et al. 2001). One chemosensor ((7-nitrobenz-2-oxa-1,3-diazol-4-yl) aminobutylamine-appended β-cyclodextrin (NC4βCD) or (7-nitrobenz-2-oxa-1,3-diazol-4-yl)aminobutylamine-appended γ-cyclodextrin (NC4γCD)) has a spacer similar to the turn-off cyclodextrin-based chemosensors and another chemosensor ((7-nitrobenz-2-oxa-1,3-diazol-4-yl)amine-appended α-cyclodextrin (NC0αCD), (7-nitrobenz-2-oxa-1,3-diazol-4-yl)amine-appended β-cyclodextrin (NC0βCD), or (7-nitrobenz-2-oxa-1,3-diazol-4-yl)amine-appended γ-cyclodextrin (NC0γCD)) has no spacer (Fig. 5.10).

5.4.1 *Skeleton-Selective Chemosensor*

The response of (7-nitrobenz-2-oxa-1,3-diazol-4-yl)aminobutylamine-appended β-cyclodextrin to guests is similar to the turn-off cyclodextrin-based chemosensors but the response of (7-nitrobenz-2-oxa-1,3-diazol-4-yl)amine-appended β-cyclodextrin to guests is quite different from those (Fig. 5.11). The fluorescence intensity of (7-nitrobenz-2-oxa-1,3-diazol-4-yl)amine-appended β-cyclodextrin increased upon addition of 1-adamantanol (1-AdOH), indicating an increase in hydrophobicity near the (7-nitrobenz-2-oxa-1,3-diazol-4-yl)amine moiety induced by accommodation of the guest (Fig. 5.12). Figure 5.11 shows the sensitivity parameters of (7-nitrobenz-2-oxa-1,3-diazol-4-yl)amine-appended β-cyclodextrin for various guests. (7-Nitrobenz-2-oxa-1,3-diazol-4-yl)amine-appended β-cyclodextrin is quite sensitive to the adamantane and borneol derivatives, which have a comparatively spherical shape that fits the β-cyclodextrin cavity. (7-Nitrobenz-2-oxa-1,3-diazol-4-yl)amine-appended β-cyclodextrin exhibits a large increase in the

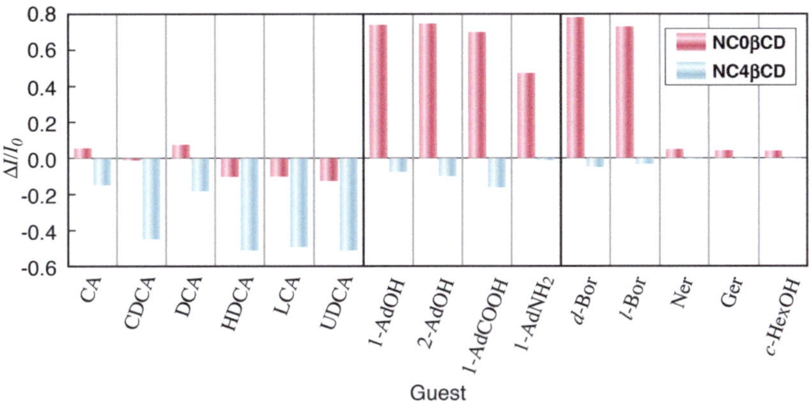

Fig. 5.11 Sensitivity parameters ($\Delta I/I_0$) of (7-nitrobenz-2-oxa-1,3-diazol-4-yl)amine-appended β-cyclodextrin (NC0βCD) and (7-nitrobenz-2-oxa-1,3-diazol-4-yl)aminobutylamine-appended β-cyclodextrin (NC4βCD): [NC0βCD] = [NC4βCD] = 5x10^{-6} M, [guest] = 1x10^{-5} M. $\Delta I/I_0 = (I - I_0)/I_0$ (where I and I_0 are the fluorescence intensities in the presence and absence of the guest, respectively) (Adapted from Ikeda et al. 2005)

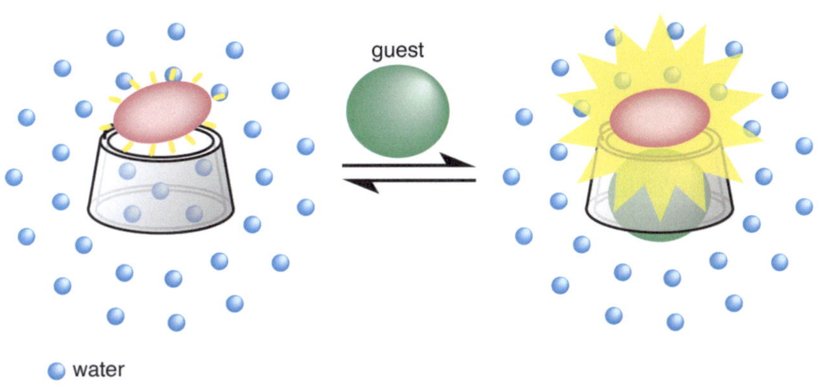

Fig. 5.12 Guest-induced conformational changes of (7-nitrobenz-2-oxa-1,3-diazol-4-yl)amine-appended cyclodextrins; Hydrophobicity near the (7-nitrobenz-2-oxa-1,3-diazol-4-yl)amine moiety increases by accommodation of the guest and its fluorescence intensity increases

fluorescence intensity in response to these guests. Notably, (7-nitrobenz-2-oxa-1,3-diazol-4-yl)amine-appended β-cyclodextrin is not sensitive to bile acids, although bile acids are strongly bound by the native β- cyclodextrin.

^{1}H NMR spectra changes upon addition of a guest suggests the fluorescence change mechanism. Upon addition of 1-adamantanol to a solution of (7-nitrobenz-2-oxa-1,3-diazol-4-yl)amine-appended β-cyclodextrin in D_2O, the motion of the (7-nitrobenz-2-oxa-1,3-diazol-4-yl)amine moiety was restricted by the van der Waals interactions with 1-adamantanol. This observation indicates that the (7-nitrobenz-2-oxa-1,3-diazol-4-yl)amine moiety is still positioned at the entrance of the

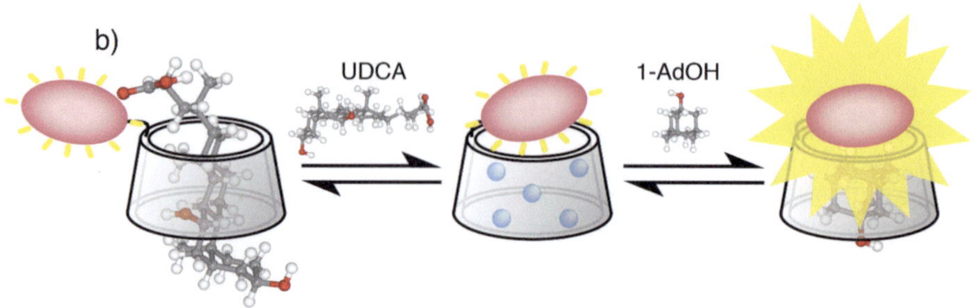

Fig. 5.13 Guest-induced conformational changes of (7-nitrobenz-2-oxa-1,3-diazol-4-yl)amine-appended β-cyclodextrins for ursodeoxycholic acid (UDCA) complex and 1-adamantanol (1-AdOH) complex; The (7-nitrobenz-2-oxa-1,3-diazol-4-yl)amine moiety moves away from the entrance of the cyclodextrin cavity after making the inclusion complex with ursodeoxycholic acid and its fluorescence intensity does not increase. The (7-nitrobenz-2-oxa-1,3-diazol-4-yl)amine moiety is still positioned at the entrance of the cyclodextrin cavity even when 1-adamantanol is accommodated and its fluorescence intensity increases (Adapted from Ikeda 2011)

cyclodextrin cavity, even when 1-adamantanol is accommodated. By contrast, upon addition of ursodeoxycholic acid to a solution of (7-nitrobenz-2-oxa-1,3-diazol-4-yl)amine-appended β-cyclodextrin in D_2O, the (7-nitrobenz-2-oxa-1,3-diazol-4-yl)amine moiety must move away from the entrance of the cyclodextrin cavity and the motion of the (7-nitrobenz-2-oxa-1,3-diazol-4-yl)amine moiety is not restricted. These different behaviors of the (7-nitrobenz-2-oxa-1,3-diazol-4-yl) amine moiety cause the difference in the fluorescence variation observed upon addition of the guest. The ^{1}H resonances for the (7-nitrobenz-2-oxa-1,3-diazol-4-yl) amine moiety of (7-nitrobenz-2-oxa-1,3-diazol-4-yl)amine-appended β-cyclodextrin alone are broader than those of (7-nitrobenz-2-oxa-1,3-diazol-4-yl)amine-appended β-cyclodextrin in the presence of ursodeoxycholic acid. This suggests that the motion of the (7-nitrobenz-2-oxa-1,3-diazol-4-yl)amine moiety of (7-nitrobenz-2-oxa-1,3-diazol-4-yl)amine-appended β-cyclodextrin is restricted in the absence of the guest, because the (7-nitrobenz-2-oxa-1,3-diazol-4-yl)amine moiety interacts with the rim of the cyclodextrin cavity.

This observation can be explained by the equation in Fig. 5.13. The bile acid derivative can be accommodated inside the β-cyclodextrin cavity, but the (7-nitrobenz-2-oxa-1,3-diazol-4-yl)amine moiety moves away from the entrance of the cyclodextrin cavity after making the inclusion complex with the bile acid derivative. The (7-nitrobenz-2-oxa-1,3-diazol-4-yl)amine moiety is still located in a hydrophilic environment in the complex with the bile acid, and its fluorescence intensity does not increase. On the other hand, the (7-nitrobenz-2-oxa-1,3-diazol-4-yl)amine moiety is still positioned at the entrance of the cyclodextrin cavity even when 1-adamantanol is accommodated.

Considering the differing guest-response patterns of (7-nitrobenz-2-oxa-1,3-diazol-4-yl)amine-appended β-cyclodextrin and (7-nitrobenz-2-oxa-1,3-diazol-4-yl)aminobutylamine-appended β-cyclodextrin, discrimination between the bile acids and adamantane derivatives at any concentration by the combined use of (7-nitrobenz-2-oxa-1,3-diazol-4-yl)amine-appended β-cyclodextrin and

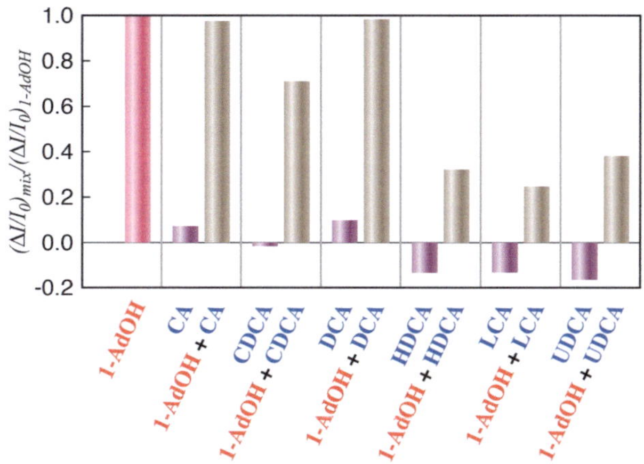

Fig. 5.14 Relative sensitivity parameters ($(\Delta I/I_0)_{mix}/(\Delta I/I_0)_{1\text{-}AdOH}$) of (7-nitrobenz-2-oxa-1,3-diazol-4-yl)amine-appended β-cyclodextrin (NC0βCD) (5×10^{-6} M) for mixtures of guests (each at 1×10^{-5} M); $(\Delta I/I_0)_{1\text{-}AdOH} = (I - I_0)/I_0$ (where I and I_0 are the fluorescence intensities in the presence and absence of 1-adamantanol, respectively), $(\Delta I/I_0)_{mix} = (I - I_0)/I_0$ (where I and I_0 are the fluorescence intensities in the presence and absence of both a bile acid derivative and 1-adamantanol, respectively)). The sensitivity parameters for a bile acid derivative alone have been normalized to the sensitivity parameters for 1-adamantanol alone (Adapted from Ikeda et al. 2005)

(7-nitrobenz-2-oxa-1,3-diazol-4-yl)aminobutylamine-appended β-cyclodextrin is now a simple matter.

Notably, 1-adamantanol can be detected even in the presence of a bile acid by using (7-nitrobenz-2-oxa-1,3-diazol-4-yl)amine-appended β-cyclodextrin (Ikeda et al. 2005). This is the first example of adamantanol being detected in the presence of a bile acid by a cyclodextrin-based chemosensor. The relative sensitivity parameters for a solution containing both a bile acid and 1-adamantanol are shown in Fig. 5.14.

5.4.2 *Bile Acids-Selective Chemosensor*

The sensitivity parameters of (7-nitrobenz-2-oxa-1,3-diazol-4-yl)amine-appended γ-cyclodextrin and (7-nitrobenz-2-oxa-1,3-diazol-4-yl)aminobutylamine-appended γ-cyclodextrin for various guests are shown in Fig. 5.15 (Ikeda et al. 2006b). (7-Nitrobenz-2-oxa-1,3-diazol-4-yl)amine-appended γ-cyclodextrin is relatively sensitive to each bile acid but has no response to other guests. (7-Nitrobenz-2-oxa-1,3-diazol-4-yl)aminobutylamine-appended γ-cyclodextrin is not sensitive to any of the guests. In most cases, the response of (7-nitrobenz-2-oxa-1,3-diazol-4-yl)amine-appended γ-cyclodextrin to the guests is an increase in the fluorescence intensity. These results suggest that the hydrophobic face of the bile acid can interact with the (7-nitrobenz-2-oxa-1,3-diazol-4-yl)amine moiety in the larger space of the entrance of the γ-cyclodextrin cavity, increasing hydrophobicity of the chromophore's

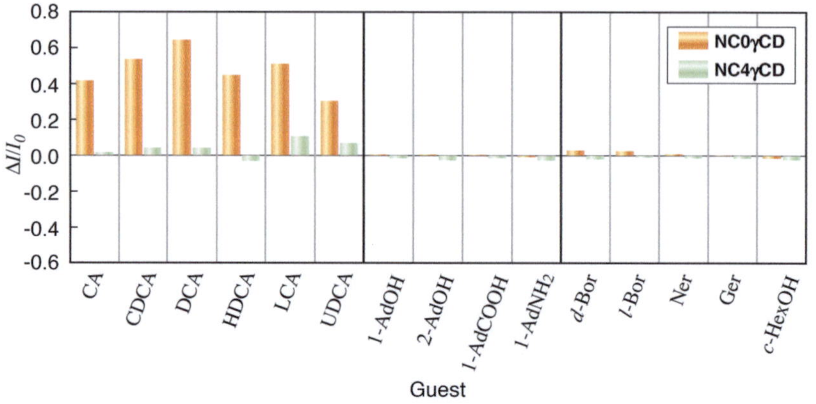

Fig. 5.15 Sensitivity parameters ($\Delta I/I_0$) of (7-nitrobenz-2-oxa-1,3-diazol-4-yl)amine-appended γ-cyclodextrin (NC0γCD) and (7-nitrobenz-2-oxa-1,3-diazol-4-yl)aminobutylamine-appended γ-cyclodextrin (NC4γCD): [NC0γCD] = [NC4γCD] = 5x10^{-6} M, [guest] = 1x10^{-5} M. $\Delta I/I_0 = (I - I_0)/I_0$ (where I and I_0 are the fluorescence intensities in the presence and absence of the guest, respectively) (Adapted from Ikeda et al. 2006b)

environment. However, the γ-cyclodextrin cavity is too large for other guests, such as adamantane derivatives, to increase hydrophobicity around the (7-nitrobenz-2-oxa-1,3-diazol-4-yl)amine moiety because the water molecules cannot be completely excluded from the cavity.

The difference in the sensitivity parameters between each bile acid is not large, although the binding affinities of the native γ-cyclodextrin are different for these bile acids. The variation in the fluorescence intensity does not correlate to the binding affinity of (7-nitrobenz-2-oxa-1,3-diazol-4-yl)amine-appended γ-cyclodextrin. The structure of the host-guest complex has a greater influence on the fluorescence intensity than the binding affinity. The $\Delta I_{max}/I_0$ of (7-nitrobenz-2-oxa-1,3-diazol-4-yl)amine-appended γ-cyclodextrin for each guest differs considerably in contrast with *N*-dansyl-leucine-appended β-cyclodextrin whose $\Delta I_{max}/I_0$ is broadly similar for each guest. When a large excess of any guest is added to a solution of *N*-dansyl-leucine-appended β-cyclodextrin, the dansyl moiety is located in a similar position outside the cyclodextrin cavity and in a similar environment. The environment around the fluorophore in the (7-nitrobenz-2-oxa-1,3-diazol-4-yl)amine-appended γ-cyclodextrin/guest complex is different for each guest. This difference in the fluorophore environment gives rise to the observed differences in sensitivity parameters for various guests. The $\Delta I_{max}/I_0$ values of (7-nitrobenz-2-oxa-1,3-diazol-4-yl)amine-appended γ-cyclodextrin for cholic acid is three times larger than that for ursodeoxycholic acid, whereas the binding constant for cholic acid is one-third of that for ursodeoxycholic acid. Therefore, the sensitivity parameter for cholic acid is similar to that for ursodeoxycholic acid. The sensitivity parameter of the turn-on chemosensor depends on both the binding affinity for the guest and the structure of the inclusion complex; the structure of the inclusion complex strongly affects the environment around the fluorophore. In contrast, the sensitivity parameter of the turn-off chemosensors generally depends on only the binding affinity for the guest.

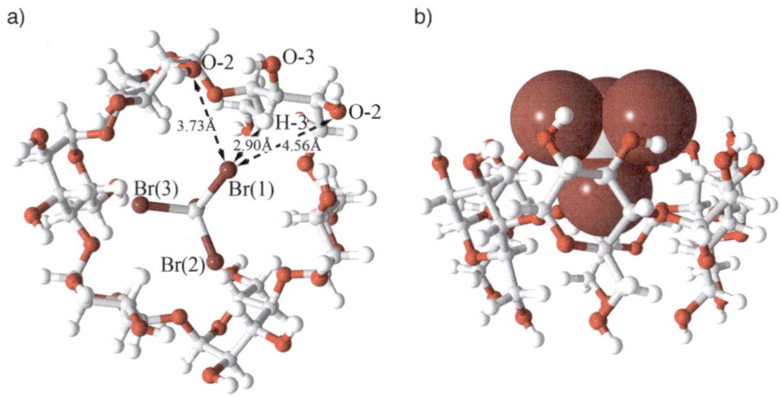

Fig. 5.16 Estimated structure of the α-cyclodextrin/CBr_4 complex (**a**) view from the secondary hydroxy side of the complex, (**b**) the side view of the complex (Adapted from Ikeda and Ueno 2009)

5.4.3 Chemosensor for Halomethanes

Although the cavity size of α-cyclodextrin is suitable for smaller guests such as halomethanes or alkanol, it is difficult to use the turn-off mechanism to construct chemosensors using α-cyclodextrin. There are few chromophores whose size and shape are suitable for the self-inclusion state in the narrow cavity of α-cyclodextrin. The chromophore in the self-inclusion state is often not excluded by guest accommodation because the binding affinity of α-cyclodextrin is not large for most guests. This disadvantage of the turn-off mechanism can be overcome by the turn-on mechanism. A fluorophore of a turn-on chemosensor is not self-included and remains at the entrance of the cyclodextrin cavity. The fluorophore does not inhibit accommodation of the guest.

Ikeda prepared a chemosensor for halomethanes using the turn-on mechanism (Ikeda and Ueno 2009). The absorption intensity of (7-nitrobenz-2-oxa-1,3-diazol-4-yl)amine-appended α-cyclodextrin decreased upon addition of CCl_4 and the fluorescence intensity of (7-nitrobenz-2-oxa-1,3-diazol-4-yl)amine-appended α-cyclodextrin increased upon addition of CCl_4. These phenomena indicate that the hydrophobicity around the (7-nitrobenz-2-oxa-1,3-diazol-4-yl)amine moiety increased upon addition of the guest. The ^{1}H NMR spectra of the α-cyclodextrin/halomethane complexes and MMFF94 molecular mechanics calculation of the α-cyclodextrin/halomethane complexes indicate that the (7-nitrobenz-2-oxa-1,3-diazol-4-yl)amine moiety remains at the entrance of the cyclodextrin cavity (Fig. 5.16).

The binding constants for halomethanes are quite a bit larger than those for alcohols having two or three carbons (Fig. 5.17). Although the reported binding constant of α-cyclodextrin for CCl_4 is only 40 M^{-1} (Fourmentin et al. 2007), the binding constant of (7-nitrobenz-2-oxa-1,3-diazol-4-yl)amine-appended α-cyclodextrin for CCl_4 is 956 M^{-1}, over 20 times larger. The (7-nitrobenz-2-oxa-1,3-diazol-4-yl)

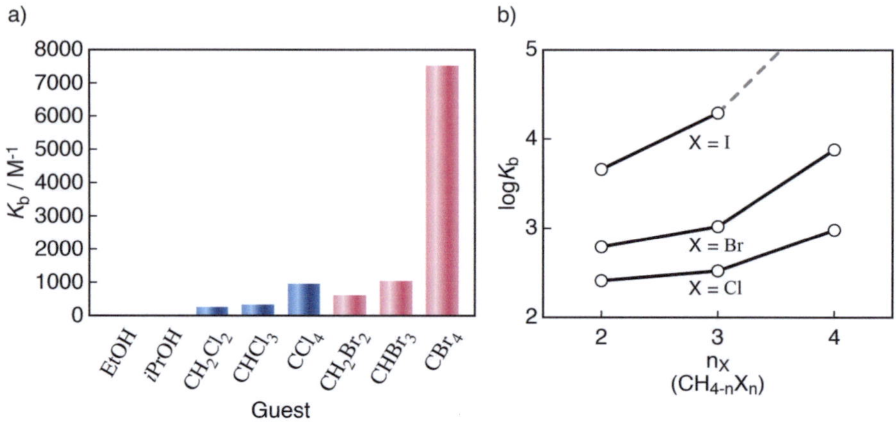

Fig. 5.17 (**a**) Binding constants (K_b) of (7-nitrobenz-2-oxa-1,3-diazol-4-yl)amine-appended α-cyclodextrin (NC0αCD) for halomethanes and alcohols and (**b**) logarithms of binding constants (logK_b) of (7-nitrobenz-2-oxa-1,3-diazol-4-yl)amine-appended α-cyclodextrin (NC0αCD) for halomethanes ($CH_{4-n}X_n$) vs number of halogen atoms bound to the methane (n_X). (Adapted from Ikeda and Ueno 2009)

amine moiety acts as a hydrophobic cap, increasing the hydrophobicity in the cavity. The binding constant for CBr_4 is eight-fold larger than that for CCl_4.

The cavity of α-cyclodextrin is too small to accommodate the entire CBr_4 molecule. One of the bromine atoms is accommodated in the α-cyclodextrin cavity, and the other bromine atoms remain at the entrance of the cyclodextrin cavity. This indicates that the selectivity for the halogen atom is due to a size effect. The van der Waals radii of chlorine and bromine are 1.69 and 1.83 Å, respectively, and the radius of the narrowest position in the α-cyclodextrin cavity is 2.2 Å (Müller and Wenz 2007).

The binding constants of (7-nitrobenz-2-oxa-1,3-diazol-4-yl)amine-appended α-cyclodextrin increase with increasing number of halogens. Of the brominated methane complexes, only the α-cyclodextrin/CBr_4 complex has a symmetric structure (C3 symmetry) which can cause the multisite interactions (Fig. 5.16) Alternate glucose units interact with bromine atoms to the same degree and this interaction causes the large binding constant.

5.5 Chemosensors for Anion or Cation Species

Inclusion complexes of cyclodextrin and chromophores can be used as chemosensors for anion or cation species. Binding of anion or cation species to the spacer or the rim of modified cyclodextrins induces change in fluorescence intensity or color change.

5.5.1 Chemosensor for Bicarbonate

Suzuki et al. prepared a pyrene-appended γ-cyclodextrin with a triamine spacer (Suzuki et al. 2006). This derivative formed an association dimer in aqueous solution and exhibited typical pyrene fluorescence around 370–400 nm together with strong excimer-like fluorescence centered at 475 nm. A new fluorescence band appeared around 390–460 nm upon addition of $NaHCO_3$. None of the anions (Cl^-, SO_4^{2-}, HPO_4^{2-}, AcO^-, ClO_4^-, NO_3^-) except bicarbonate (HCO_3^-) induced the new fluorescence band, and metal cations (Zn^{2+}, Na^+, or K^+) showed no ability to cause the new fluorescence band. The proposed association behavior is shown in Fig. 5.18. Binding of bicarbonate causes conformational change for the association dimer from a well-stacked conformation to an imperfectly stacked conformation, inducing the new fluorescence band. This selective fluorescence change can act as a chemosensor for bicarbonate, which is a physiologically important anion that plays a vital role in maintaining the pH of biological fluids and in signal transduction in intracellular events.

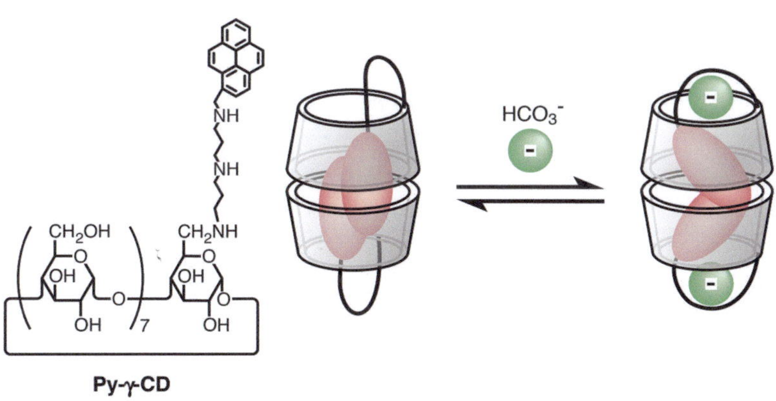

Fig. 5.18 Structure of *N*-[9-(1-pyrenyl)-4,8-diazanonyl]-6-amino-6-deoxy-γ-cyclodextrin (Py-γ-CD) as a bicarbonate sensor and its guest-induced conformational change; Binding of bicarbonate causes conformational change for the association dimer from a well-stacked conformation to an imperfectly stacked conformation, inducing the new fluorescence band. (Adapted from Suzuki et al. 2006)

5.5.2 Chemosensors for Fe^{3+} or Ru^{3+} Cations and Phosphate or Pyrophosphate Anions

Pitchumani et al. elucidated that the 1:1 inclusion complex of per-6-amino-β- cyclodextrin and *p*-nitrophenol can be used as a color-change chemosensor for Fe^{3+} and Ru^{3+} in water (Suresh et al. 2010). Binding of these cations causes an appreciable color change from intense yellow to colorless. They also showed that the 1:2 inclusion complex of per-6-amino-β-cyclodextrin and *p*-nitrophenol can be used as a color-change chemosensor for phosphate or pyrophosphate anions (Azath et al. 2011). Binding of these anions causes an appreciable color change from colorless to intense yellow.

5.6 Conclusion

Chromophore-appended cyclodextrins are effective for detecting molecules in water via fluorescence and absorption changes. The response pattern of cyclodextrin chemosensors for guests can be changed by varying the spacer between the cyclodextrin cavity and the chromophore. The cavity size of cyclodextrin is also an important factor for the response pattern. Many types of cyclodextrin chemosensors having different responses for guests are expected to be used for new pattern recognition systems to detect molecules.

References

Anslyn EV (2007) Supramolecular analytical chemistry. J Org Chem 72:687–699. https://doi.org/10.1021/jo0617971

Aoyagi T, Nakamura A, Ikeda H, Ikeda T, Mihara H, Ueno A (1997) Alizarin yellow-modified β-cyclodextrin as a guest-responsive absorption change sensor. Anal Chem 69:659–663. https://doi.org/10.1021/ac960727z

Azath IA, Suresh P, Pitchumani K (2011) Per-6-ammonium-β-cyclodextrin/*p*-nitrophenol complex as a colorimetric sensor for phosphate and pyrophosphate anions in water. Sens Actuators B: Chem 155:909–914. doi.org/10.1016/j.snb.2011.01.041

Czarnik AW (ed) (1993) Fluorescent chemosensors for ion and molecule recognition. American Chemical Society, Washington, DC

de Silva AP, Gunaratne HQN, Gunnlaugsson T, Huxley AJM, McCoy CP, Rademacher JT, Rice TE (1997) Signaling recognition events with fluorescent sensors and switches. Chem Rev 97:1515–1566. https://doi.org/10.1021/cr960386p

Desvergne JP, Czarnik AW (eds) (1997) Chemosensors of ion and molecule recognition, NATO ASI Series C492. Kluwer Academic, Dordrecht

Dunbar RA, Bright FV (1994) Comparison of inter- and intramolecular cyclodextrin complexes. Supramol Chem 3:93–99. doi.org/10.1080/10610279408029847

Fourmentin S, Outirite M, Blach P, Landy D, Ponchel A, Monflier E, Surpateanu G (2007) Solubilisation of chlorinated solvents by cyclodextrin derivatives: a study by static headspace gas chromatography and molecular modelling. J Hazard Mater 141:92–97. doi.org/10.1016/j.jhazmat.2006.06.090

Hall KS, Nissan RA, Quintana RL, Hollins RA (1988) Inclusion complexes of diisopropylfluorophosphate with cyclodextrins. J Catal 112:464–468. doi.org/10.1016/0021-9517(88)90161-3

Hashimoto S, Thomas JK (1985) Fluorescence study of pyrene and naphthalene in cyclodextrin-amphiphile complex systems. J Am Chem Soc 107:4655–4662. https://doi.org/10.1021/ja00302a010

Huang J, Bright FV (1990) Unimodal Lorentzian lifetime distributions for the 2-anilinonaphthalene-6-sulfonate-β-cyclodextrin inclusion complex recovered by multifrequency phase-modulation fluorometry. J Phys Chem 94:8457–8463. https://doi.org/10.1021/j100385a019

Ikeda H (2011) Fluorescent cyclodextrins as chemosensors for molecule detection in water. In: Mirsky VM, Yatsimirsky AK (eds) Artificial receptors for chemical sensors. Wiley-VCH, Weinheim, pp 113–134

Ikeda H, Ueno A (2009) Fluorescent α-cyclodextrin as a chemosensor for halomethanes. Chem Commun 4281–4283. https://doi.org/10.1039/B907629H

Ikeda H, Nakamura M, Ise N, Oguma N, Nakamura A, Ikeda T, Toda F, Ueno A (1996) Fluorescent cyclodextrins for molecule sensing: fluorescent properties, NMR characterization, and inclusion phenomena of *N*-dansylleucine-modified cyclodextrins. J Am Chem Soc 118:10980–10988. https://doi.org/10.1021/ja960183i

Ikeda H, Nakamura M, Ise N, Toda F, Ueno A (1997) NMR studies of conformations of *N*-dansyl-L-leucine-appended and *N*-dansyl-D-leucine-appended β-cyclodextrin as fluorescent indicators for molecular recognition. J Org Chem 62:1411–1418. https://doi.org/10.1021/jo960425x

Ikeda H, Murayama T, Ueno A (2005) Skeleton-selective fluorescent chemosensor based on cyclodextrin bearing a 4-amino-7-nitrobenz-2-oxa-1,3-diazole moiety. Org Biomol Chem 3:4262–4267. https://doi.org/10.1039/B508477F

Ikeda H, Li Q, Ueno A (2006a) Chiral recognition by fluorescent chemosensors based on *N*-dansyl-amino acid-modified cyclodextrins. Bioorg Med Chem Lett 16:5420–5423. doi.org/10.1016/j.bmcl.2006.07.069

Ikeda H, Murayama T, Ueno A (2006b) Bile acids-selective chemosensors based on NBD-amine-modified cyclodextrins. J Incl Phenom Macrocycl Chem 56:101–105. https://doi.org/10.1007/s10847-006-9069-2

Kuwabara T, Matsushita A, Nakamura A, Ueno A, Toda F (1993) A novel color-change indicator for molecules. Guest-induced color-to-colorless change of *p*-nitrophenol-modified β-cyclodextrin. Chem Lett 22:2081–2084. https://doi.org/10.1246/cl.1993.2081

Kuwabara T, Nakamura A, Ueno A, Toda F (1994) Inclusion complexes and guest-induced color changes of pH-indicator-modified β-cyclodextrins. J Phys Chem 98:6297–6303. https://doi.org/10.1021/j100076a011

Kuwabara T, Takamura M, Matsushita A, Ueno A, Toda F (1996) A novel color-changeable host for molecules. Guest-induced colorless-to-color change of phenolphthalein-modified β-cyclodextrin. Supramol Chem 8:13–15. doi.org/10.1080/10610279608233963

Kuwabara T, Takamura M, Matsushita A, Ikeda H, Nakamura A, Ueno A, Toda F (1998) Phenolphthalein-modified β-cyclodextrin as a molecule-responsive colorless-to-color change indicator. J Org Chem 63:8729–8735. https://doi.org/10.1021/jo980613i

Kuwabara T, Nakajima H, Nanasawa M, Ueno A (1999) Color change indicators for molecules using methyl red-modified cyclodextrins. Anal Chem 71:2844–2849. https://doi.org/10.1021/ac9814041

Lavigne JJ, Anslyn EV (2001) Sensing a paradigm shift in the field of molecular recognition: from selective to differential receptors. Angew Chem Int Ed 40:3118–3130. https://doi.org/10.1002/1521-3773(20010903)40:17<3118::AID-ANIE3118>3.0.CO;2-Y

Li YH, Chan LM, Tyer L, Moody RT, Himel CM, Hercules DM (1975) Study of solvent effects on the fluorescence of 1-(dimethylamino)-5-naphthalenesulfonic acid and related compounds. J Am Chem Soc 97:3118–3126. https://doi.org/10.1021/ja00844a033

Matsushita A, Kuwabara T, Nakamura A, Ikeda H, Ueno A (1997) Guest-induced colour changes and molecule-sensing abilities of *p*-nitrophenol-modified cyclodextrins. J Chem Soc Perkin Trans 2:1705–1710. https://doi.org/10.1039/A701220I

Mirsky VM, Yatsimirsky AK (eds) (2011) Artificial receptors for chemical sensors. Wiley-VCH, Weinheim

Müller A, Wenz G (2007) Thickness recognition of bolaamphiphiles by α-cyclodextrin. Chem Eur J 13:2218–2223. https://doi.org/10.1002/chem.200600764

Nelson G, Patonay G, Warner IM (1988) Effects of selected alcohols on cyclodextrin inclusion complexes of pyrene using fluorescence lifetime measurements. Anal Chem 60:274–279. https://doi.org/10.1021/ac00154a018

Prodi L, Montalti M, Zaccheroni N (eds) (2011) Luminescence applied in sensor science. Springer, Heidelberg

Saenger W (1980) Cyclodextrin inclusion compounds in research and industry. Angew Chem Int Ed Engl 19:344–362. https://doi.org/10.1002/anie.198003441

Suresh P, Azath IA, Pitchumani K (2010) Naked-eye detection of Fe^{3+} and Ru^{3+} in water: colorimetric and ratiometric sensor based on per-6-amino-β-cyclodextrin/*p*-nitrophenol. Sens Actuators B: Chem 146:273–277. doi.org/10.1016/j.snb.2010.02.047

Suzuki I, Ui M, Yamauchi A (2006) Supramolecular probe for bicarbonate exhibiting anomalous pyrene fluorescence in aqueous media. J Am Chem Soc 128:4498–4499. https://doi.org/10.1021/ja055772f

Uchiyama S, Santa T, Okiyama N, Fukushima T, Imai K (2001) Fluorogenic and fluorescent labeling reagents with a benzodfurazan skeleton. Biomed Chromatogr 15:295–318. https://doi.org/10.1002/bmc.75

Ueno A, Minato S, Suzuki I, Fukushima M, Ohkubo M, Osa T, Hamada F, Murai K (1990) Host-guest sensory system of dansyl-modified β-cyclodextrin for detecting steroidal compounds by dansyl fluorescence. Chem Lett 19:605–608. https://doi.org/10.1246/cl.1990.605

Ueno A, Kuwabara T, Nakamura A, Toda F (1992) A modified cyclodextrin as a guest responsive colour-change indicator. Nature 356:136–137. https://doi.org/10.1038/356136a0

You L, Zha D, Anslyn EV (2015) Recent advances in supramolecular analytical chemistry using optical sensing. Chem Rev 115:7840–7892. https://doi.org/10.1021/cr5005524

Chapter 6
Silica Materials Containing Cyclodextrin for Pollutant Removal

Nadia Morin-Crini, Marc Fourmentin, Sophie Fourmentin, Giangiacomo Torri, and Grégorio Crini

Contents

N. Morin-Crini · G. Crini (✉)
UFR Sciences et Techniques, Laboratoire Chrono-environnement, Université de Bourgogne Franche-Comté, Besançon, France
e-mail: nadia.crini@univ-fcomte.fr; gregorio.crini@univ-fcomte.fr

M. Fourmentin
Laboratoire de Physico-Chimie de l'Atmosphère (LPCA, EA 4493), Université du Littoral Côte d'Opale, Dunkerque, France
e-mail: marc.fourmentin@univ-littoral.fr

S. Fourmentin
Unité de Chimie Environnementale et Interactions sur le Vivant (UCEIV, EA 4492), Dunkerque, France
e-mail: sophie.fourmentin@univ-littoral.fr

G. Torri
Istituto di Chimica e Biochimica G. Ronzoni, Milano, Italy
e-mail: torri@ronzoni.it

S. Fourmentin et al. (eds.), *Cyclodextrin Applications in Medicine, Food, Environment and Liquid Crystals*, Environmental Chemistry for a Sustainable World, https://doi.org/10.1007/978-3-319-76162-6_6

Abstract This chapter reviews the use of cyclodextrin-silica hybrid systems and cyclodextrin-functionalized silica used as adsorbents or filters for the removal of inorganic and organic pollutants from aqueous solutions in solid-phase extraction and adsorption-oriented processes. Actually, there is a need to develop efficient processes for the synthesis and application of multifunctional silica-based materials for pollutant removal by adsorption or filtration, and for sample purification and concentration using solid-phase extraction.

On one hand, of silica-based adsorbents are low-cost, robust inorganic solids having large surface areas, high porosity, and excellent mechanical, physical and chemical properties, and wide possibilities of functionalization due to silanol reactivity. On the other hand, cyclodextrins are natural molecules obtained from the enzymatic degradation of starch. They belong to the family of cage molecules due to their structure which is composed of a hydrophobic cavity that can encapsulate other molecules. Cyclodextrin-functionalized silicas usually display improved access to the binding sites because the moieties are located on the external surface of the material. In cyclodextrin-silica hybrid systems prepared through sol-gel or self-assembly process, cyclodextrin molecules are located within the framework of nanoporous silicas. Here, both high cyclodextrin loadings, robust structures and higher surface area are observed. Cyclodextrin-based silica materials have strong binding affinities for chemical substances such as metal ions, dyes, pesticides, and drugs.

Abbreviations

AAm	Acrylamide
APTES	3-aminopropyltriethoxysilane
BET	Brunauer-Emmett-Teller
BPA	Bisphenol A
CD	Cyclodextrin
CPMAS	Cross-polarization magic angle spinning
CTAB	Cetyltrimethylammonium bromide
EDS	Energy-dispersive X-ray spectroscopy
EPI	Epichlorohydrin
FSM	Folded sheets mechanism
FT-IR	Fourier transform infrared
GPTS	Glycidoxypropyl trimethoxysilane
HMDI	Hexamethylene diisocyanate
HMS	Hexagonal mesoporous silica
MCM	Mobil crystalline materials
MCT-CD	Monochlorotriazinyl-cyclodextrin
MS	Mass spectrometry
MSU	Michigan State University

NMR	Nuclear magnetic resonance
PAAM	Polyacrylamide
PAH	Polycyclic aromatic hydrocarbons
PCB	Polychlorobiphenyls
SBA	Santa Barbara amorphous
SDS	Sodium dodecylsulfate
SEM	Scanning electron microscopy
TEM	Transmission electron microscopy
TEOS	Tetraethylorthosilane
TGA	Thermogravimetric analysis
Triton X-45	A nonionic surfactant
VOC	Volatile organic compounds
XPS	X-ray photoelectron spectroscopy
XRD	X-ray diffraction

6.1 Introduction

Amongst the numerous techniques for pollutant removal of filtration/separation, liquid-solid adsorption using conventional adsorbents is the procedure of choice and gives the best results as it can be used to remove different types of pollutants (Volesky and Holan 1995; McKay 1996; Yang 2003; Crini and Badot 2010). As adsorption is a surface phenomenon, any porous solid having a large surface area may be an adsorbent. In general, the selection of an adsorbent in a suitable adsorption process is based on the following criteria: low-cost and readily available, granular type with a good particle size distribution, a well-developed structure, porous material with a large total surface area, presence of surface charge due to functional groups considered as adsorption sites, high physical strength in solution, suitable mechanical properties, efficient for a large range of pollutants, a long life, and able to be regenerated if required (Crini 2005; Crini and Badot 2010). For an excellent adsorbent to remove a large amount of pollutant in a short period of time, it should also possess short adsorption equilibrium times, fast adsorption rates, high capacity, and selectivity.

What is the best adsorbent? There is no direct answer to this question because each solid material has advantages and drawbacks (Volesky 1990; McKay 1996; Wase and Forster 1997). However, there is no doubt that activated carbons are one of the oldest and most widely used adsorbents in industry because systems using carbons are both technologically simple, efficient, and also economically feasible although the initial cost of the carbon can be high (Morin-Crini and Crini 2012). Other conventional adsorbents can be also used and the list includes commercial ion-exchange resins and inorganic-based materials such as activated aluminas, silica, zeolites and molecular sieves.

Amongst these inorganic-based materials, silica beads are good adsorbents due to their high adsorption capabilities because of their large surface areas and high porosity with highly uniform pore distribution and tunable pore size, excellent physical and chemical properties such as water stability, thermal and mechanical stability. Their versatility is also another interesting aspect: for instance, there are several types of contacting systems available for industrial applications such as batch methods, fixed-bed type processes or fluidized beds.

Silica is also non-toxic and a low-cost inorganic material with extraordinarily wide possibilities of functionalization due to the silanol reactivity. Indeed, the presence of silanol groups on the surface makes silica a better solid for the immobilization of a wide range of inorganic and organic ligands. Numerous three-dimensional network structures can be obtained such as amorphous silica, fumed silica, silica gels, and mesoporous silica (Bergna 1994; Cooney 1999; Mocanu et al. 2001; Berggren et al. 2005; Morin-Crini and Crini 2012).

Among them, literature data clearly indicate that well-prepared mesoporous materials show excellent pollutant adsorption capabilities compared to their amorphous counterparts. In general, these materials can be prepared by the hydrolysis of alkoxysilane precursor (tetraethoxysilane for example) in the presence of suitable surfactant, the template, and catalyst, to form condensed polymerized networks of siloxanes (Beck et al. 1992; Kresge et al. 1992, 1995; Davis 2002; Sohmiya et al. 2015; Vunain et al. 2016). The sol-gel chemistry leads to the synthesis of ordered mesoporous silicas, e.g. the well-known mobile crystalline materials such as MCM-41 or SBA-15, with large surface area, high porosity, and high amount of silanol groups.

Mesoporous silica can be further modified by immobilization of various functional groups at the particle surface as well as at the pore wall surface to form organic-inorganic hybrid materials (Goyal et al. 2011; Samiey et al. 2014). The modification provides new opportunities for fine-tuning the chemical, physical and mechanical properties of these novel materials. The modified materials are used not only in electrochemical detection, electronic devices, controlled drug delivery, and catalysis but also in separation technology.

The application of silica beads containing cyclodextrin (CD) molecules as adsorbents in solving environmental problems such as pollutant removal or filtration from water, wastewater, and atmosphere, e.g. indoor air and gas treatment, has recently received a lot of attention (Sharma and Sanghi 2012; Gibson 2014). In particular, the past decade has seen an explosive interest in ordered mesoporous silica-based materials. Their unique physical and chemical properties make them more superior and useful in various fields than conventional carbons. In these compounds, the high mechanical properties and physical strength of inorganic substrates, the silica beads, is combined with advantages of a complexing substrate, the cyclodextrin, resulting in strong binding affinities toward target pollutants and relatively high pollutant adsorption capacities (Morin-Crini and Crini 2012; Sharma and Sanghi 2012).

The main objectives of this chapter are to provide a summary of the recent information concerning the innovative synthesis of silica containing cyclodextrin and to describe the developments in the use of these materials for the removal of pollutants by reviewing some selected studies reported in the literature.

6.2 Cyclodextrins and Silicas, Raw Materials to Produce Innovative Complexing Networks

The most characteristic feature of cyclodextrins, natural substances obtained from starch, is their ability to form inclusion compounds with various molecules, ions and polymers. This remarkable property combined with their relatively low-cost and non-toxic character to humans have led to their use in pharmaceuticals, cosmetics, as food additives, as well as in the complexation of pollutants. Numerous books and chapters books can be consulted (Bender and Komiyama 1978; Szejtli 1982, 1988, 1998; Atwood et al. 1984; Duchêne 1987, 1991; Cram 1988; Frömming and Szejtli 1994; Robyt 1998; Dodziuk 2006; Bilensoy 2011; Crini 2014; Morin-Crini et al. 2015). Cyclodextrin molecules themselves, however, are highly water-soluble, and must therefore be processed into solid forms or grafted on solid supports (such as silica) before they can be implemented into usable separation technology (Mocanu et al. 2001; Crini and Morcellet 2002; Landy et al. 2012; Guo and Wilson 2013; Sharma 2015).

As already mentioned, silica-based materials are robust inorganic solids displaying both high specific surface area (200–1500 m^2/g) and a three-dimensional structure made of highly open spaces interconnected to each other via SiO_4 tetrahedra, giving rise to highly porous structures, up to 1 cm^3/g or even more (Bergna 1994). Most of them can be manufactured quite easily at room temperature by sol-gel processing for example, involving hydrolysis of silicon alkoxide precursors such as tetramethoxysilane or tetraethoxysilane, and catalytic polycondensation to produce a macromolecular network of siloxane bonds. Accurate tuning of the experimental parameters affecting the production steps, i.e. polymerization, gelation, aging, drying and heating steps, allows control over the microstructure of the final materials. Porosity control can be achieved by the surfactant template route, giving rise to novel ordered mesoporous materials that possess large uniform pore sizes (1.5–10 nm), highly ordered nanochannels, large surface areas (>1000 m^2/g), and tunable liquid crystal-like structures. The chemical reactivity of silicas is essentially governed by their surface properties, especially via the weakly acid silanol groups (Bergna 1994). Originally used mainly in chromatography, silica-based materials are actually used in numerous industrial applications including catalysis, pharmacy, medicine, dentistry, cosmetology, paper, paints, coatings, and also in environment (Descalzo et al. 2006; Pagliaro 2009).

6.3 Incorporation of Cyclodextrin Molecules into Silica-Based Materials

In the literature, there have been numerous studies on the preparation, characterization, properties and applications of cyclodextrin-based silica materials in separation technology. There are several synthetic procedures and numerous materials have been characterized and proposed as adsorbents. Here, we propose to classify these adsorbents in two main class: cyclodextrin-functionalized silicas prepared through grafting or coating reactions and cyclodextrin-silica hybrid systems prepared through sol-gel or self-assembly process. Fig. 6.1 shows three schematic representations of silica networks containing cyclodextrin molecules: (a) grafted matrices, (b) coated materials and (c) nanoporous frameworks.

The first approach involves the grafting or coating of cyclodextrin moieties onto silica gel by using generally cross-linking reactions (Crini and Morcellet 2002). Cyclodextrin-grafted or cyclodextrin-coated silicas offer improved access to the binding sites because the moieties are located on the external surface of the material, and the structure has improved thermal, mechanical and chemical stability owing to the robustness of the bead. However, the disadvantage of this synthetic approach is often uneven distribution of cyclodextrin molecules. In addition, it is known that these materials have low cyclodextrin loading which can limit adsorption capacities. In the second approach, the cyclodextrin molecules are located within the framework of nanoporous silicas (Bibby and Mercier 2003). These materials possess both high cyclodextrin group loadings and robust structures.

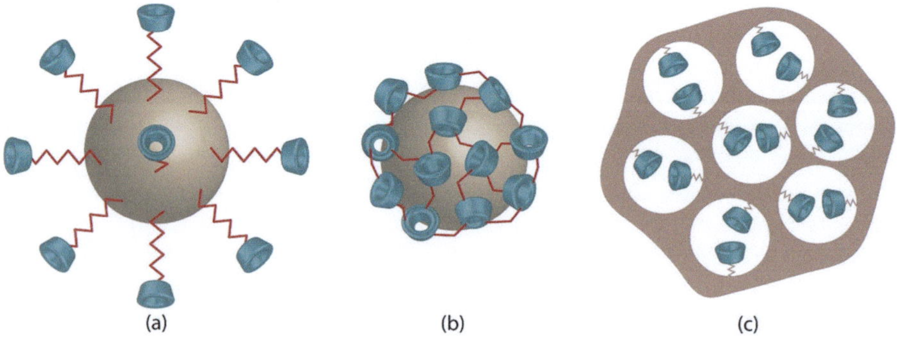

Fig. 6.1 Schematic representations of three silica networks containing cyclodextrin molecules: (**a**) cyclodextrin-functionalized silicas prepared through grafting, (**b**) cyclodextrin-functionalized silicas prepared through coating and (**c**) incorporation of cyclodextrins into nanoporous frameworks

6.4 Cyclodextrin-Silica Hybrid Systems

Recent advances in the development of innovative functionalized materials such as organic-inorganic hybrid systems are having a major impact on analytical and environmental chemistry, catalysis, and biorefinery production (Descalzo et al. 2006; Walcarius and Mercier 2010; Rahmat et al. 2010; Al Othman 2012; Han and Zhang 2012; Samiey et al. 2014; Lee and Park 2015; Dinker and Kulkarni 2015). Numerous fields are concerned such as water analysis for metal detection, pollutant removal from liquid or gas phases, adsorbents for solid-phase extraction, electrochemical detection of metals, and also chemical processes using immobilized catalysts.

The organic-inorganic hybrid systems have been obtained through the coupling of inorganic and organic components by template synthesis. The main methods include sol-gel process, self-assembly process, assembling or dispersion of nano-building blocks or interpenetrating networks (hierarchical structures). The incorporation of functionalities onto material pore surfaces or into the frameworks can be achieved by post-synthesis grafting or co-condensation method (Hoffman et al. 2006; Vartuli et al. 2008; Samiey et al. 2014). Organic-inorganic hybrid systems are classes of materials whose structure includes both organic and inorganic units that interact with each other at the molecular level. The inorganic part provides mechanical strength and the organic part shows functional activities. These materials are divided into two classes, class I and class II, on the basis of interaction between organic and inorganic components. In class I, organic and inorganic are embedded and there are weak interactions, such as hydrogen bonding, van der Waals, π–π or electrostatic interactions between them, and in class II, the two components are bonded together through strong covalent bonds such as coordinative bonds.

At the beginning of the 1990s, the formation of mesostructured silica using surfactants as templates (Myers 1992) was reported by researchers from the Mobil Oil Corporation and by Kuroda's group (Hench and West 1990; Yanagisawa et al. 1990a, b; Beck et al. 1992, 1994; Kresge et al. 1992, 1995; Chen et al. 1993; Inagaki et al. 1993, 1996a, b; Fukushima et al. 1995; Wen and Wilkes 1996; Lim et al. 1998; Lim and Stein 1999; Ciesla and Schüth 1999; Ying et al. 1999; Polarz et al. 2001). These materials found potential applications such as heterogeneous catalysis, electrochemical chemistry, host-guest chemistry, and separation processes (Walcarius 2001; Pan et al. 2009). This new route of synthesis immediately attracted attention in the material science community.

Numerous works based on mesoporous silica and others oxides such as alumina were published (Ying et al. 1999; Patarin et al. 2002; Sanchez et al. 2005; Simsek et al. 2012; Yamamoto and Kuroda 2016). At the same period, a novel generation of mesoporous hybrid silicas containing organic groups on the surfaces of ordered mesoporous silicas was developed by grafting or anchoring of organic guests onto the mesopore channel surface or by the direct incorporation of organic groups through co-condensation of organotrialkoxysilanes with tetraalkoxysilanes (Wen and Wilkes 1996; Ogoshi and Chujo 2005; Hoffman et al. 2006; Al Othman and Apblett 2009). As documented in a recent comprehensive review (Samiey et al. 2014), sol-gel chemistry is a versatile tool for the synthesis of ordered organic-inorganic hybrid materials with advanced properties that are often difficult to achieve either from totally inorganic or from totally organic materials. The previous reviews of Hench and West (1990), Wen and Wilkes (1996), Hoffman et al. (2006), Vartuli et al. (2008), and Walcarius and Mercier (2010) can be also consulted.

Usually, this process involves the hydrolysis and condensation of a tetraalkoxysilane ($Si(OR)_4$ with R = Me or Et) in the presence of a supramolecular template such as surfactant or more recently water-soluble polymer. The silica framework can be formed around preformed liquid crystal mesophases but, in many cases, the organized architectures are obtained via a self-assembly cooperative process taking place *in situ* between the templates and the silica network precursors. The organic template is then removed by calcination or solvent extraction to give the resulting mesoporous open structure. Both ionic (cationic form such as long chain quaternary ammonium) or non-ionic (neutral amine, water-soluble copolymers) surfactants can be used to produce ordered mesoporous silicas by the sol-gel process.

Mesoporous silicate M41S (Mobil Oil Corporation), a well-known family of nanostructured mesoporous materials is synthetized using tetraethylorthosilane $Si(OC_2H_5)_4$, known as TEOS, as a silica source in the presence of long-chain alkyltrimethylammonium halide surfactants (Valtchev et al. 2009). MCM-41 having a hierarchical structure is undoubtedly the best known and most widely studied of this family of materials. MCM is an abbreviation of Mobil Crystalline Materials. Most common materials are based on 2D and 3D structures but lamellar and less ordered mesostructures, e.g. wormlike, can be also obtained. The other members are the cubic MCM-48 and lamellar MCM-50 forms. Other types of ordered mesoporous silicas are MSU-2, SBA-15 and SBA-16, FSM-16 and HMS. MSU, SBA, FSM and HMS are abbreviations of Michigan State University, Santa Barbara Amorphous, Folded Sheets Mechanism and Hexagonal Mesoporous Silica, respectively (Zhao et al. 1996; Rahmat et al. 2010).

The interaction between the inorganic precursor and the template is a key factor in the control of the mesostructured materials. Ordered silica-based mesoporous materials are solids displaying a periodic and regular arrangement of well-defined and controllable mesopores with size ranging between 2–10 nm, and amorphous inorganic framework structures. They are highly porous (pore volume > 0.7 mL/g), having high specific surface areas up to 1500 m^2/g, and they share characteristics of both silica gels and molecular sieve zeolites, showing however larger pore sizes.

Compared to their non-ordered homologues, these mesostructured porous solids also offer other significant advantages: (i) exceptionally good accessibility to active centers due to highly ordered nanochannels of uniform pore size; (ii) very high number of functional groups that can be attached to the (mostly internal) surface of mesoporous silica, as a consequence of very large surface areas; (iii) fast mass transport rates inside the porous structure due to the regular spatial arrangement of mesopore channels of monodisperse dimensions; and (iv) good mechanical and hydrothermal stabilities. This last point is controversial because these properties strongly dependent on mesostructure types and post-synthesis treatments. All these

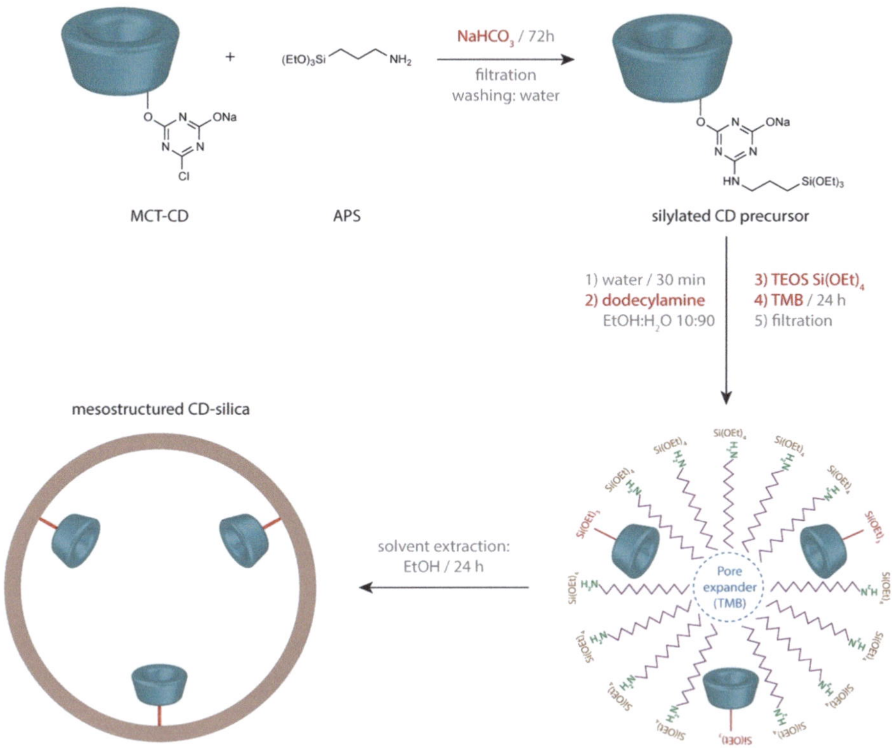

Fig. 6.2 Incorporation of cyclodextrin molecules into mesostructured silica (adapted from Huq et al. 2001)

attractive features make such innovative materials of interest for the adsorptive removal of pollutants from waters.

Mercier's group previously reported cyclodextrin-silica hybrid systems with uniform framework mesoporosity as a new class of efficient materials (Huq et al. 2001) for adsorption and separation of water-soluble aromatic molecules (Bibby and Mercier 2003), phenols or pesticides such as dichlorodiphenyldichloroethylene and dichlorodiphenyltrichloroethane (Sawicki and Mercier 2006). This group utilized the concept of surfactant-directed mesostructured oxide assembly to produce cyclodextrin-silica hybrid materials with well-defined nanometer-scale porosity. The materials were prepared using an entirely aqueous/ethanolic one-step procedure involving incorporation of ethoxysilane precursors (silylated-cyclodextrin) in the presence of structure-directing surfactants solutions including alkylamine agents such as dodecylamine and pore expanding additives such as trimethylbenzene.

A silylated-cyclodextrin derivative was first prepared by reaction between monochlorotriazinyl-cyclodextrin (MCT-CD) and 3-aminopropyltriethoxysilane (APS or APTES). The second step was the co-polymerization of this derivative with tetraethylorthosilane in the presence of a structure-directing template. Fig. 6.2 shows the synthetic procedure proposed by Huq et al. (2001). In preliminary experiments, the incorporation of cyclodextrin molecules inside mesoporous silica frameworks was attempted using two methods: by the grafting of the cyclodextrin into the channels of preformed mesoporous silica hosts and by the direct synthesis of cyclodextrin-containing mesoporous materials by a one-step process. While the grafting procedure failed to incorporate cyclodextrin inside the pore channels of mesoporous silica substrates, the direct synthesis approach yielded materials whose physical and chemical characteristics as determined by numerous techniques such as X-ray diffraction XRD, nitrogen adsorption analysis, transmission electron microscopy (TEM), and elemental analysis were indicative of the presence of cyclodextrin inside the pore channels of the materials. Materials with cyclodextrin loadings up to 0.39 mmol/g and uniform pore channels with diameters in the range of 38–42 Å were prepared by this method. Preliminary adsorption experiments using aqueous monocontaminated solutions showed that the materials were promising in environmental applications.

Indeed, in another work, Mercier's results demonstrated that the materials could selectively separate organic molecules by using the shape- and function-specific inclusion properties of receptor binding sites (Bibby and Mercier 2003). The materials exhibited high adsorption capacities due to inclusion complex formation: up to 0.33 mmol/g for *p*-nitrophenol. These adsorption capacities were superior to those achieved with pure silica materials. This can be also attributed to their higher surface areas, good accessibility to active centers, which increase selectivity, and higher mass transport rates inside the porous structure. Mercier' group also found that materials with higher cyclodextrin loadings generally possessed greater performances for aromatic pollutants. However, the authors demonstrated that adsorption capacity varied not only as a function of the amount of cyclodextrin molecules in the material but also as a function of the complexation thermodynamics of the pollutants with cyclodextrin. Thus, *p*-nitrophenol was found to have the highest adsorption capacity compared to the other pollutants (phenol, *m*-nitrophenol, *p*-chlorophenol and *p*-nitroaniline). In another study (Sawicki and Mercier 2006), the same materials were used for the removal of 14 pesticides from aqueous solutions (initial concentrations in the range 0.06–0.27 μg/L). The results showed that the complete uptake of all pesticides was observed when 250 mg of material were used for adsorption. The cyclodextrin receptor sites in the materials were very effective toward the binding of pesticide molecules to the adsorbents. Fig. 6.3 depicts bidentate complexation of cyclodextrin receptors with *p,p'*-dichlorodiphenyl-based pesticide within a cyclodextrin-silica hybrid system pore channel proposed by the authors.

The Mercier's group studies also highlighted the importance of molecular-scale engineering of target-specific materials in which optimal adsorptivity was not necessarily obtained by maximizing the number of binding sites, but rather by controlling their placement and orientation within the structure of mesosilica. Indeed, only adsorbents containing low to intermediate amounts of cyclodextrin molecules (2–4% with respect to total Si in sample) were found to have optimal affinity toward the pesticides target (Sawicki and Mercier 2006). Materials produced with

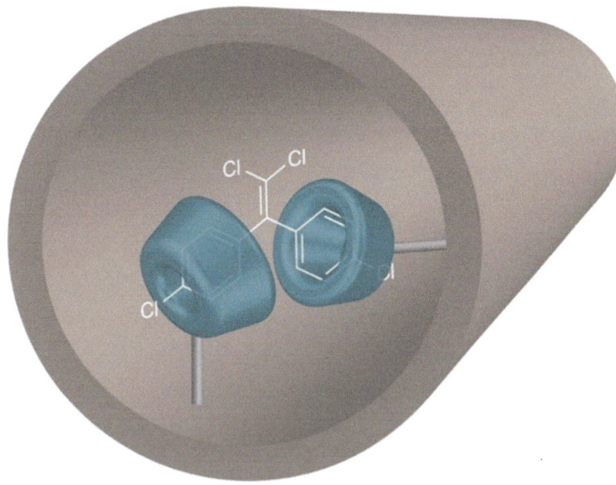

Fig. 6.3 Depiction of bidentate complexation of cyclodextrin receptors with a *p,p'*-dichlorodiphenyl-based pesticides within a cyclodextrin-silica hybrid system pore channel (adapted from Sawicki and Mercier 2006)

cyclodextrin loadings higher than 8% had poor performance. This can be explained by the destruction of the periodic mesoporous structure with high cyclodextrin amount. Two interesting results were reported by Walcarius and Mercier (2010): (i) the location of cyclodextrin moieties within the structure of a siliceous framework may reduce the materials' susceptibility to chemical and biological degradation; (ii) cyclodextrin-silica hybrid materials exhibited higher adsorption properties toward pollutants after multiple uses, indicating high chemical and mechanical stability. Mercier's group concluded that the incorporation of cyclodextrin molecules into mesostructured silica was an innovative and promising tool for environmental protection.

Using similar synthetic procedure, Bacquet's group also reported mesoporous silicas containing both cyclodextrin and amino groups for *p*-nitrophenol removal from aqueous solutions (Willai et al. 2008; Degoutin and Bacquet 2013). However, their results revealed an adsorption capacity of only 61 μmol/g, far inferior to what was reported by Mercier's group (Bibby and Mercier 2003). The synthetic procedure is schematized in Fig. 6.4. The strategy was based on a classical direct co-condensation between the components via a sol-gel pathway but templated by three different surfactants, namely an anionic template (sodium dodecylsulfate SDS), a cationic one (cetyltrimethylammonium bromide CTAB) or a neutral surfactant (Triton X-45, a nonionic polyethylene oxide containing an aromatic part). Each of them presents difference on the pore size and the repartition of the chemical groups. However, in the three cases, the use of the surfactant gave materials with smaller pores. Adsorption mechanism was interpreted by the presence of two main interactions between *p*-nitrophenol and the porous organo-silica depending on the surfactant used: hydrogen bonds and inclusion complex with Triton X-45, ionic interactions and inclusion complex with SDS and CTAB. No significant differences were observed for the three bifunctional materials.

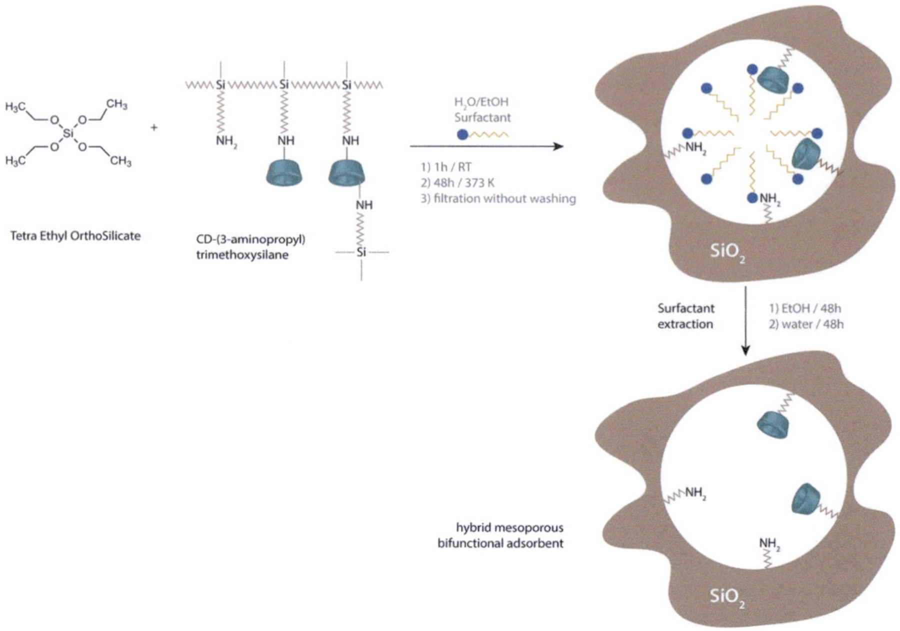

Fig. 6.4 Preparation of templated hybrid mesoporous bifunctional organo-silica proposed by Degoutin and Bacquet (2013)

Others researchers also reported that samples with high cyclodextrin molecules loading (up to 6%) were generally much less effective than those with lower cyclodextrin content (Liu et al. 2004; Gibson 2014; Mahmud and Wilson 2016). Liu et al. (2004) proposed a new mesoporous organo-silica material (cyclodextrin-silica-4%) containing microporous cyclodextrin prepared by the co-polymerization of a silylated cyclodextrin monomer with tetraethylorthosilane in the presence of cetyltrimethylammonium bromide template. Fig. 6.5 depicts the synthesis procedure. Surfactant extraction resulted in an adsorbent containing covalently bound microporous cyclodextrin moieties capable to adsorb efficiently humic acid from water. Nitrogen adsorption experiments showed that cyclodextrin-silica-4% material had a BET surface area of 460 m^2/g and an average mesopore diameter of 2.52 nm. Small-angle powder XRD pattern of cyclodextrin-silica-4% material revealed the lack of highly ordered mesoporous structure. Solid-state C-13 and Si-29 nuclear magnetic resonance (NMR) studies provided evidence for the presence of covalently attached cyclodextrins in the mesoporous material. Adsorption experiments showed that new material removed up to 99% of humic acid from an aqueous solution containing 50 ppm of humic acid at a solution-to-solid ratio of 100 mL/g. The authors claimed that the new mesoporous cyclodextrins-containing organo-silica material was potentially useful both in environmental remediation and chromatographic separation.

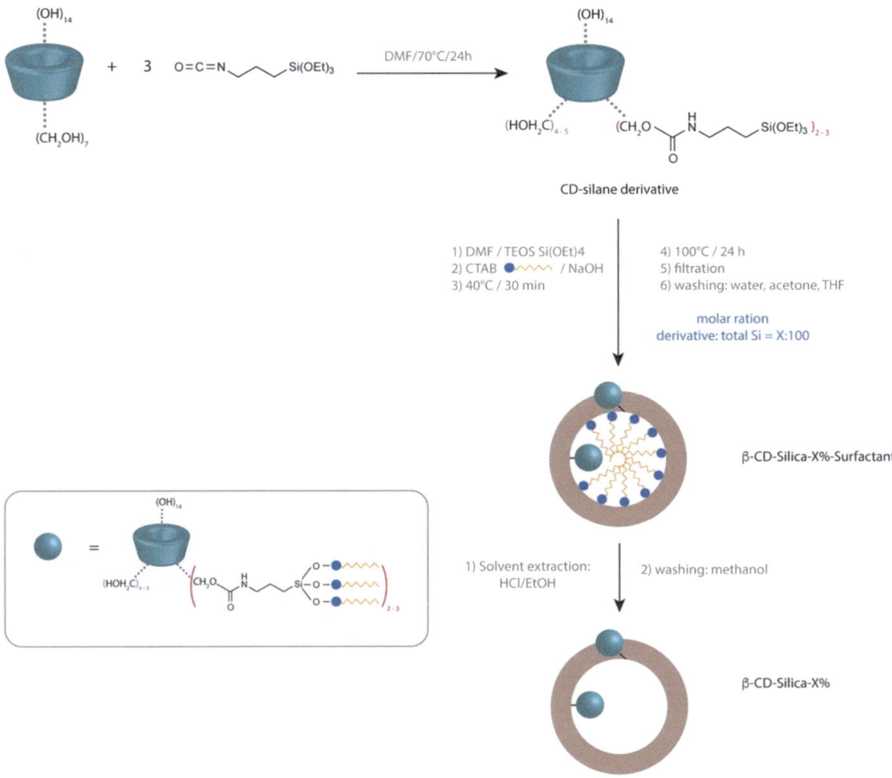

Fig. 6.5 Synthesis of mesoporous organo-silica material containing microporous cyclodextrin molecules (adapted from Liu et al. 2004)

Gibson (2014) recently published an interesting tutorial review on the removal of organic pollutants from the aqueous phase by mesoporous silica. After a discussion about mesosilica formation (MCM-41 and SBA-15) and silica surface modification, the review focused on the use of mesosilica for the removal of organic compounds such as dyes and phenols, and emerging contaminants, e.g. pharmaceutical, from aqueous solutions. High extraction capacities could be obtained, although results were cyclodextrin loading-dependent. Indeed, extraction capacities were significantly more important with increased cyclodextrin loading (between 2–8%) into mesosilica but materials produced with cyclodextrin loadings higher than 8% had poor performance. Gibson (2014) concluded that mesoporous cyclodextrin-silica nanocomposites were efficient adsorbents. However, cyclodextrin-loading, accessibility, molecular size and structure were crucial factors when assessing extraction performance of cyclodextrin-loaded mesosilica.

Wilson's group reported two comprehensive studies on the synthesis and characterization of surface-modified mesoporous silica materials with cyclodextrin molecules and their adsorption properties toward two types of gas phase (nitrogen and methyl chloride) and *p*-nitrophenol in aqueous solution (Wilson and Mahmud 2015; Mahmud and Wilson 2016). This group synthesized silica containing microporous cavities provided by surface-bound cyclodextrin by co-condensation of a cyclodextrin-functionalized triethoxysilane with tetraethylorthosilane using alkylamine-based surfactants such as dodecylamine, tetradecylamine, and hexadecylamine as structure directing agents. The materials had an ordered silica mesostructure framework that depended on the type of surfactant template and the level of loading of cyclodextrin. The incorporation of cyclodextrin within the mesoporous framework was supported by spectroscopic techniques such as Fourier transform infrared (FT-IR) spectroscopy, Raman, mass spectrometry (MS) and NMR, and thermogravimetric analysis (TGA) data (Mahmud and Wilson 2016). The MALDI-TOF MS characterization and adsorption data provided corroborating support that cyclodextrin molecules were incorporated within and onto the surface of the silica framework. Small-angle XRD and nitrogen adsorption also provided evidence of ordered silica mesostructured frameworks. C-13 solid cross-polarization magic angle spinning (CPMAS) NMR data were interesting to show the presence of residual pore expander such as 1,3,5-trimethylbenzene in the material. Mahmud and Wilson (2016) demonstrated that the use of a pore expander for the synthesis of such composites may result in entrapped impurities during the formation of the silica framework, in contrast of the results published by Huq et al. (2001). The authors supposed that 1,3,5-trimethylbenzene containing aromatic group was entrapped in the silica network during the condensation process. Small-angle X-ray diffraction data showed that greater loading of cyclodextrin was accompanied by a concomitant decrease in the tetraethylorthosilane content, resulting in a reduction in the stability of the silica framework, in agreement with the loss of long-range ordering. TGA results confirmed that cyclodextrin was covalently bound to the silica network. For materials with similar cyclodextrin loading, the textural properties (surface area and pore volume) doubled as the surfactant changed from C12 (dodecylamine) to C16 (hexadecylamine). The textural properties decreased with cyclodextrin loading (2 to 6%). The surface area also decreased by *ca.* 1.5-fold as the cyclodextrin loading varied from 2% to 6%. The adsorption capacity of gas phase with polar and apolar species (CH_3Cl and N_2) varied along with the adsorption properties in aqueous solution toward *p*-nitrophenol according to the CD loading (2–6%) and surfactant template employed. Indeed, incremental variations in the uptake of gas phase adsorbates and *p*-nitrophenol from an aqueous solution were observed, according to the composition of materials (Wilson and Mahmud 2015). The adsorption capacity of *p*-nitrophenol increased from 61% to 84% as the cyclodextrin loading increased from 2% to 6% and as the alkyl chain length of the surfactant template varied from C12 to C16. The adsorption properties of materials with CH_3Cl in the gas phase and for *p*-nitrophenol in aqueous solution adopted a multi-layer adsorption profile, as described by the BET isotherm model. Their results revealed the structural contribution of surface modification and framework incorporation of cyclodextrin with mesoporous silica framework materials. Mahmud and Wilson (2016) concluded that the general concept 'more cyclodextrin, more adsorption' may not necessarily apply due to changes in the mechanical stability of the framework, potential steric effects due to loading of the cyclodextrin moiety in mesopore channels (that reduce the accessibility of pollutants), and limitations on the surface immobilization of cyclodextrin for the synthetic conditions.

Alahmadi et al. (2014) studied the covalent attachment of cyclodextrin on MCM-41 through two different reactions using toluene diisocyanate as linker. The FT-IR spectra and TGA analysis demonstrated that the cyclodextrin molecules were covalently attached to the mesoporous silica while the preservation of the MCM-41 channel system was checked by XRD and nitrogen adsorption analysis. These materials were used to evaluate the adsorption properties of organotin compounds such as tributyltin, triphenyltin and dibutyltin. The results showed that mesoporous silica MCM-41 functionalized with cyclodextrin, using toluene diisocyanate, was a more highly effective adsorbent for organotin compounds especially for triphenyltin, compared to mesoporous silica MCM-41 functionalized with cyclodextrin using 3-chloropropyltriethoxysilane and toluene diisocyanate as linkers.

The main drawbacks in the synthesis of cyclodextrin-containing silicas are the use of aggressive and toxic solvents and activating agents in multistep procedures of organic reactions at elevated temperatures. These conditions may affect the structure of final product. New approaches for ordered cyclodextrin-containing silicas synthesis under mild conditions were proposed by Trofymchuk et al. (2016). MCM-41 materials with hexagonally ordered mesoporous structure were prepared by post-synthesis grafting and by co-condensation methods. Cyclodextrin molecules activated by a *N,N′*-carbonyldiimidazole were employed for post-synthesis treatment of 3-aminopropyl-modified MCM-41 support (3-aminopropyltriethoxysilane was used as silica source) as well as for sol-gel synthesis with cyclodextrin-containing organosilane and tetraethylorthosilane, as silica source, participation in the presence of surfactant CTAB. The successful incorporation of cyclic oligosaccharide moieties in silica surface layer was verified by means of FT-IR spectroscopy and elemental analysis. Obtained cyclodextrin-containing materials were characterized by XRD, TEM, and low-temperature adsorption-desorption of nitrogen. In spite of commensurable loading of cyclodextrin groups attained by both proposed approaches (up to 0.028 $\mu mol/m^2$), it was found that co-condensation procedure provided uniform distribution of cyclodextrin functionalities in silica framework, whereas post-synthesis grafting resulted in modification of external surface of silica surface. Adsorption of benzene from aqueous solutions onto the surface of cyclodextrin-containing materials prepared by co-condensation method was studied as the function of time and equilibrium concentration. Langmuir and Freundlich models were used to evaluate adsorption processes and parameters. Calculated maximum adsorption capacity for benzene was 111 mg/g. Adsorption experiments showed that ordered cyclodextrin-containing silicas could be promising for the trace amount removal of aromatics from water.

Kawamura et al. (2015) prepared novel organic-inorganic hybrid nanoparticles with a bisphenol A (BPA)-responsive hydrogel layer on the surface of SiO_2 nanoparticles via surface-initiated atom transfer radical polymerization of acrylamide (AAm), acryloyl-modified cyclodextrin and *N,N′*-methylenebisacrylamide. The resulting CD-PAAm/SiO_2 nanoparticles underwent a change in size in response to

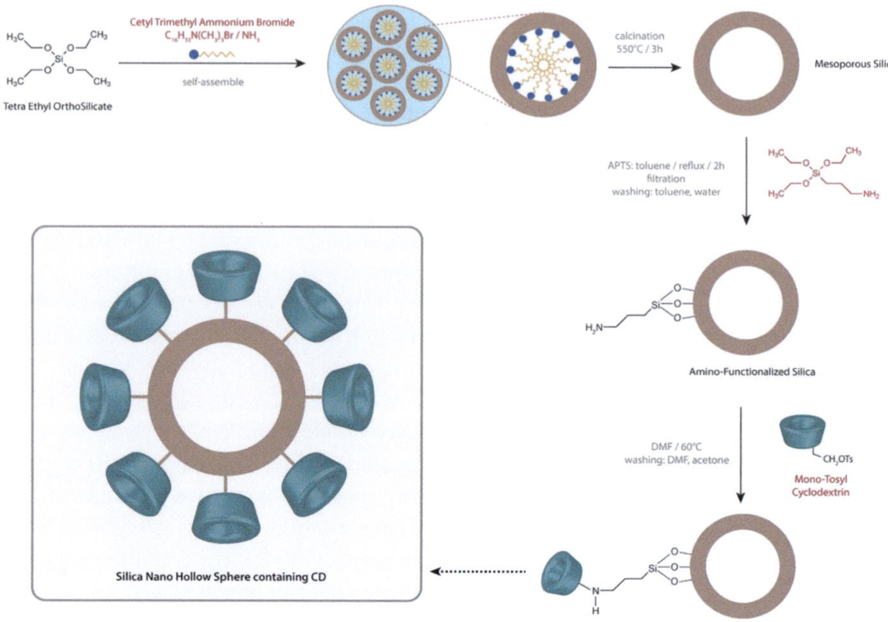

Fig. 6.6 Modification of silica nano-hollow sphere containing cyclodextrin molecules (adapted from Ebadi and Rafati 2015)

BPA. The BPA-responsive shrinkage of the CD-PAAm/SiO_2 nanoparticles was caused by an increase in the crosslinking density of the CD-PAAm hydrogel layer, which resulted from the formation of CD-BPA-CD complexes acting as dynamic crosslinks. The authors concluded that the smart functions of BPA-responsive hybrid nanoparticles could provide useful tools for constructing molecular sensors and adsorption materials.

Ebadi and Rafati (2015) prepared new silica mesoporous nanoparticles functionalized by cyclodextrin molecules for methylene blue removal from aqueous solution. Modified nano-hollow sphere silica were proposed by sol-gel method using amino-functionalized silica and mono-tosyl-cyclodextrin derivatives according the procedure described in Fig. 6.6. Structure of nanoparticles was identified by scanning electron microscopy (SEM) images and FT-IR spectroscopy. Nanoparticles showed regular sphere particles with radius less than 100 nm, with a surface area of 754 m^2/g (from BET data). FT-IR data also indicated both the grafting and crosslinking of cyclodextrin molecules onto nanoparticles. Adsorption behaviors of methylene blue onto nanoparticles were studied from equilibrium and kinetic viewpoints. Experimental data showed that high adsorption capacities were obtained with a maximum adsorption capacity of 99.22 mg/g at pH = 10.5. The results were however strongly pH-dependent. The equilibrium data could be well described by several isotherms but Toth isotherm model was designed for investigation of heterogeneous adsorption systems. Host-guest interactions between cyclodextrin and organic molecules had a great contribution to dye adsorption. These nanoparticles could be applied in the elimination, enrichment and detection of some environmental pollutants. The authors concluded that the new nanoparticles could be a potential material for *in situ* remediation of contaminated surface and ground water.

Mesoporous silicas containing cyclodextrin were also useful for pollutants present in air samples and for drug complexation. Mauri-Aucejo et al. (2012, 2015, 2016) published a series of works on the determination of phenolic compounds, polycyclic aromatic hydrocarbons (PAH) and volatile organic compounds (VOC) in air and in water samples by using cyclodextrin-silica hybrid composites. The preparation of these materials was very easy and inexpensive. Proposed samplers compared with other conventional solid phases present the advantages of a wider range of operative conditions for VOC desorption (Mauri-Aucejo et al. 2012). Samplers were tested based on results for the determination of BTEX, i.e. benzene, toluene, ethylbenzene, *o*-, *m*- and *p*-xylene, in air. Operational parameters were optimized and quantitative recovery was obtained using a solid phase from 2-hydroxypropyl-beta- cyclodextrin and acetonitrile as the extraction solvent. The recoveries obtained were 89 ± 4% for benzene, 90 ± 6% for toluene, 91 ± 2% for ethylbenzene, and 87.0 ± 0.9%, 88 ± 4%, 88 ± 4% for *o*-, *m*- and *p*-xylene, respectively. Moreover, results indicated a good reproducibility with a coefficient of variation below 6% and no significant difference between the reproducibility intra-synthesis and inter-synthesis. The proposed procedure has been applied to the determination of BTEX in several contaminated air samples and compared with results provided by a reference method. The authors also showed that cyclodextrin-silica composite samplers were particularly suitable for the sampling of phenol, cresol isomers, eugenol, guaiacol, 4-ethylguaiacol, and *p*-ethylphenol in air samples. The proposed method constituted an alternative to other methods due to low-cost and higher recoveries.

In another work (Mauri-Aucejo et al. 2015), the method has been applied to the assessment of exposure in different areas of a farm and regarding the quantification of these compounds in the vapors generated by burning incense sticks and an essential oil marketed as air fresheners. The acquired results were comparable with those provided from a reference method for a 95% of confidence level. The possible use of these samplers for other toxic compounds such as phthalates was also evaluated by qualitative analysis of extracts from incense sticks and essential oil samples. In a recent work (Mauri-Aucejo et al. 2016), the authors used the materials as adsorbent in solid-phase extraction combined with high-performance liquid chromatography to determine PAH in water samples. The experimental results indicated that the material exhibited high adsorption capacities toward PAH. Under optimum conditions, the quantification limits of the method were in the range of 0.09–2.4 μg/L and fine linear correlations between peak height and concentration were found around 1.3–70 μg/L. The method had good repeatability and reproducibility, with coefficients of variation under 8%. Due to the concentration results, this material might represent an alternative for trace analysis of PAH in water trough solid phase extraction.

Results on drug complexation by mesoporous silicas containing cyclodextrin were reported by Pasqua et al. (2013). Ordered silicas SBA-15 was hybridized using two different synthetic procedures to produce a covalent bond with cyclodextrin molecules as a drug delivery device for progesterone. In the first approach, SBA-15 silica (S_{BET} = 766 m^2/g) was first let to react with 3-glycidoxypropyltrimethoxysilane (GPTS) to produce an epoxide ring functional group on mesoporous silica. The latter was then reacted under basic conditions with mono-6-deoxy-6-mercapto-cyclodextrin, prepared in its turn in two steps from cyclodextrin through monotosylation followed by thiolation with thiourea. In the second approach, a silica suitably functionalized with a terminal thiol group, obtained by the reaction of SBA-15 silica with 3-mercaptopropyltrimethoxy-silane, was reacted with monotosyl-cyclodextrin (Fig. 6.7). The obtained materials (denoted SC1 and SC2) were characterized by XRD, nitrogen adsorption, SEM and C-13 CP/MAS NMR. The BET surface area obtained were 150 m^2/g and 190 m^2/g for SC1 and SC2, respectively. NMR solid state spectroscopy permitted a complete characterization for both materials. SEM observations did not evidence substantial variations in morphological properties. Progesterone was loaded on both the materials producing complete filling of mesopores and cyclodextrin cavities. Its release was studied at different pH values. The results showed that only one of the two progesterone-loaded delivery device (SC2) was, however, able to retain the drug in the system during the first period at acid pH (2 h) and release it after pH increase. A possible explanation for the immediate release from SC1 was the increase of solubility in acidic solution of progesterone adsorbed on the external surface and included in the hybrid material.

6.5 Cyclodextrin-Functionalized Silica Materials

Liu et al. (2016b) proposed a novel and low-cost cyclodextrin-functionalized silica gel prepared with grafting cyclodextrin to silica gel containing salicylamide as adsorbent for UO_2^{2+} removal (Fig. 6.8). Salicylamide (*o*-hydroxybenzamide) is known as drug for its analgesic and antipyretic properties but also as organic ligand for metal chelation. At the following conditions: pH 4.5, equilibrium time 60 min and initial UO_2^{2+} concentration 25 mg/L, the equilibrium adsorption capacity was found to be 6.45 mg/g. Adsorption performances were strongly pH-dependent. Modelling indicated that Langmuir and Freundlich models were suitable to describe the adsorption process. Thermodynamic data also indicated that the process was endothermic, spontaneous, and dominated by entropy rather than enthalpy change. The new material also adsorbed UO_2^{2+} in aqueous solution in presence of interfering ions such as Na^+, Fe^{3+}, Cu^{2+}, Mg^{2+}, La^{3+}, Mn^{2+}, Zn^{2+}, Pb^{2+} and Hg^{2+}. Apparently, the performance of the adsorbent was not appreciably deteriorated after repeated use and regeneration for five cycles. This new and efficient adsorbent for uranium(VI) was considered as low-cost since raw materials were cheap and commercially available, and the synthetic procedure was easy.

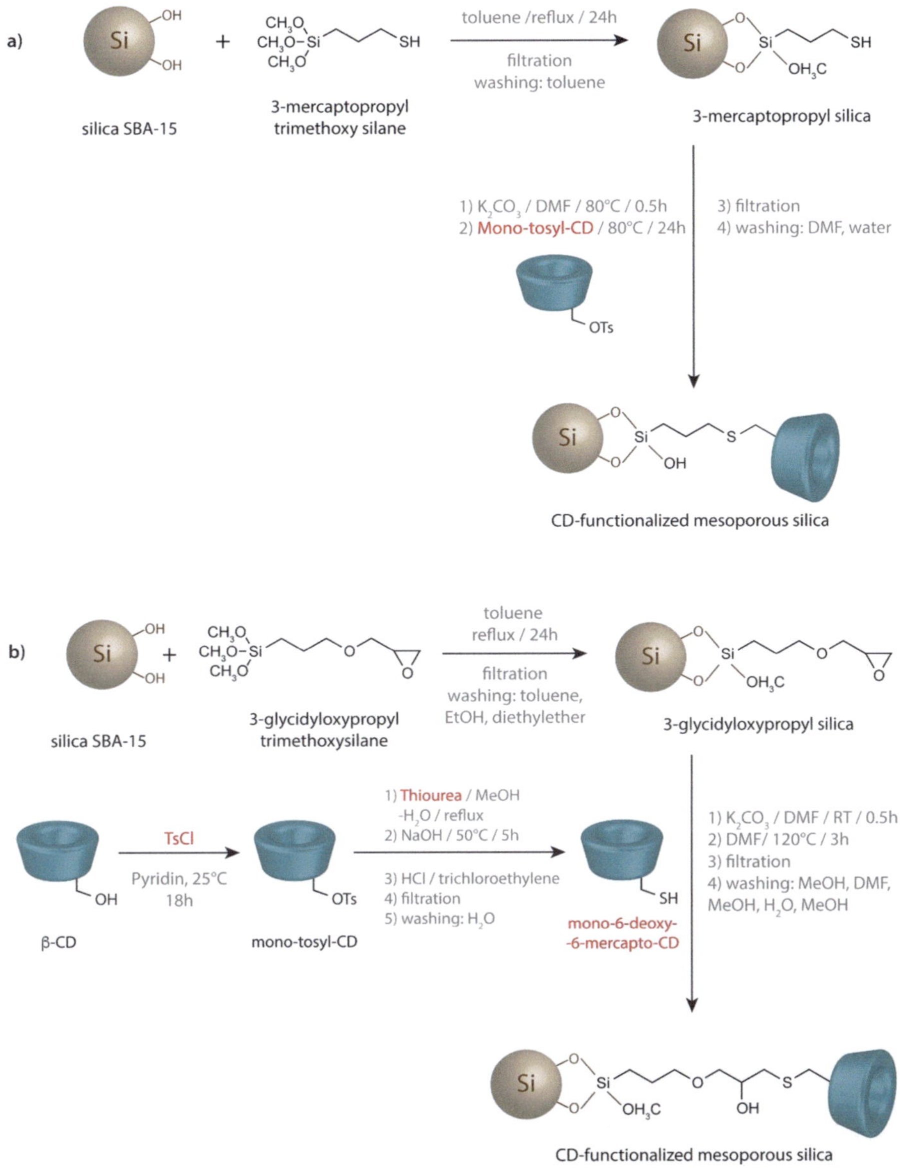

Fig. 6.7 Synthetic strategies proposed by Pasqua et al. (2013) to prepare cyclodextrin-functionalized mesoporous silica

De Carvalho et al. (2014) proposed a cyclodextrin-functionalized silica adsorbent for the removal of methylene blue in aqueous media. Modification of the silica surface using cyclodextrin (~44%) was performed in one-step by refluxing using citric acid as a linking agent (Fig. 6.9). The material obtained was characterized

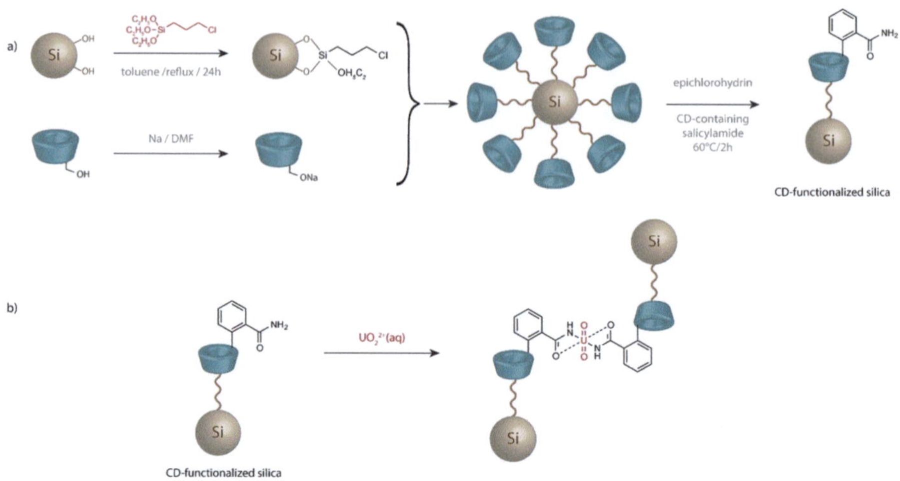

Fig. 6.8 (a) Synthesis of cyclodextrin functionalized silica gel and (b) its application for adsorption of uranium(VI) (adapted from Liu et al. 2016b)

using SEM, FT-IR spectroscopy, XRD and TGA. Based on these data, the authors showed that two mechanisms might occur in solution during the synthesis, as shown in Fig. 6.9. In the first route (A), an intermediary was formed between the linking agent and the cyclodextrin molecules which became both more active and capable of being functionalized to the surface of the silica. In a second route (B), the cyclodextrin molecules might be modified with groups from citric acid and became covalently linked to the surface of the silica without the direct intervention of the linking agent. However, it was difficult to choose between these two mechanisms and more studies should be conducted to improve knowledge of the functionalization mechanism. The authors reported interesting adsorption capacities. Indeed, the maximum capacity of the nano-adsorbent for adsorption of the dye was 212 mg/g and the best adsorption was achieved at pH values higher than 3.5. This value was in agreement with adsorption performances obtained employing commercial activated charcoal. The correlation coefficients obtained using the Langmuir isotherm enabled elucidation of the adsorption mechanism. Thermodynamic data also showed that the mechanism was spontaneous and temperature-dependent (adsorption decreased with increasing temperature), with adsorption following the pseudo-second order kinetic model and being fastest during the early stages, with equilibrium achieved after around 3 h. Advantages of these new materials include not only good adsorption properties but also ease of preparation and relatively low cost. The authors claimed that cyclodextrin-silica adsorbents can be used for removal of dyes from aqueous media, and could therefore substitute other more expensive adsorbents.

Belyakova's group proposed new functional organosilicas for adsorption of toxic metals such as mercury(II), cadmium(II) and zinc(II) from aqueous solution (Belyakova et al. 2014; Belyakova and Lyashenko 2014; Shvets and Belyakova 2015). The materials were synthesized using a nanoporous amorphous highly

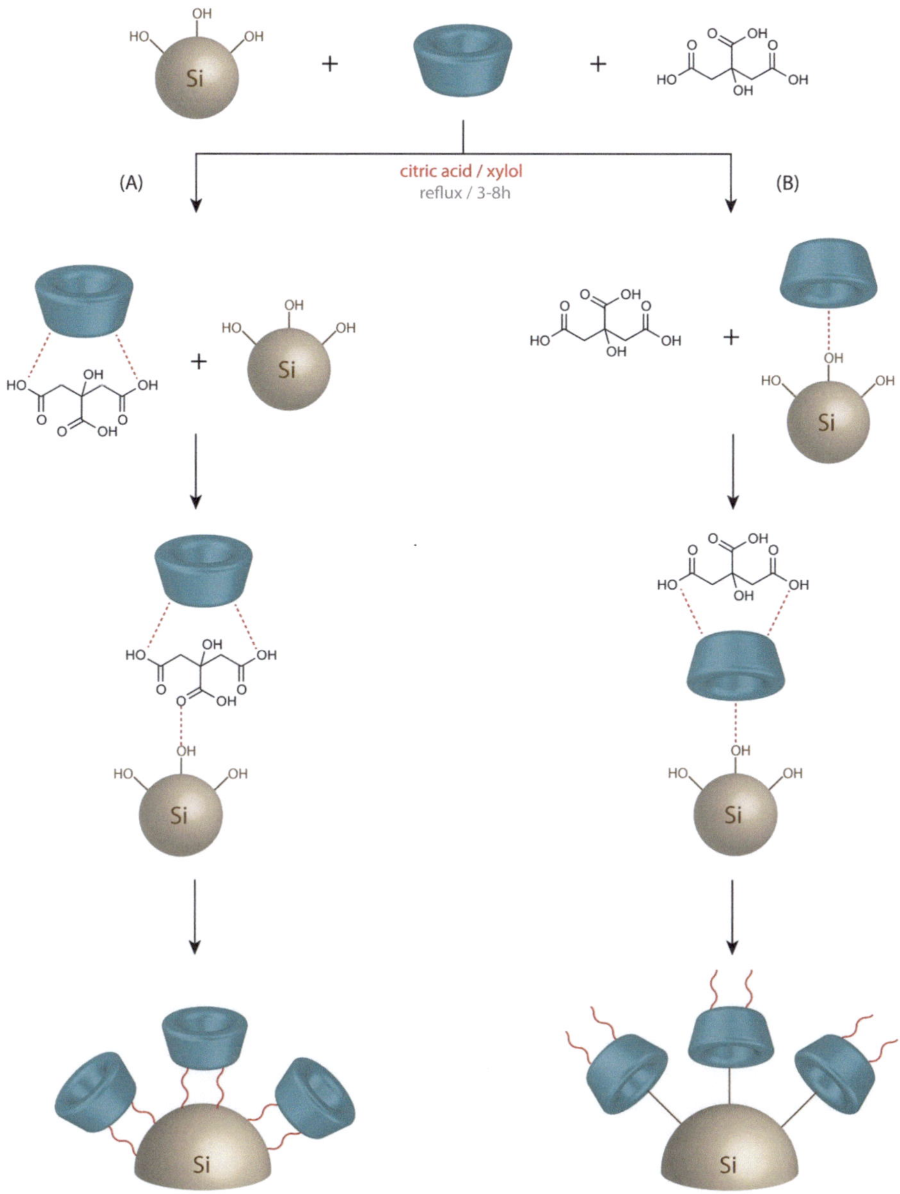

Fig. 6.9 Functionalization of silica gel with cyclodextrin molecules using citric acid as a bonding agent (adapted from de Carvalho et al. 2014)

disperse silica with the following characteristics: specific surface area 133 m^2/g, particle size 0.3–0.5 mm, concentration of silanol groups 0.4 mmol/g, adsorption pore volume 0.8 cm^3/g and mean pore diameter 46 nm. For chemical functionalization, the authors used aminopropylsilica preliminary obtained by the electrophilic substitution of aminopropyl radicals for the proton of silanol groups and the grafting reaction with three cyclodextrin-derivatives (mono-tosyl, bromoacetyl and thiosemicarbazidoacetyl functional derivatives) to introduce in the material secondary alcohol, bromoacetyl and thiosemicarbazidoacetyl functional groups of the wide edge of cyclodextrin molecules. The material obtained by grafting of mono-tosyl-cyclodextrin was denoted CD-1-SiO_2, the one obtained with bromoacetyl-cyclodextrin, CD-2-SiO_2 and the one obtained with thiosemicarbazidoacetyl-cyclodextrin, CD-3-SiO_2. They were characterized by elemental analysis, adsorption/desorption of nitrogen, potentiometric titration, SEM, TGA and FT-IR spectroscopy. In all cases, the chemical modification of the surface of the silica by cyclodextrin molecules leaded a decrease in the surface area (by 26–32%), adsorption volume (25–48%) and diameter of pores (37–63%). Nevertheless, cyclodextrin-silicas remained highly disperse materials with nanosized pores. Cyclodextrin content were similar in the three types of materials: 0.02, 0.01 and 0.1 mmol/g for CD-1-SiO_2, CD-2-SiO_2 and CD-3-SiO_2, respectively. Adsorption data demonstrated that cyclodextrin-silicas were found to have high affinity to mercury(II), cadmium(II), and zinc(II) cations, indicating strong interactions between the chemical groups grafted in the cyclodextrin molecules and metal cations. This was confirmed by a detailed FT-IR analysis (Belyakova and Lyashenko 2014). The authors clearly indicated that only the side functional groups of the grafted cyclodextrin were the real adsorption sites for metal removal. The adsorption ability of cyclodextrin-silicas with respect to mercury(II) and cadmium(II) increased in the series CD-1-SiO_2 < CD-2-SiO_2 < CD-3-SiO_2. For zinc (II), the series was CD-1-SiO_2 < CD-3-SiO_2 < CD-2-SiO_2. The maximum monolayer capacities for mercury(II) were: 34 ± 2, 12.6 ± 0.6 and 22 ± 1 mg/g for CD-1-SiO_2, CD-2-SiO_2, and CD-3-SiO_2, respectively. These results were consistent with the chemical and elemental analysis data for supramolecular compounds formed on the surface of nanoporous CD-silicas. The data were also in agreement with Pearson's theory. The softness of the side functional groups of the wide edge of cyclodextrin increased in the series secondary alcohol < bromoacetyl < thiosemicarbazidoacetyl groups. Belyakova and Lyashenko (2014) concluded that supramolecular surface structures formed whose chemical composition depended on the nature of the adsorbed cations and the functional substituents in the grafted cyclodextrin molecules. This research group also investigated the impact of hardness salts modeling soft and hard waters on adsorption of trace amounts of cadmium(II) by cyclodextrin-silicas, which differed by chemical nature of adsorption centers (Belyakova et al. 2014; Shvets and Belyakova 2015). The driving force of cadmium(II) adsorption on the surface of functional cyclodextrin-containing silica was the formation of inclusion complexes cyclodextrin-nitrate-anion. The results demonstrated high affinity of nanoporous organo-silica to cadmium(II) in its adsorption from multicomponent solutions with rapid kinetics (equilibrium was reached in 30 min). The adsorption of trace amounts of cadmium (II) from multi-component solutions did not decrease, but even increased in the presence of hardness salts, simulating soft and hard water. Equilibrium adsorption of cations from aqueous nitrate solutions within a broad interval of concentrations has been described by means of the Langmuir and Freundlich models and was interpreted in terms of the theory of hard and soft acids and Pearson bases. It has been proved the formation of supramolecular structures on the surface of synthesized organosilicas as a result of cadmium (II) adsorption.

Shen's group (Shen et al. 2014a, b; Shen et al. 2015a, b; Han et al. 2016) proposed a new cyclodextrin-grafted silica gel, denoted CD@Si, for *p*-nitrophenol removal. Cyclodextrin was grafted onto the surface of silica using (3-chloropropyl)-trimethoxysilane and ethylenediamine as linking groups according to the procedure described in Fig. 6.10 (Shen et al. 2015a). The obtained CD@Si adsorbent was characterized through FT-IR spectroscopy, X-ray photoelectron spectroscopy (XPS), contact angle measurement, TGA, solid-state C-13 NMR, SEM, and XRD analyses. FT-IR spectroscopy, XPS and NMR data demonstrated the successful graft of cyclodextrin molecules on silica surface. From TGA data, the amount of cyclodextrin grafted was estimated about 20% (w/w). SEM analysis indicated that the morphologies of the different materials, i.e. activated gel, silica-grafted (3-chloropropyl)-trimethoxysilane and CD@Si, were uniform in both shape and size, and the graft of (3-chloropropyl)-trimethoxysilane and cyclodextrin molecules did not affect the morphology of silica gel. Adsorption data indicated that (i) the adsorption of *p*-nitrophenol onto CD@Si was a very fast process: the equilibrium can be reached in 5 s with a maximum adsorption capacity of 41.5 mg/g at pH > 8.5, much faster than many reported adsorbents based on CD; (ii) the adsorption process followed the pseudo-second-order and Freundlich models; (iii) thermodynamic data indicated that the process was spontaneous and exothermic but temperature-dependent (adsorption was more favorable in lower temperature); and (iv) CD@Si can be recycled and reused at least five runs with acceptable adsorption capacity, showing the chemical stability of the material. A systematic study of the adsorption mechanism using spectroscopic data showed that the two main interactions were inclusion complex formation and hydrogen bond interactions as illustrated in Fig. 6.10. In another recent work, the authors provide a new strategy to increase the adsorption rate of CD@Si adsorbents using different experimental conditions during synthetic procedure (Shen et al. 2015b). The authors claimed that these new cyclodextrin-grafted silica gels were promising for pollutant removal not only due to their efficiency but also from an economic point of view (Han et al. 2016).

Bhattarai et al. (2014) developed silica coated cyclodextrin polymeric adsorbents for the removal of several emerging pollutants such as steroid hormones (17-β-estradiol), bisphenol A and perfluorooctanoic acid. Three different approaches were used to functionalize cyclodextrin onto silica (40x100 mesh size). In the first approach, cyclodextrin molecules can be polymerized onto surface silica by a reaction between the hydroxyl groups with a coupling agent such as epichlorohydrin (EPI) or hexamethylene diisocyanate (HMDI). In the second method, cyclodextrin were supported on silica with copolymers glycidoxypropyltrimethoxysilane (GPTS) and 3-aminopropyltriethoxysilane (APTES). In the third approach, cyclodextrin were coated on silica with both cross-linking agents (EPI or HMDI) and copolymers (GPTS or APTES). Fourteen different adsorbents were synthesized under different experimental conditions, e.g. solvent, temperature, and reactant concentration. All the adsorbents were characterized using FT-IR, TEM and TGA techniques, and

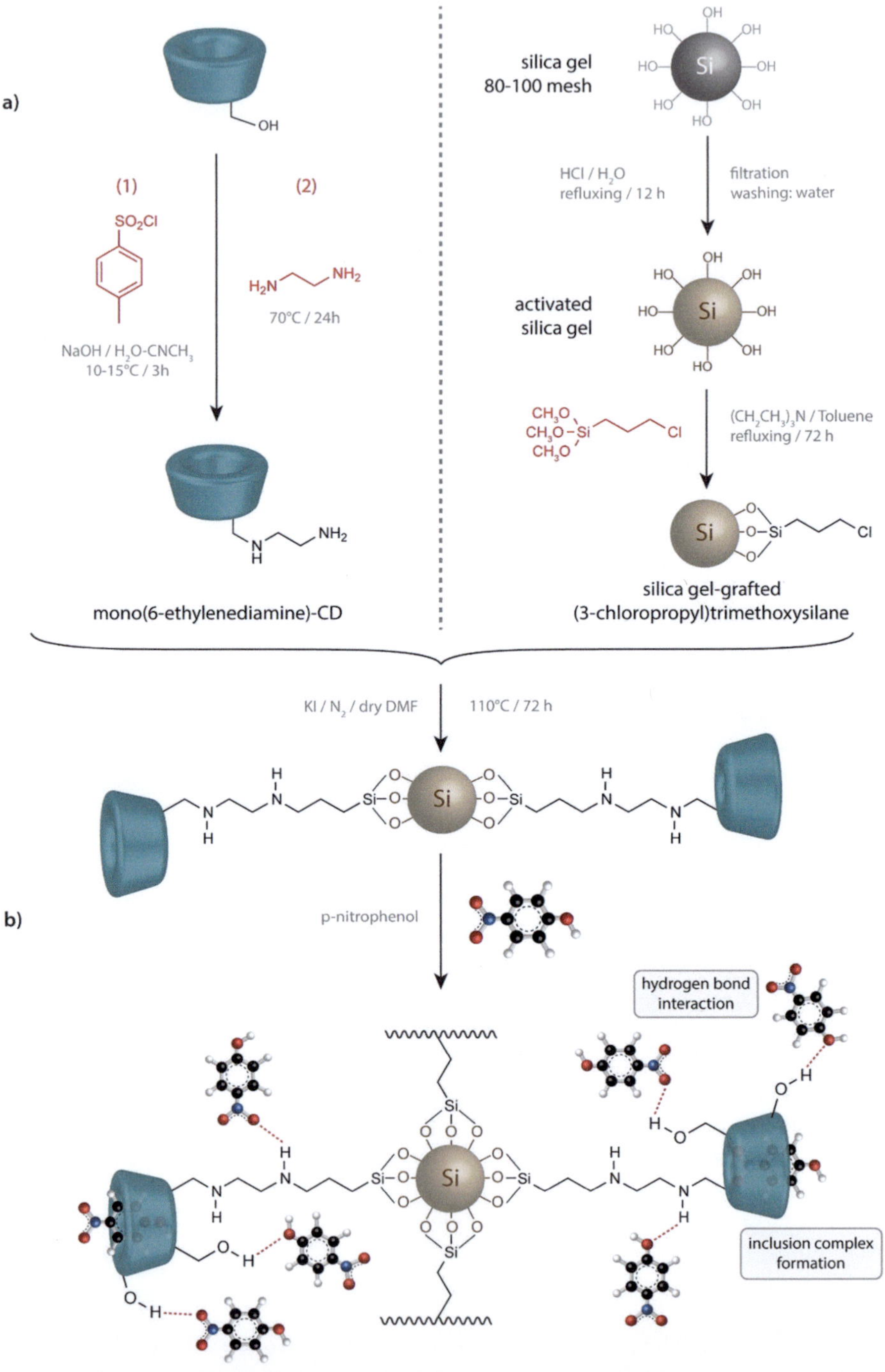

Fig. 6.10 (a) Preparation of cyclodextrin-grafted silica according to Shen et al. (2015a) and (b) adsorption mechanism proposed for *p*-nitrophenol removal from aqueous solution

nitrogen adsorption analysis. Adsorption results showed that the material prepared by using HMDI as crosslinking agent with DMSO as solvent was the best in term of performance for the removal of 17-β-estradiol, perfluorooctanoic acid, and bisphenol A with more than 90% abatement in all case. This material had thermal stability of up to 300 °C. This adsorbent was also resynthesized in seven batches and its performance was reproducible for the removal of ten steroid hormones at 1.5 g/L adsorbent dosage. It showed very good regeneration potential for four successive adsorption-regeneration cycles to remove steroid hormones and perfluorooctanoic acid without significant loss in its performances. The new adsorbent showed higher adsorption capacity (38.8 ± 5.6 μg/g) for 17-β-estradiol when compared with that of commercially available activated carbon (23.6 ± 6.3 μ/g). Furthermore, the cyclodextrin loading on pollutant removal was studied which showed that the adsorbate removal increased with increase in loading of cyclodextrin on the substrate (a loading of 0.23 g/g was recommended). The results suggested the main role of the cyclodextrin molecules in the adsorption process (physisorption was negligible since the BET surface area of the materials were less than 1 m^2/g), and also possible numerous inter-molecular hydrogen bond interactions between the cyclodextrin host molecules and the guest pollutants.

3,5-dimethylphenylcarbamoylated cyclodextrin bonded silica gel was used as adsorbent in solid-phase extraction to selectively enrich forchlorfenuron and thidiazuron, a plant growth regulators used on kiwi fruits and grapes for example, followed by determination with surface-enhanced Raman spectroscopy (Chen et al. 2016). This new adsorbent exhibited high adsorption capacities, 40 and 30 μg/g for chlorfenuron and thidiazuron, respectively, and showed excellent selectivity.

A highly effective clean-up adsorbent was developed by Liu et al. (2016a) for eliminating matrix interferences, especially main organochlorine pesticide residues during the determination of highly chlorinated polychlorinated biphenyls in seafood. The multifunctional adsorbent was prepared by grafting carboxymethyl-cyclodextrin on the surface of amino functionalized mesoporous nanoparticles. The amino group functionalized mesoporous SiO_2 can remove most of matrix interference in samples. Moreover, carboxymethyl-cyclodextrin had stronger host-guest complexation with 1,1,1-trichloro-2,2-bis(*p*-chlorophenyl) ethane, 2,2-bis(*p*-chlorophenyl)-1,1-dichloro-ethylene, and 1,1-dichloro-2,2-bis(*p*-chlorophenyl) ethane. However, it showed weaker adsorption ability toward highly chlorinated polychlorinated biphenyls due to a steric hindrance effect. A gas chromatography-mass spectrometry method coupled with the multifunctional adsorbent as a clean-up adsorbent for dispersive solid phase extraction was developed for the analysis of several highly chlorinated polychlorinated biphenyls in seafood samples. The results indicated that the multifunctional adsorbent as a purification material could easily and effectively remove matrix interferences in seafood samples within a short time. The recoveries for polychlorinated biphenyls were in the range of 88.4–103.2%, with relative standard deviations varying between 1.3 and 5.7%.

6.6 Other Materials

Matias et al. (2015) proposed porous silica-based aerogels/xerogels containing cyclodextrin with different degrees of hydrophobicity/hydrophilicity as adsorbents for phenolic compounds removal. The materials were synthesized by using a combination of methyltrimethoxysilane (MTMS) and tetramethylorthosilicate precursors in different proportions, in order to provide a variable number of methyl and hydroxyl groups in the aerogels structure, and by cyclodextrin functionalization of the aerogels to introduce possible inclusion complex formation. The sol-gel synthesis followed a one-step NH_4OH-catalyzed procedure and the drying of the gels was accomplished by supercritical fluid drying/extraction with CO_2, to obtain aerogels, and evaporative drying, to produce xerogels. The characterization of materials showed that silica-based aerogels/xerogels containing cyclodextrin had high surface area (800 m^2/g) and tailored levels of hydrophobicity/hydrophilicity. The aerogels with intermediate and high contact angles showed higher adsorption capacities, with favorable isotherms, for the phenolic compounds, such as phenol, *p*-cresol and *p*-chlorophenol, with higher hydrophobicity, which showed the important role of hydrophobic interactions in the adsorption process. The best results were obtained for the aerogel functionalized with cyclodextrin when adsorbing the *p*-chlorophenol (adsorption capacity 117.8 mg/g), proving the positive effect of the presence of the cyclodextrin hydrophobic cavity in the adsorption of this highly hydrophobic pollutant.

Chen et al. (2014) synthetized cyclodextrin-modified cellulose nanocrystals (CNC)@Fe_3O_4@SiO_2 superparamagnetic nanorods for the removal of two model drugs (procaine hydrochloride and imipramine hydrochloride). Cellulose nanocrystals and silica served as supporting template and coating materials, respectively. The silica shell was previously functionalized with primary amine and then reacted with mono-chlorotriazine-cyclodextrin derivative. Cellulose nanocrystals are attracting increasing interest due to their abundance, uniform nanorod shape, good mechanical strength, liquid crystalline character, high specific surface area, biocompatibility, biodegradability, and sustainability. Cellulose nanocrystals can be produced at an industrial scale by acid hydrolysis of pulp fibers. During the synthetic process proposed by the authors, sustainable natural materials and low-cost chemicals were used, and mild reaction conditions were adopted. TEM and SEM images indicated good dispersion of Fe_3O_4 nanoparticles with uniform silica coating on Cellulose nanocrystals. The thickness of the silica coating was controlled by manipulating the amounts of precursor solution used. TGA data confirmed that the silica coating significantly enhanced the thermal stability of cellulose nanocrystals. The onset decomposition temperature of CNC@Fe_3O_4@SiO_2 hybrids increased by 60 °C compared to pure cellulose nanocrystals. XRD, energy-dispersive X-ray spectroscopy (EDS), and FT-IR analyses confirmed the structure of CNC@Fe_3O_4@SiO_2 and the successful grafting of CD. The CNC@Fe_3O_4@SiO_2@CD hybrids displayed good adsorption toward the two pharmaceutical compounds. The adsorption capacities were determined to be 13 ± 0.09 mg/g and 14.8 ± 0.16 mg/g for procaine hydrochloride and imipramine hydrochloride, respectively. Compared with conventional membrane separations that generally suffer from fouling, poor permeate quality and low flux enhancement, the magnetic separation proposed by the authors demonstrated several advantages including simple cleanup, good magnetic properties, effective adsorption characteristic toward two model drugs with fast kinetics, and easy recovery of materials. The authors concluded that the procedure reported can be extended to prepare other inorganic-organic nanocomposites using cellulose nanocrystals as the template material.

Superparamagnetic Fe_3O_4 core could be rapidly separated from matrix to simplify time-consuming washing extraction as reported by Wang et al. (2015). These authors proposed core-shell superparamagnetic Fe_3O_4@CD composites for host-guest adsorption of polychlorinated biphenyls (PCB). Their protocol involved the synthesis of Fe_3O_4 particles through a solvothermal reaction and the covering of a silica layer bonded cyclodextrin over Fe_3O_4 via a sol-gel process to produce core-shell Fe_3O_4@CD composites. Cyclodextrin molecules were linked covalently to Fe_3O_4 nanoparticles to generate the binding sites, enhancing the stability of Fe_3O_4 nanoparticles in water. The adsorption capacity of composites to PCB28 and PCB52 in aqueous solutions was investigated. To estimate the theoretical binding site number of Fe_3O_4@CD, the obtained binding data were replotted according to Scatchard equation. The host-guest interaction between cyclodextrin and PCB were further examined with density functional theory calculations. It provided theoretical evidence of cyclodextrin as host molecule had a higher binding amount towards PCB-28 than PCB-52 on the basis of their optimized geometries and calculated complexation energies. The authors concluded that the new nanomaterial was an ideal candidate for various applications, including the recognition and removal of environmentally deleterious substances.

Although cholesterol plays significant biochemical function in the human body, excess of it leads to various disorders, and thus, its control/separation is important in medical science and food industries. However, efficient and selective separation of cholesterol is challenging because cholesterol often exists in microheterogeneous or insoluble forms in remote organ and exists with other substances. Sinha et al. (2015) have described a colloidal magnetic mesoporous silica-based approach for efficient separation of cholesterol in different forms. Magnetic mesoporous silica was functionalized with cyclodextrin for selective binding with cholesterol via host-guest interaction. The colloidal form of magnetic mesoporous silica offered effective interaction with cholesterol of any form, and magnetic property of magnetic mesoporous silica offered easier separation of bound cholesterol. Functionalized material was efficient in separating cholesterol crystals, water-insoluble cholesterol, and the microheterogeneous form of cholesterol from milk or a cellular environment. The authors concluded that the new material could be used to remove cholesterol from a complex environment and extended for large-scale cholesterol separation from food.

Ennist et al. (2014) synthesized novel dendronized silica substrates that adsorbed targeted analytes. First- and second- generation poly(arylether) dendrons were appended to silica surfaces. Using Cu(I) mediated cycloaddition click chemistry, cyclodextrin was tethered to the dendronized surfaces and to a nondendronized surface for comparison purposes. This synthesis strategy afforded a modular, versatile method for surface functionalization in which the density of functional groups could be readily varied by changing the generation of dendron used. The surfaces have been characterized and studied using XPS and vibrational sum frequency spectroscopy. Fluorescence spectroscopy was used to study the surfaces' ability to retain coumarin 152 (C152). These studies indicated that the cyclodextrin-functionalized surfaces not only adsorbed C152 but also retained it through multiple aqueous washes. Furthermore, these observations were quantified and showed that substrates functionalized with first-generation dendrons had a more than 6 times greater capacity to adsorb C152 than slides functionalized with monomeric cyclodextrin. The first-generation dendrons also had 2 times greater the capacity than the larger generation dendrons. This result was explained by describing a dendron that had an increased number of cyclodextrin monomers but, when covalently bound to silica, had a footprint too large to optimize the number of accessible monomers. Overall, both dendronized surfaces demonstrated an increased capacity to adsorb targeted analytes over the slides functionalized with monomeric cyclodextrin. Their study provided a methodology for characterizing and evaluating the properties of novel, highly functional surfaces.

6.7 Conclusion

The field of silica containing cyclodextrin molecules has been studied for about 25 years but is still an area attracting a lot of attention. These kinds of materials often present the best properties of each of its components in a synergic way and have high performances of physical, chemical and mechanical properties. The recent state-of-the-art in their synthesis is reviewed in this chapter, based on a substantial number of relevant references published recently. Of course, this is an ambitious project since a direct comparison of data obtained using different materials is difficult to make. Nevertheless, as demonstrated in this chapter, cyclodextrin-silica material such as cyclodextrin-silica hybrid systems prepared through sol-gel or self-assembly process and cyclodextrin-functionalized silicas prepared through grafting or cross-linking reactions are regarded as effective adsorbents for pollutant removal. Indeed, from a point of view of their efficiency, there is no doubt that these materials exhibited interesting adsorption capacities toward pollutants and will find industrial environmental applications. However, it is important to note that research is mainly focused on the performances of these innovative materials, while their economic aspect is neglected. Cost is an important parameter for comparing adsorbents. An important question is the following: could mesosilica adsorbents compete with conventional activated carbon (that are produced on a significantly larger scale) in price, durability and efficiency? To date, there is no systematic and comparative study. Mesosilica is not yet produced on an industrial scale. In addition, starting materials such as alkoxides as the silica source are relatively expensive and the product made from them can only find use in applications such as catalysis or CO_2 capture where price is not a major issue. Recently, the use of inexpensive inorganic silicates (water glass) has been proposed in the synthesis of mesostructured materials (Han and Zhang 2012; Al Othman 2012).

In addition, although extensive work has been done, future research needs to look into some of the following aspects. Numerous progresses have been realized on the control of the structure and at the same time control of the size and the size distribution of the particles, and the cyclodextrin content of different materials. The chemical procedures involving multistep reactions are also well-known. However, several issues related to the synthesis using eco-friendly procedures are still poorly investigated. Adsorption processes are basically at the stage of laboratory-scale study in spite of unquestionable progresses. Much work is necessary to demonstrate the possibilities on an industrial scale. The experimental conditions should be chosen to simulate real effluents on the basis of thermodynamics and reaction kinetics studies. Indeed, a major challenge is the design of mesosilica in a form compatible with treatment of large volumes of polycontamined real effluents. The use of these materials is still underdeveloped in real-world environmental remediation applications. For example, only a few reports have focused on their use for the removal of emerging pollutants from water such as pharmaceuticals products. Other questions concern their long-term stability and effective reusability after regeneration, and adsorption mechanisms. Significant decreases in crystallinity, average pore diameter and pore volume can be observed when the materials were used in alkaline conditions. So, it is necessary to continue to search for and select the most promising types of cyclodextrin-silica materials. Finally, the works reviewed in this chapter indicated that adsorption onto these new materials is becoming a promising alternative to replace conventional materials although the mechanisms involved are not fully understood.

References

Al Othman ZA (2012) A review: fundamental aspects of silicate mesoporous materials. Materials 5:2874–2902

Al Othman ZA, Apblett AW (2009) Synthesis and characterization of a hexagonal mesoporous silica grafted with 3-glycidoxypropyltrimethoxy-silane. Mater Lett 6:2331–2334

Alahmadi SM, Mohamad S, Maah MJ (2014) Organic-inorganic hybrid materials based on mesoporous silica MCM-41 with beta-cyclodextrin and its applications. Asian J Chem 26(14):4323–4329

Atwood JL, Davies EE, MacNicol DD (1984) Inclusion compounds. Academic Press, Michigan

Beck JS, Vartuli JC, Roth WJ, Leonowicz ME, Kresge CT, Schmitt KD, Chu CTW, Olson DH, Sheppard EW, McCullen SB, Higgins JB, Schlenker JL (1992) A new family of mesoporous molecular-sieves prepared with liquid-crystal templates. J Am Chem Soc 114(27):10834–10843

Beck JS, Vartuli JC, Kennedy GJ, Kresge CT, Roth WJ, Schramm SE (1994) Molecular or supramolecular templating – defining the role of surfactant chemistry in the formation of microporous and mesoporous molecular-sieves. Chem Mater 6(10):1816–1821

Belyakova LA, Lyashenko DY (2014) Nanoporous functional organosilicas for sorption of toxic ions. Russ J Phys Chem A 88(3):489–493
Belyakova LA, Lyashenko DY, Shvets AN (2014) Sorption of Cd(II) from multicomponent nitrate solutions by functional organosilicas. J Water Chem Technol 36(2):56–61
Bender ML, Komiyama M (1978) Cyclodextrins chemistry. Springer Verlag, Berlin
Berggren A, Palmqvist AEC, Holmberg K (2005) Surfactant-templated mesostructured materials from inorganic silica. Soft Matter 1:219–226
Bergna HE (1994) The colloid chemistry of silica, Advances in Chemistry Series, vol 234. American Chemical Society, Washington, DC
Bhattarai B, Muruganandham M, Suri RPS (2014) Development of high efficiency silica coated beta-cyclodextrin polymeric adsorbent for the removal of emerging contaminants of concern from water. J Hazard Mater 273:146–154
Bibby A, Mercier L (2003) Adsorption and separation of water-soluble aromatic molecules by cyclodextrin-functionalized mesoporous silica. Green Chem 5(1):15–19
Bilensoy E (2011) Cyclodextrin in pharmaceutics, cosmetics, and biomedicine. Wiley, Hoboken, p 395
Chen CY, Burkett SL, Li HX, Davis ME (1993) Studies on mesoporous materials. II. Synthesis mechanism of MCM-41. Microporous Mater 2:27–34
Chen L, Berry RM, Tam KC (2014) Synthesis of beta-cyclodextrin-modified cellulose nanocrystals (CNCs)@Fe_3O_4@SiO_2 superparamagnetic nanorods. ACS Sustain Chem Eng 2(4):951–958
Chen XM, Yan KL, Xiao XH, Li GK (2016) Analysis of forchlorfenuron and thidiazuron in fruits and vegetables by surface-enhanced Raman spectroscopy after selective solid-phase extraction with modified beta-cyclodextrin. J Sep Sci 39(12):2340–2346
Ciesla U, Schüth F (1999) Ordered mesoporous materials. Microporous Mesoporous Mater 27:131–149
Cooney DO (1999) Adsorption design for wastewater treatment. Lewis Publishers, Boca Raton
Cram DJ (1988) The design of molecular hosts, guests and their complexes. Angew Chem Int Ed 27(8):1009–1020
Crini G (2005) Recent developments in polysaccharide-based materials used as adsorbents in wastewater treatment. Prog Polym Sci 30:38–70
Crini G (2014) Review: a history of cyclodextrins. Chem Rev 114:10940–10975
Crini G, Badot PM (2010) Sorption processes and pollution. PUFC, Besançon
Crini G, Morcellet M (2002) Synthesis and applications of adsorbents containing cyclodextrins. J Sep Sci 25:789–813
Davis ME (2002) Ordered porous materials for emerging applications. Nature 417:813–821
de Carvalho LB, Carvalho TG, Magriotis ZM, Ramalho TD, Pinto LDA (2014) Cyclodextrin/silica hybrid adsorbent for removal of methylene blue in aqueous media. J Incl Phenom Macrocycl Chem 78(1–4):77–87
Degoutin S, Bacquet M (2013) Novel porous organosilica containing amino and beta-cyclodextrin groups. J Porous Mater 20(4):663–671
Descalzo AB, Martinez-Manez R, Sancenon R, Hoffmann K, Rurack K (2006) The supramolecular chemistry of organic-inorganic hybrid materials. Angew Chemie Int Ed 45(36):5924–5948
Dinker MK, Kulkarni PS (2015) Recent advances in silica-based materials for the removal of hexavalent chromium: a review. J Chem Eng Data 60:2521–2540
Dodziuk H (2006) Cyclodextrins and their complexes. Wiley-VCH GmbH & Co, Weinheim, p 489
Duchêne D (1987) Cyclodextrins and their industrial uses. Éditions de Santé, Paris
Duchêne D (1991) New trends in cyclodextrins and derivatives. Éditions de Santé, Paris
Ebadi A, Rafati AA (2015) Preparation of silica mesoporous nanoparticles functionalized with beta-cyclodextrin and its application for methylene blue removal. J Mol Liq 209:239–245
Ennist JH, Gobrogge EA, Schlick KH, Walker RA, Cloninger MJ (2014) Cyclodextrin-functionalized chromatographic materials tailored for reversible adsorption. ACS Appl Mater Interfaces 6(20):18087–18097

Frömming KH, Szejtli J (1994) Cyclodextrins in pharmacy. Kluwer Academic Publishers, Dordrecht

Fukushima Y, Inagaki S, Kuroda K (1995) Mesoporous materials with uniform pore-size - synthesis and their applications. Nippon Kagaku Kaishi 5:327–333

Gibson LT (2014) Mesosilica materials and organic pollutant adsorption: part B - removal from aqueous solution. Chem Soc Rev 43(15):5173–5182

Goyal AK, Johal ES, Rath G (2011) Nanotechnology for water treatment. Curr Nanosci 7:640–654

Guo R, Wilson LD (2013) Cyclodextrin-based microcapsule materials - their preparation and physiochemical properties. Curr Org Chem 17:14–21

Han Y, Zhang D (2012) Ordered mesoporous silica materials with complicated structures. Curr Opin Chem Eng 1:129–137

Han YQ, Zhou WJ, Shen HM, Liu QP, Yu WY, Ji HB, She YB (2016) Progress in the immobilization of beta-cyclodextrin and their application in adsorption of environmental pollutants. Chin J Org Chem 36(2):248–257

Hench LL, West JK (1990) The sol-gel process. Chem Rev 90:33–72

Hoffman F, Cornelius M, Morell J, Fröba M (2006) Silica-based mesoporous organic-inorganic hybrid materials. Angew Chem Int Ed 45:3216–3251

Huq R, Mercier L, Kooyman PJ (2001) Incorporation of cyclodextrin into mesostructured silica. Chem Mater 13(12):4512–4519

Inagaki S, Fukushima Y, Kuroda K (1993) Synthesis of highly ordered mesoporous materials from a layered polysilicate. J Chem Soc Chem Commun 8:680–682

Inagaki S, Fukushima Y, Kuroda K, Kuroda K (1996a) Adsorption isotherm of water vapor and its large hysteresis on highly ordered mesoporous silica. J Colloid Interface Sci 180(2):623–624

Inagaki S, Koiwai A, Suzuki N, Fukushima Y, Kuroda K (1996b) Syntheses of highly ordered mesoporous materials, FSM-16, derived from kanemite. Bull Chem Soc Jpn 69(5):1449–1457

Kawamura A, Katoh T, Uragami T, Miyata T (2015) Design of molecule-responsive organic-inorganic hybrid nanoparticles bearing cyclodextrin as ligands. Polym J 47(2):206–211

Kresge CT, Leonowicz ME, Roth WJ, Vartuli JC, Beck JS (1992) Ordered mesoporous molecular-sieves synthetized by a liquid-crystal template mechanism. Nature 359:710–712

Kresge CT, Vartuli RWJ, Leonowicz ME, Beck JS, Schmitt KD, Chu CTW, Olson DH, Sheppard EW, McCullen SB, Higgins JB, Schlenker JL (1995) M41S: a new family of mesoporous molecular-sieves prepared with liquid-crystal templates. Sci Technol Catalysis, Book Series 92:11–19

Landy D, Mallard I, Ponchel A, Monflier E, Fourmentin S (2012) Remediation technologies using cyclodextrins: an overview. Environ Chem Lett 10:225–237

Lee SY, Park SJ (2015) A review on solid adsorbents for carbon dioxide capture. J Ind Eng Chem 23:1–11

Lim MH, Stein A (1999) Comparative studies of grafting and direct synthesis of inorganic-organic hybrid mesoporous materials. Chem Mater 11(11):3285–3295

Lim MH, Blanford CF, Stein A (1998) Synthesis of ordered microporous silicates with organosulfur surface groups and their applications as solid acid catalysts. Chem Mater 10(2):467–470

Liu CQ, Naismith N, Economy J (2004) Advanced mesoporous organosilica material containing microporous beta-cyclodextrins for the removal of humic acid from water. J Chromatogr A 1036(2):113–118

Liu HB, Zhang JB, Gan N, Chen YJ, Huang J, Cao YT, Li TH, Lan HZ (2016a) Application of a multifunctional magnetic mesoporous material for seafood sample clean-up prior to the determination of highly chlorinated polychlorinated biphenyls. RSC Adv 6(1):183–189

Liu HJ, Jing PF, Liu XY, Du KJ, Sun YK (2016b) Synthesis of beta-cyclodextrin functionalized silica gel and its application for adsorption of uranium(VI). J Radioanal Nucl Chem 310(1):263–270

Mahmud ST, Wilson LD (2016) Synthesis and characterization of surface-modified mesoporous silica materials with beta-cyclodextrin. Cogent Chem 2:1132984. https://doi.org/10.1080/23312009.2015.1132984

Matias T, Marques J, Quina MJ, Gando-Ferreira L, Valente AJM, Portugal A, Duraes L (2015) Silica-based aerogels as adsorbents for phenol-derivative compounds. Colloids Surf A Physicochem Eng Asp 480:260–269

Mauri-Aucejo AR, Llobat-Estelles M, Egea MG, Guillem C, Amoros P (2012) Samplers for VOCs in air based on cyclodextrin-silica hybrid microporous solid phases. Analyst 137(5):1275–1283

Mauri-Aucejo AR, Ponce-Catala P, Belenguer-Sapina C, Amoros P (2015) Determination of phenolic compounds in air by using cyclodextrin-silica hybrid microporous composite samplers. Talanta 134:560–567

Mauri-Aucejo A, Amoros P, Moragues A, Guillem C, Belenguer-Sapina C (2016) Comparison of the solid-phase extraction efficiency of a bounded and an included cyclodextrin-silica microporous composite for polycyclic aromatic hydrocarbons determination in water samples. Talanta 156:95–103

McKay G (1996) Use of adsorbents for the removal of pollutants from wastewaters. CRC Press, Boca Raton

Mocanu G, Vizitiu D, Carpov A (2001) Cyclodextrins polymers. J Bioact Compat Polym 16:315–342

Morin-Crini N, Crini G (2012) Review of current physical and chemical technologies for the decontamination of industrial wastewater. Int J Chem Eng 5:143–186

Morin-Crini N, Fourmentin S, Crini G (2015) Cyclodextrines (in French). Presses Universitaires de Franche-Comté, Besançon. 370 p

Myers D (1992) Surfactant science and technology. VCH, New-York

Ogoshi T, Chujo Y (2005) Organic-inorganic polymer hybrids prepared by the sol-gel method. Compos Interfaces 11:539–566

Pagliaro M (2009) Silica-based materials for advanced chemical applications. RSC Publishing, Royal Society of Chemistry. 192 p

Pan BJ, Pan BC, Zhang MW, Lv L, Zhang QX, Zheng SR (2009) Development of polymeric and polymer-based hybrid adsorbents for pollutant removal from waters. Chem Eng J 151:19–29

Pasqua L, Veltri L, Gabriele B, Testa F, Salerno G (2013) Progesterone inclusion into cyclodextrin-functionalized mesoporous silica. J Porous Mater 20(4):917–925

Patarin J, Lebeau B, Zana R (2002) Recent advances in the formation mechanisms of organized mesoporous materials. Curr Opin Colloid Interface Sci 7(1–2):107–115

Polarz S, Smarsly B, Bronstein L, Antonietti M (2001) From cyclodextrin assemblies to porous materials by silica templating. Angew Chem Int Ed 40(23):4417–4421

Rahmat N, Abdullah AZ, Mohamed AR (2010) A review: Mesoporous Santa Barbara Amorphous-15, types, synthesis and its application towards biorefinery productin. Am J Appl Sci 7:1579–1586

Robyt JF (1998) Chapter 8: Cyclodextrins. In: Essentials of carbohydrate of chemistry. Springer, New York, pp 245–261

Samiey B, Cheng CH, Wu JG (2014) Organic-inorganic hybrid polymers as adsorbent for removal of heavy metal ions from solutions: a review. Materials 7:673–726

Sanchez C, Julian B, Belleville P, Popall M (2005) Applications of silica-based organic-inorganic hybrid materials. J Chem Mater 15:3559–3592

Sawicki R, Mercier L (2006) Evaluation of mesoporous cyclodextrin-silica nanocomposites for the removal of pesticides from aqueous media. Environ Sci Technol 40(6):1978–1983

Sharma SK (2015) Green chemistry for Dyes removal from wastewater. Scrivener Publishing LLC/Wiley, Beverley

Sharma SK, Sanghi R (2012) Advances in water treatment and pollution prevention. Springer, Dordrecht

Shen HM, Ji HB, Wu HK, Shi HX (2014a) Recent advances in the immobilization of beta-cyclodextrin and their application. Chin J Org Chem 34(8):1549–1572

Shen HM, Wu HK, Ji HB, Shi HX (2014b) Progress in the construction of beta-cyclodextrin-Fe_3O_4 supramolecular systems and their application. Chin J Org Chem 34(4):630–646

Shen HM, Zhu GY, Yu WB, Wu HK, Ji HB, Shi HX, She YB, Zheng YF (2015a) Fast adsorption of p-nitrophenol from aqueous solution using beta-cyclodextrin grafted silica gel. Appl Surf Sci 356:1155–1167
Shen HM, Zhu GY, Yu WB, Wu HK, Ji HB, Shi HX, Zheng YF, She YB (2015b) Surface immobilization of beta-cyclodextrin on hybrid silica and its fast adsorption performance of p-nitrophenol from the aqueous phase. RSC Adv 5(103):84410–84422
Shvets O, Belyakova L (2015) Synthesis, characterization and sorption properties of silica modified with some derivatives of beta-cyclodextrin. J Hazard Mater 283:643–656
Simsek EB, Duranoglu D, Beker U (2012) Heavy metal adsorption by magnetic hybrid-sorbent: an experimental and theoretical approach. Sep Sci Technol 47:1334–1340
Sinha A, Basiruddin SK, Chakraborty A, Jana NR (2015) Beta-cyclodextrin functionalized magnetic mesoporous silica colloid for cholesterol separation. ACS Appl Mater Interfaces 7(2):1340–1347
Sohmiya M, Saito K, Ogawa M (2015) Host-guest chemistry of mesoporous silicas: precise design location, density and orientation of molecular guests in mesopores. Sci Technol Adv Mater 16:054201. https://doi.org/10.1088/1468-6996/16/5/054201
Szejtli J (1982) Cyclodextrins and their inclusion complexes. Akadémiai Kiadó, Budapest
Szejtli J (1988) Cyclodextrin technology. Kluwer Academic Publishers, Dordrecht
Szejtli J (1998) Introduction and general overview of cyclodextrin chemistry. Chem Rev 98:1743–1754
Trofymchuk IM, Roik N, Belyakova L (2016) Sol-gel synthesis of ordered beta-cyclodextrin-containing silicas. Nanoscale Res Lett 11:1–12
Valtchev V, Mintova S, Tsapatsis M (2009) Ordered porous solids – recent advances and prospects. Elsevier, Amsterdam, p 800
Vartuli JC, Roth WJ, Degnan TF (2008) Mesoporous materials (M41S): from discovery to application. In: Scharwz JA, Contescu CI, Putyera K (eds) Dekker encyclopedia of nanoscience and nanotechnology. Taylor and Francis, New-York, pp 1797–1811
Volesky B (1990) Biosorption of metals. CRC Press, Boca Raton
Volesky B, Holan ZR (1995) Biosorption of heavy metals. Biotechnol Prog 11:235–250
Vunain E, Mishra AK, Mamba BB (2016) Dendrimers, mesoporous silicas and chitosan-based nanosorbents for the removal of heavy-metal ions: a review. Int J Biol Macromol 86:570–586
Walcarius A (2001) Electrochemical applications of silica-based organic-inorganic hybrid nanocomposites. Mater Chem 13:3351–3372
Walcarius A, Mercier L (2010) Mesoporous organosilica adsorbents: nanoengineered materials for removal of organic and inorganic pollutants. J Mater Chem 20:4478–4511
Wang ML, Liu P, Wang Y, Zhou DM, Ma C, Zhang DJ, Zhan JH (2015) Core-shell superparamagnetic Fe_3O_4@beta-CD composites for host-guest adsorption of polychlorinated biphenyls (PCBs). J Colloid Interface Sci 447:1–7
Wase J, Forster C (1997) Biosorbents for metal ions. Taylor & Francis, Bristol
Wen J, Wilkes GL (1996) Organic/inorganic hybrid network materials by the sol-gel approach. Chem Mater 8:1667–1681
Willai S, Bacquet M, Morcellet M (2008) Organosilica mesoporous materials with double functionality: Amino groups and β-cyclodextrin synthesis and properties. In: Ganachaud F (ed) Silicon based polymers, Springer Science, Dordrecht, pp 213–221
Wilson LD, Mahmud ST (2015) The adsorption properties of surface-modified mesoporous silica materials with beta-cyclodextrin. Int J Technol 6(4):533–545
Yamamoto E, Kuroda K (2016) Colloidal mesoporous silica nanoparticles. Bull Chem Soc Jpn 89(5):501–539
Yanagisawa T, Shimizu T, Kuroda K, Kato C (1990a) Trimethylsilyl derivatives of alkyltrimethylammonium-kanemite complexes and their conversion to microporous SiO_2 materials. Bull Chem Soc Japan 63(5):1535–1537

Yanagisawa T, Shimizu T, Kuroda K (1990b) The preparation of alkyltrimethylammonium-kanemite complexes and their conversion to microporous materials. Bull Chem Soc Jpn 63(4):988–992

Yang TR (2003) Adsorbents: fundamentals and applications. Wiley-Interscience, Hoboken

Ying JY, Mehnert CP, Wong MS (1999) Synthesis and applications of supramolecular-templated mesoporous materials. Angew Chem Int Ed 38(1–2):56–77

Zhao XS, Lu GQ, Millar GJ (1996) Advances in mesoporous molecular sieve MCM-41. Ind Eng Chem Res 35:2075–2090

Chapter 7
Supramolecular Liquid Crystals Based on Cyclodextrins

Pier-Luc Champagne, Rajesh Kumar, and Chang-Chun Ling

Contents

Abstract Cyclodextrin-based materials represent an environmentally friendly alternative to toxic liquid crystalline materials. Cyclodextrins are well-known for their cavity and inclusion properties. They are scaffold molecules that can be chemically modified to obtain functional materials for various applications. For instance, amphiphilic cyclodextrins have attracted tremendous interests from researchers of different fields because of their ability to self-assemble and to encapsulate medicines. They can also be designed to form supramolecular liquid crystals. Since the first report of a class of 6-alkylthiolated β-cyclodextrin derivatives that exhibit ther-

P.-L. Champagne · R. Kumar · C.-C. Ling (✉)
Alberta Glycomics Centre, Department of Chemistry, University of Calgary, Calgary, Alberta, Canada
e-mail: ccling@ucalgary.ca

S. Fourmentin et al. (eds.), *Cyclodextrin Applications in Medicine, Food, Environment and Liquid Crystals*, Environmental Chemistry for a Sustainable World, https://doi.org/10.1007/978-3-319-76162-6_7

motropic liquid crystalline properties in 1993, there has been only few developments in this area. But there is an increasing interest to develop cyclodextrin-based liquid crystalline materials, owing to their potential utilities in different areas.

In this chapter, we review cyclodextrins-based molecular designs, their synthesis, as well as characterization of their thermotropic and lyotropic liquid crystalline properties. The presence of numerous hydroxyl groups and a face-to-face pseudo-symmetry in native cyclodextrins create numerous opportunities for the design of smart materials. It has been shown that not only the nature of the substituent, but also its location highly influences the self-assembly behavior of the cyclodextrin derivatives. After an introduction on cyclodextrins and liquid crystals, we summarize various approaches used to chemically modify cyclodextrins for the development of thermotropic and lyotropic liquid crystals, such as generating amphiphilic derivatives, or appending mesogenic groups to both monomeric and polymeric cyclodextrin backbones. The last section presents examples of applications of cyclodexrin-based liquid crystals for bio-sensing and liquid crystal displays.

7.1 Introduction

7.1.1 Cyclodextrins Structures

Cyclodextrins (CDs) are polyhydroxylated macrocycles based on D-glucose. The three most commonly encountered members are α-, β- and γ-CDs which contain 6, 7 and 8 D-glucopyranosyl units, respectively giving birth to a characteristic truncated cone-shaped cavity of different sizes (Fig. 7.1).

The inner cavity volume of these cyclodextrins vary between 0.17–0.42 nm^3 and their diameters from 0.49 to 0.79 nm. The cavities of cyclodextrins are relatively hydrophobic compared to outer surface, which provides them with ability to form inclusion complexes with organic guest molecules that are characterized with improved water-solubility, air-stability, reduced tissue-irritability, and other benefits. The electrochemical potential map of β-cyclodextrin shown in Fig. 7.2, greatly

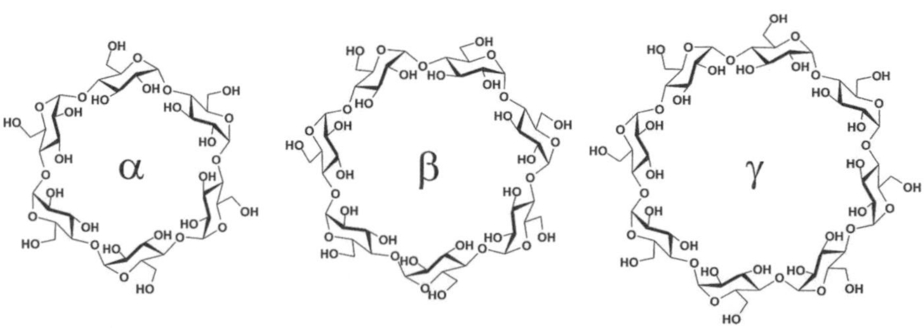

Fig. 7.1 Graphical representation of the three most commonly encountered cyclodextrins; α, β, and γ cyclodextrins

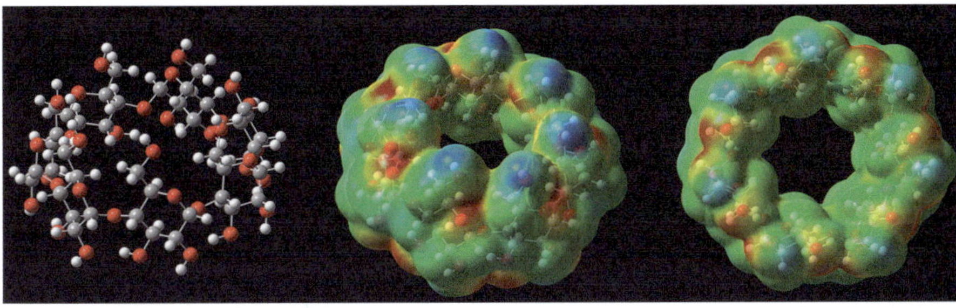

Fig. 7.2 Electrochemical potential map of the truncated cone-shaped structure of β- cyclodextrin

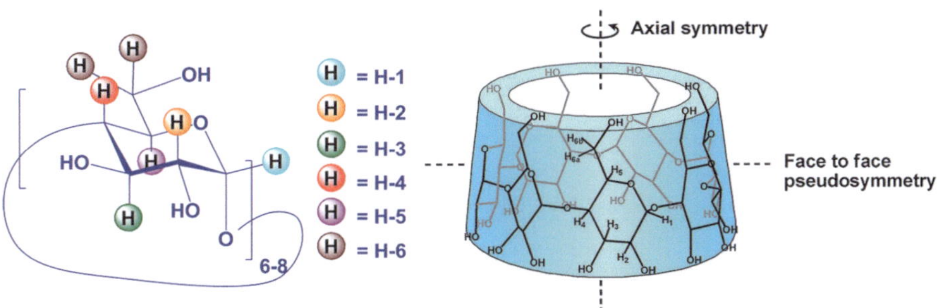

Fig. 7.3 Commonly used nomenclature for hydrogens in native cyclodextrins as well as their axial symmetry and face to face pseudosymmetries

represents the hydrophobic character of cyclodextrins cavity. The red, blue and green colors illustrate the negative, positive, and neutral electrochemical regions, respectively. Such representation emphasizes the contrast between the hydrophobic inner cavity and hydrophilic outer-surface of cyclodextrins.

For these reasons, cyclodextrins have attracted considerable attention from pharmaceutical and other industries (Hirayama and Uekama 1999; Rasheed 2008). Structurally, the D-glucose units are linked together via α(1,4)-glycosidic linkages, which arranges all the primary hydroxyl groups (OH-6's) at the narrower face (primary face) of the cavity, and all the secondary hydroxyl groups (OH-2's and OH-3's) at the wider face (secondary face) (Fig. 7.3).

Consequently, all cyclodextrins possess an axial symmetry, as well as a face-to-face pseudo symmetry. Although cyclodextrins improve aqueous solubility of hydrophobic molecules, they are unfortunately restrained by their own solubility in aqueous media. α, β, and γ- cyclodextrins have a water solubility of 14.5%, 1.85%, and 23.2% (w/v) respectively at room temperature (Davis and Brewster 2004). Their lower solubility is partly attributed to the intramolecular hydrogen bond network which lessens their ability to form H-bonding with water molecules (Rasheed 2008). Enhanced solubility can be achieved by reducing the strength of this network. For instance, partial methylations of cyclodextrins have shown to increase water-solubility tremendously (Rasheed 2008). Because of their high polarity,

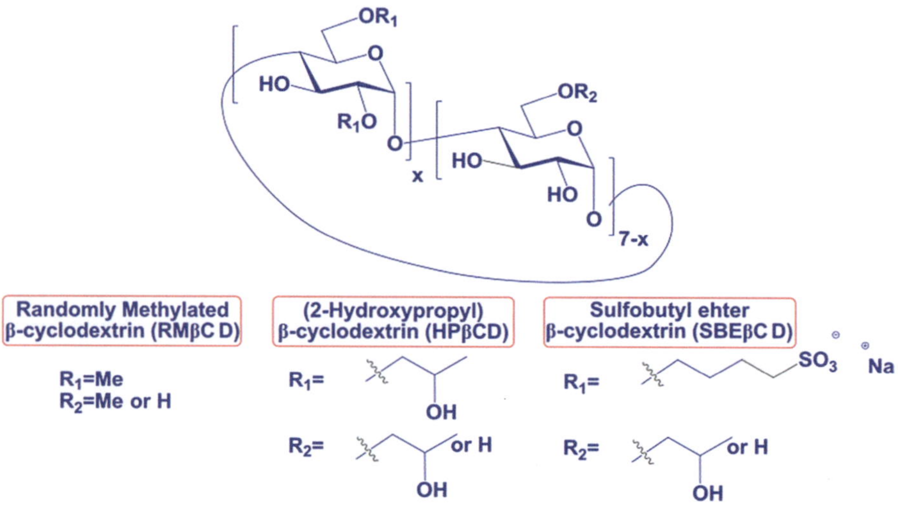

Fig. 7.4 Structures of commercially available CD derivatives; randomly methyl-β-cyclodextrin (RMβ-CD), 2-hydroxypropyl β-CD (HPβCD), and sulfobutyl ether (SBEβ-CD)

native cyclodextrins also have very limited solubility in most organic solvents. To expand their applications in pharmaceutical industries, various chemical modifications have been developed. The three most widely used and commercially viable strategies are to partially alkylate the β-cyclodextrin scaffold with methyl, 2-hydroxypropyl, or sulfobutyl groups, to afford respectively RMβ-CD, HPβ-CD and SBEβ-CD (Fig. 7.4). Unfortunately, due to coexistence of multiple types of hydroxyl groups and limited differences in their reactivity, all chemically modified cyclodextrin derivatives available on the market are sold as mixtures that contain different degrees of substitutions (Rasheed 2008).

7.1.2 Cyclodextrin Modifications

At first sight, since the hydroxyl groups are the only available functional group present at the two rims of the cavity, one might believe cyclodextrin chemical modifications to be simple but they are in fact very challenging, due to their presence in large number. Three types of hydroxyl groups can be identified within a native cyclodextrin molecule: those respectively attached to 2, 3, and 6-positions of glucopyranosyl units (Szejtli and Huber 2012). In the case of β-cyclodextrin, each type of hydroxyl groups consists of 7 identical copies. Due to the circular geometry, the hydroxyl groups of the seven glucose moieties form an intramolecular H-bonding network, resulting in a relatively rigid structure. At the primary face, the primary hydroxyl groups attached to the 6 positions have an additional degree of rotational freedom, making the hydrogen bond network weaker. More importantly, they experience lower steric hindrance, resulting in higher reactivity (Khan et al. 1998). On the other hand, all secondary hydroxyl groups located at the 2- and 3-positions experience reduced rotational freedom allowing the formation of a stronger hydrogen bond network. Furthermore, the increase in steric hindrance contributes to their lower reactivity.

X-ray experiments revealed that the OH-3 groups act as H-bond donors while the OH-2 as H-bond acceptors; this makes OH-2's the most acidic (Saenger et al. 1976). Since the protons of OH-3 groups are involved in intramolecular hydrogen bonding, they are generally less reactive, thus most challenging to modify. Consequently, by controlling the reactivity of the chemical reagent used, one can achieve regioselective modification of cyclodextrins. For instance, the OH-6's can be selectively protected with bulky groups such as *tert*-butyldimethylsilyl, while less bulky group such as trimethysilyl can be introduced to all 3 positions (Cramer et al. 1969; Takeo et al. 1988). The most acidic OH-2 groups can be deprotonated first to allow selective OH-2 monoalkylation. However, the proton exchange between the OH-2 and OH-6 positions often reduces the regioselectivity which produces mixtures of O2- and O6-substituted derivatives (Khan et al. 1998).

Since cyclodextrins possess the capability to form complexes, it is possible in few cases to achieve regioselective modifications by taking advantage of their inclusion properties (Ueno and Breslow 1982). This process is greatly influenced by the solvent effect and the size of the substituted cyclodextrin cavity. For instance, the 6-tosylated-α-cyclodextrin product is obtained in pyridine, while the reaction in aqueous media yields the mono-2-tosylate α-cyclodextrin. On the other hand, if β-cyclodextrin is subjected to a monotosylation in water, the 6-tosylate is obtained (Fujita et al. 1984).

In general, since the OH-6's in cyclodextrins are the most nucleophilic and least sterically hindered, per-substitutions at the primary face are less troublesome. On the other hand, since all primary hydroxyl groups have identical chemical reactivities, partial substitutions such as mono-, di-, and tri-substitutions are usually much more challenging, as they usually lead to the formation of many regioisomers (also referred to as positional isomers) which are extremely difficult to separate. Consequently, partial substitutions usually result in low yields of the desired product.

Persubsitutions at the secondary face are especially difficult because of the inherited lower chemical reactivity of secondary hydroxyl groups. Moreover, there are twice the amounts of –OH groups at the secondary face compared to primary face; as the reaction progresses, the reactive sites become increasingly more hindered, making the approach of electrophiles more challenging.

7.1.3 Synthesis of Amphiphilic Cyclodextrins

The face-to-face pseudosymmetry (Zhang et al. 1992) can be taken advantage of by chemical modifications, to introduce hydrophobic groups to only one face of the molecule. This results in the formation of amphiphilic molecules. These amphiphilic cyclodextrins act like detergent in water, thus possess the ability to form self-assembled systems (Bonini et al. 2006). Depending on the size and geometry of the introduced hydrophobic groups, different nano-materials can be obtained through their self-assembly (Nagarajan 2002) such as spherical micelles, liposomes and other rod-like objects (Nagarajan 2002). Once assembled into such self-organized structures, the inherent cyclodextrin cavity and the hydrophobic regions found within the apolar chains can be used to host the hydrophobic guest molecules, creating efficient drug delivery systems for the treatment of disease. This has attracted considerable interest from pharmaceutical researchers (Rajewski and Stella 1996; Challa et al. 2005; Hoare and Kohane 2008). Such amphiphilic systems are advantageous compared to non-amphiphilic cyclodextrins which can only encapsulate the desired cargo into the hydrophobic cyclodextrin cavity, for enhanced distribution of drug to the disease model. By forming self-assembled nanostructures, a concentrating effect may be achieved, with many replicas of the drug trapped in one nanostructure. Upon administration, self-assembled nanoparticles can reach the targeted disease cells, and deliver replicas of drug molecules at the same time, thus augmenting the therapeutic potency and efficacy (Yallapu et al. 2010). Consequently, this could greatly improve drug safety while reducing toxicity, as lower dosage could be administered without diminishing the overall therapeutic effect of the medicine (Rajewski and Stella 1996).

Amphiphilic cyclodextrin complexes can be designed to control the average release rate of a drug. In fact, cyclodextrins have been modified to offer a desired drug-releasing mechanism to improve the oral bioavailability of steroids, cardiac glycosides, non-steroidal anti-inflammatory drugs, benzodiazepines etc. (Rasheed 2008). For instance, when a drug has to be delivered to the enteric region, a delayed release is required to prevent premature leaking of the cargo into the stomach, which presents a harsh acidic environment. In order to reach the upper small intestine, a cyclodextrin-based drug delivery system (DDS) bearing weak acidic groups at its surface was designed. Such formulations display reduced solubility in the stomach while increased solubility in neutral and alkaline regions, owing to ionization of the acidic groups. This pH dependent behavior allows a delayed release of the drug into the enteric region. As an example, 6-*O*-(carboxymethyl)-β-cyclodextrin derivatives have been developed as a drug delivery system for time-controlled release of oral medicine; the system exhibited restricted solubility in acidic conditions but improved solubility with increasing pH (Uekama 2004).

7.1.4 Liquid Crystals

Self-assembly of simple systems into well-ordered nano-structures is a fascinating topic. The understanding of the fundamentals of soft-matter self-organization has stimulated tremendous research interest throughout the last decade (Hoeben et al. 2005; Yagai and Kitamura 2008). Liquid crystal represents a state of the matter between solid and liquid phase. It displays the fluidity of liquids while possessing the molecular alignment of crystalline solid. Liquid crystalline phases are referred to as mesophases and the molecules responsible for such behavior, as mesogens. In order to obtain spontaneous self-assembly of specific material into highly-ordered molecular structures at nanoscales, molecular building blocks must be precisely engineered to display the desired characteristics. In an attempt to improve our understanding of such phenomena, some background information about the transitions from crystalline solid to isotropic liquid will be discussed.

Commonly, two main parameters are used to report molecular arrangement in different phases, the orientational order which describes the alignment of molecules to a director (n), and the positional order which refers to the center of mass of each molecule with respect to each other. A material in crystalline solid phase is composed of molecules having both positional and orientational order with periodically repeating patterns. Such nanoscale organization is possible because of non-covalent intermolecular forces. Within the crystalline phase, the sole movement of the molecules is the thermal vibration within the crystal lattice (Gray 1962). As the temperature increases, these vibrations become stronger and eventually overcome the intermolecular interactions. At the transition temperature (melting), the solid material enters an isotropic liquid phase. It can be rationalized that at such temperature, the long range positional order is destroyed, providing freedom to the molecules for random traveling.

On the other hand, some materials possess liquid crystalline phases, associated with transitions observed between the crystalline and isotropic liquid phases, characterized by the presence of either positional or orientational order or both. Higher degrees of orientational order give the mesophase enhanced anisotropic properties while lower degrees of positional order provide greater fluidity (McArdle 1990).

Liquid crystals are classified into two main families: thermotropic and lyotropic. Thermotropic liquid crystals are formed from pure products which self-organize into mesophases upon temperature variation. In contrast, lyotropic liquid crystals are formed by the addition of solvent to an amphiphilic molecule, resulting in long range positional orders due to the self-organization properties of the molecule.

Throughout this chapter several examples of cyclodextrin-based liquid crystal designs will be discussed. First, cyclodextrin-based thermotropic liquid crystals will be presented and rationalized based on the interactions and substituents responsible for their self-assembly. In the second section, lyotropic liquid crystalline systems containing cyclodextrins will be discussed, as well as some of their potential applications.

7.1.5 Thermotropic Liquid Crystals

Thermotropic liquid crystals are materials whose self-assembly behavior is influenced by temperature variation in the solid states. They can exist in many different phases. The three most common phases are: nematic (Fig. 7.5), smectic (Fig. 7.6) and discotic (Fig. 7.7) liquid crystalline phases.

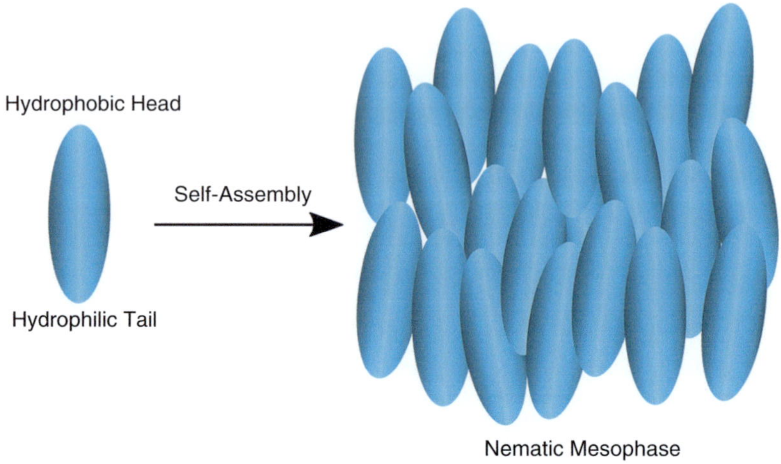

Fig. 7.5 Pictogram representation of molecular arrangement in the nematic liquid crystalline mesophases

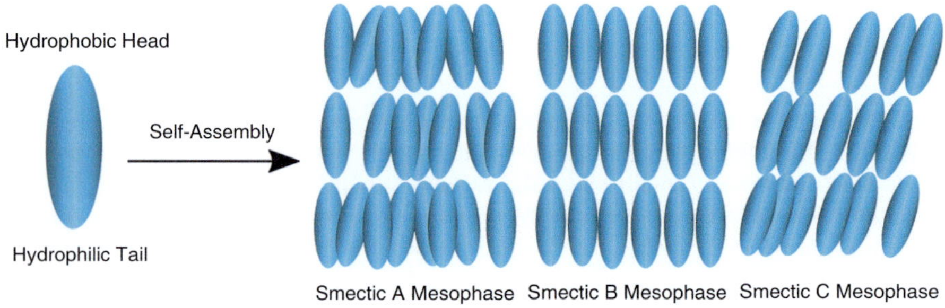

Fig. 7.6 Pictogram representations of molecular arrangement in the smectic A, B, and C liquid crystalline mesophases

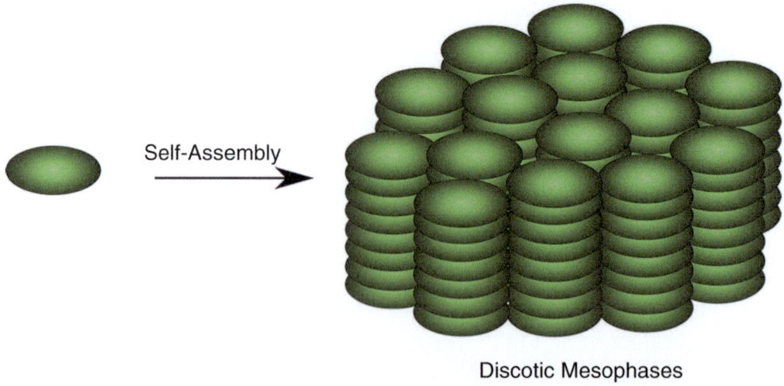

Fig. 7.7 Pictogram representations of molecular arrangement in the discotic liquid crystalline mesophases

7.1.5.1 Nematic Phase

The nematic phase (N) is the thermotropic liquid crystalline phase with the least order. This phase is characterized by self-organization of molecules into ordered structures in which the long axis of all molecules (director **n**) aligns with each other, procuring orientational order to the material. However, no positional order is found in nematic liquid crystalline phase since the center of mass of the molecules is isotropically distributed (Linlin et al. 2015).

7.1.5.2 Smectic Phases

Smectic mesophases are often observed when decreasing the temperature of nematic phase. They possess some degree of positional order in addition to the orientational order. The center of mass of the molecules aligns more or less with each other, forming layered structures (Collings and Hird 1997). Within the same layer, the molecules can move around, gain some degree of freedom but overall all molecules maintain a degree of translational order. Several types of smectic phases are known depending on the type and degree of order (Fig. 7.6). The least ordered smectic mesophase is called Smectic A (Sm A). Sm A mesophases consist of layered structures in which the director **n** is parallel to the layer normal but possess no order within the layers. Other smectic mesophases, such as Smectic C, differ from Smectic A by the tilted angle of their director **n**, with respect to the layer normal. Several other smectic mesophases are also known such as Smectic B, I, and F, which are generally characterized by the intra-layer short range positional order. They are also referred to as hexatic smectic phases. Within these phases, the center of mass of the molecules arranges in hexagonal pattern. The three different hexatic smectic phases (Sm B, Sm I, Sm F) differ in their respective director **n** orientation. Sm B director **n** aligns parallel to the layer normal, while in the case of Sm I, **n** is tilted towards the corner of the hexagon (apex), contrasting with Sm F whose director **n** is tilted towards the side of the hexagon.

7.1.5.3 Discotic Liquid Crystal

Discotic mesophases are usually formed by disk-shaped molecules which orient themselves in a layer-like fashion and packed into stacks (called discotic columns, Fig. 7.7). Similarly to smectic mesophases, the columns formed can possess random or ordered positions by arranging themselves into cubic or hexagonal arrays (Kumar 2010).

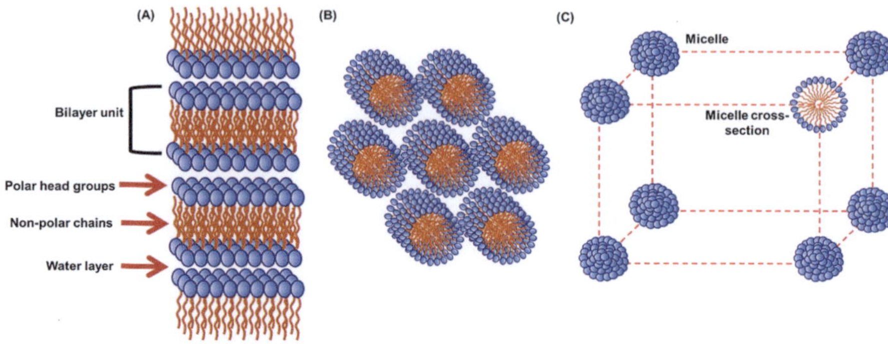

Fig. 7.8 Various types of lyotropic liquid crystals; (**A**) Lamellar, (**B**) Hexagonal, and (**C**) Cubic

7.1.6 Lyotropic Liquid Crystals

Lyotropic liquid crystals, consist of two-component systems mostly displayed by amphiphilic molecules in solvent producing ordered mesophases (Hiltrop 1994). Lyotropic liquid crystals morphologies are highly dependent on concentration and temperature of the amphiphilic systems bearing hydrophilic head and hydrophobic tail (Tolédano and Neto 1998). At critical micelle concentration (CMC) in aqueous solutions, amphiphilic molecules aggregate into micelles with polar head groups directed outward and hydrophobic tails inward. Upon increasing the concentrations, the micelles can fuse to each other forming various types of lyotropic liquid crystals such as cubic micelles, hexagonal phase, cubic phase and lamellar bilayer phases. The lamellar phase consists of amphiphilic molecules arranged in bilayer structures (Fig. 7.8A). The non-polar tails of two oppositely directed molecules are intertwined into the inner membrane and polar head groups found at the surface. The bilayer structures resulting from this self-assembly are separated by layers of water at the interface. Generally, lamellar lyotropic liquid crystal phases are obtained in solution of amphiphilic molecules concentrations higher than 50%.

In hexagonal phase, micelles are fused together; forming long ranged hexagonal cylinder arrays that exhibit birefringent textures (Fig. 7.8B). The distance between adjacent micellar cylinders is approximately 1 to 5 nm depending upon the concentration of amphiphilic and solvent molecules. Typically, hexagonal lyotropic liquid crystals are highly viscous with water content of 30 to 60% by weight. Cubic lyotropic liquid crystal phase, on the other hand, shows cubical arrangements of micelles that lead to viscous isotropic phases (Fig. 7.1C).

7.2 Cyclodextrin-Based Thermotropic Liquid Crystals

In the following section, several cyclodextrin-based thermotropic liquid crystals will be discussed. Based on their designs, they can be classified into three families: (a) cyclodextrin liquid crystalline systems that combine simple aliphatic chains and polar groups; (b) cyclodextrin liquid crystals that have incorporated mesogenic groups; (c) polymer-based cyclodextrin liquid crystals.

7.2.1 Thermotropic Liquid Crystals Based on Simple Amphiphilic Cyclodextrin Derivatives

7.2.1.1 H-Bond Mediated Mesophase Formation Based on Amphiphilic Cyclodextrin Derivatives

The early generations of amphiphilic cyclodextrin derivatives were prepared by introducing hydrophobic alkyl groups such as alkylamino, alkylthio, or alkylsulfonyl to the primary face of β-cyclodextrin. (Tanaka et al. 1987; Taneva et al. 1989). Unfortunately, these derivatives were a commonly prepared via the "per-6-tosylate of β-CD" through S_N2 substitutions. It was later found that the "per-6-tosylate of β-cyclodextrin" could only be isolated in pure form and in low yields by repeated high-performance liquid chromatography using reversed phase C18 column. This confirmed that the amphiphilic cyclodextrins prepared earlier were impure as they contained mixtures with different degrees of substitutions. (Ashton et al. 1991) Despite the heterogeneity in these materials, they were found to possess excellent properties in self-assembly such as forming mono-layer and Langmuir-Blodgett films at the water-air interface, micelles and liposomes in water etc. However, their thermotropic properties remained unknown until 1993 when Ling and Darcy (Ling et al. 1993) published the first series of cyclodextrin-amphiphiles that showed thermotropic liquid crystalline properties. Their successful synthesis began with the use of per-6-bromo-6-deoxy-β-cyclodextrin, which was prepared from native β-cyclodextrin **1** using Gadelle and Defaye's protocol (Gadelle et al. 1991) which employs Br_2/Ph_3P as a reagent in anhydrous dimethylformamide. This afforded the per-6-deoxy-6-bromo-β-cyclodextrin **2** in excellent regioselectivity and high purity. The subsequent per-substitution by alkylthiolate of different lengths allowed them to synthesize per-6-alkylthio-substituted amphiphilic β-cyclodextrin derivatives **3a-e** in excellent yields (>90%) and purities (Fig. 7.9).

The incorporated chain lengths varied from C_2-C_{18}, which provided them opportunities to comprehensively study the thermotropic properties of this family of amphiphilic cyclodextrins. Both cross-polarized optical microscopy (POM) and differential scanning calorimetry (DSC) were used to determine the phase transitions of the materials upon heating and cooling, and cross-polarized optical microscopy further provided information on the birefringence of the material during phase tran-

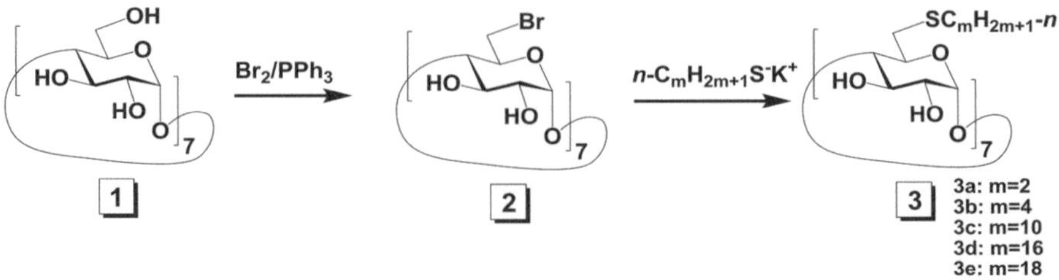

1 2 3

Fig. 7.9 Synthetic scheme of the first family of amphiphilic cyclodextrins that displayed mesogenic properties

sitions and formed textures. It was discovered that although all synthesized compounds showed some degrees of thermotropic liquid crystallinity, compounds **3d** and **3e** that contain chain lengths of 16 and 18 carbons respectively, displayed the best mesogenic properties over a wide temperature range (215 to 280 °C). However, upon heating further, both compounds decomposed before reaching their isotropic liquid phase. By comparing the textures of the two cyclodextrin compounds to monosaccharide mesogens bearing a single chain, the authors assigned the formed mesophases by compounds **3d** and **3e** to be smectic; thus, bilayer structures were proposed in the formed liquid crystalline phases. When analyzed by X-ray powder diffraction, derivative **3e** revealed a sharp peak in the low angle region which was attributed to a bilayer depth of 38 Å, less than twice the calculated molecular length. The authors suggested that the CD molecules may exist at some degrees of inclination in the layers. In the later work, Ling and co-workers (Ward et al. 2014) explained the results using an interdigitation model by the amphiphilic CD molecules in the bilayer (Fig. 7.10).

H-bonding was believed to be the major intermolecular force that drives the amphiphilic cyclodextrin molecules to self-assemble into highly ordered liquid crystalline mesophases. Because of their cyclic bidimensional geometry, amphiphilic cyclodextrin molecules can establish complex inter- and intra-molecular H-bond network in solid state. To elucidate the important role that H-bond played during the molecular self-assembling, Ling and co-workers (Ward et al. 2014) prepared a series of per-6-substituted β-cyclodextrin derivatives (**4–9**), all modified with the *n*-octadecylthio group at the 6-positions (Fig. 7.11).

Compounds **4** and **5** were per-substituted at all O2 and O3 positions with methyl and benzyl groups respectively, thus they contained no hydroxyl groups, therefore were incapable of inducing molecular self-assembly via H-bonding in solid state. However, in principle, molecules of compound **5** could interact with each other via π-π interaction in solid state because of their phenyl groups. Compounds **6–9** are all amphiphilic as they were regioselectively modified at all O2 positions with alkyl groups, namely methyl **6**, ethyl **7**, allyl **8** or benzyl **9**, but all possess seven hydroxyl groups (OH-3's). Thus, all compounds **6–9** should be capable of intermolecular H-bonding in solid state. On the other hand, since the alkyl groups at O2 position increase in size from **6** to **9**, their physical dimensions could effectively determine

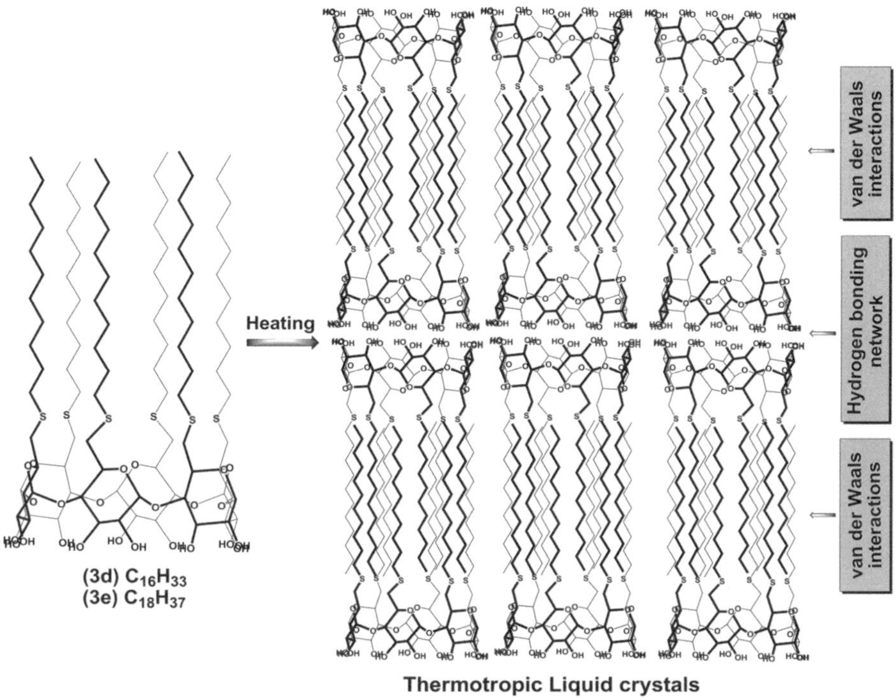

Fig. 7.10 Schematic structures of synthesized β-cyclodextrin derivatives exhibiting thermotropic liquid crystal properties by Ling et al.

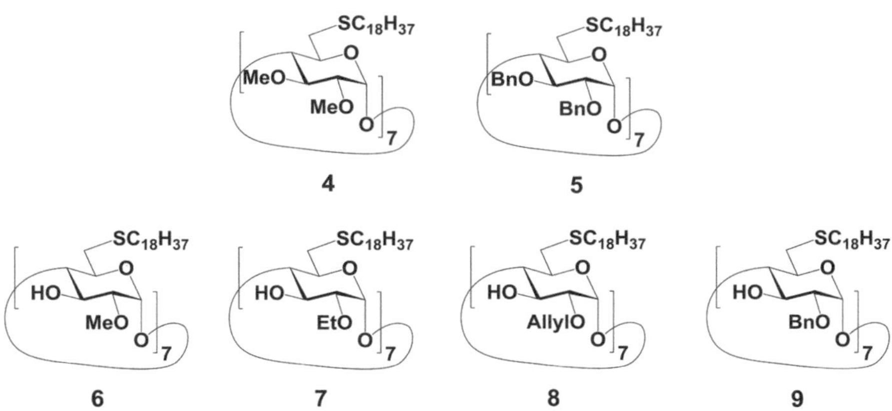

Fig. 7.11 Synthesized per-6-substituted β-cyclodextrin derivatives containing *n*-octadecylthio groups

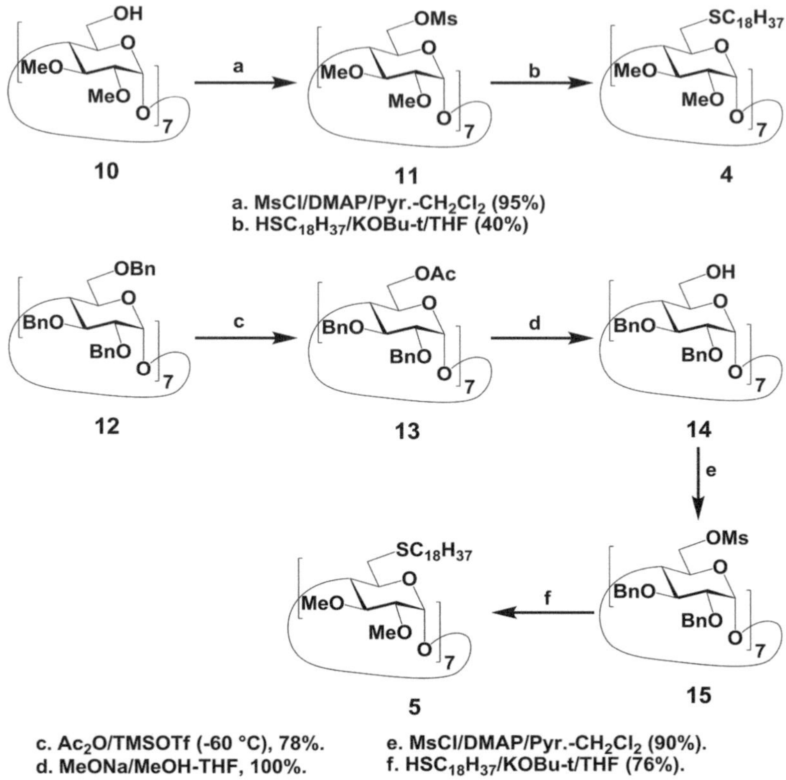

Fig. 7.12 Schematic synthesis of per-6-octadecylthiolated β-cyclodextrin derivatives **4** and **5**

how close the molecules in the bilayers could get to each other in solid state. Logically, the ability to engage intermolecular H-bonding network in solid state should decrease from compounds **6** to **9**.

Compound **4** was synthesized by first mesylating the per-2,3-di-*O*-methylated β-cyclodextrin **10** in a mixture of anhydrous pyridine-dichloromethane to obtain desired per-6-mesylate **11** in 95% yield. Subsequent nucleophilic substitutions of the 6-mesylates with potassium *n*-octadecylalkylthiolate, generated in situ by mixing *n*-octadecylalkylthiol with potassium tert-butoxide (KOBu-t) in THF afforded the desired β-cyclodextrin derivative **4** in 40% yield. The synthesis of the analogous compound **5** started from the per-2,3,6-*O*-benzylated compound **12**, which was first subjected to an acetolysis at −60 °C in acetic anhydride using trimethylsilyl trifluoromethanesulfonate (TMSOTf) as the catalyst; this elegantly converted the less sterically hindered 6-*O*-benzyl groups to the corresponding 6-*O*-acetates in a highly regioselective manner to afford compound **13** (78%). Subsequent Zemplén *O*-transesterification in anhydrous methanol-tetrahydrofuran mixture using sodium methoxide as a catalyst, produced the heptol **14** quantitatively. Compound **14** was then activated in a similarly manner as compound **11** to afford the per-6-mesylate **15** (90%, yield), which was ultimately converted to target compound **5** (76% yield).

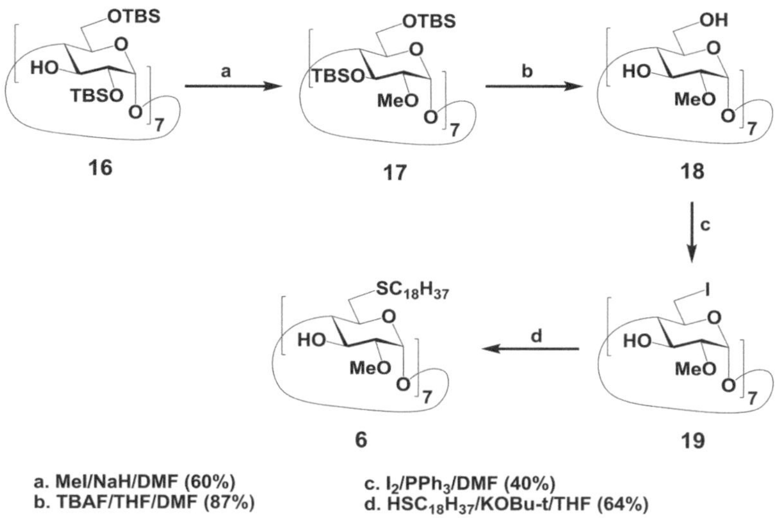

Fig. 7.13 Synthetic scheme of per-6-*n*-octadecylthio-2-*O*-methyl-β-cyclodextrin derivative **6**

Compound **6** was synthesized by reacting a per-2,6-di-*O*-silylated derivative of β-cyclodextrin **16** with methyl iodide in the presence of sodium hydride in anhydrous dimethylformamide (Fig. 7.12) (Fügedi 1989). After the deprotonation, the O2-silyl groups underwent a clean migration to O3-positions of β-cyclodextrin, which provided the opportunity for O2 to be methylated under Williamson etherification conditions. The corresponding compound **17** was isolated in 60% yield (Ashton et al. 1995). The protecting tert-butyldimethylsilyl groups were then removed using tetra-*n*-butylammonium fluoride (TBAF) to obtain polyhydroxylated β-cyclodextrin **18** (87% yield). A regioselective per-6-iodination was then performed using I_2/PPh_3 as a reagent to afford the corresponding per-6-iodide **19** in 40% yield. Finally, all the 6-iodides were substituted by *n*-octadecylthiolate to provide the desired β-cyclodextrin derivative **6** in 64% yield (Fig. 7.13).

The remaining three compounds **7**, **8** and **9** were synthesized according to Fig. 7.14 using compound **20** as a common starting material. First, a regioselective O2-alkylation was performed using NaH as a base, and either ethyl iodide, or allyl bromide or benzyl bromide as an electrophile. The corresponding intermediates **21**–**23** were obtained in 30–52% yields. Then the 6-*O*-*tert*-butyldimethylsiloxy groups were directly converted to the corresponding 6-bromide using homologuous brominating conditions as reported for **19** (bromine/triphenylphosphine in anhydrous dichloromethane); this afforded the corresponding per-6-bromides **24**–**26** in 46–94% yields. Final nucleophilic substitutions by *n*-octadecylthiolate afforded the desired targets **7**–**9** in 39–80% yields (Fig. 7.14). All synthesized β-cyclodextrin derivatives were characterized by 1D and 2D NMR experiments as well as MALDI-mass spectrometry.

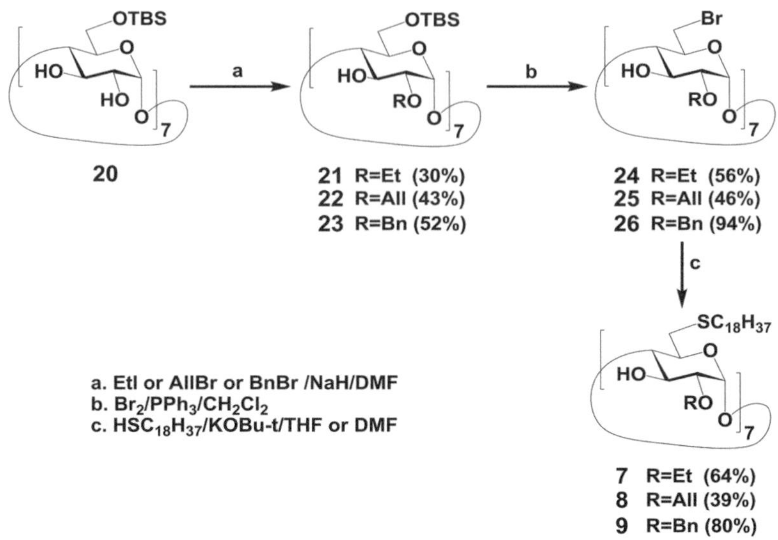

Fig. 7.14 Synthetic scheme of per-6-octadecylthio-2-*O*-alkylated β-cyclodextrin derivatives 7-9

Mesomorphic Properties The mesomorphic properties of β-cyclodextrin derivatives **4–5** were first investigated using differential scanning microscopy to reveal the phase-transitions as well as their associated enthalpic changes. It was found that both compounds **4–5** did not display liquid crystallinity, since only a single phase-transition was observed upon heating, which was attributed to the crystalline-isotropic liquid transition. Thus, clearly, the possible π-π interactions of benzene units present in **5** were not enough to induce long-range ordering in solid states. Interestingly, the two amphiphilic compounds **8–9** with larger groups (allyl for **8** and benzyl for **9**) at O2-positons also displayed a single phase transition, as observed by differential scanning calorimetry. This observation was further confirmed by cross-polarized optical microscopy as no birefringence was detected. Thus, both compounds **8** and **9** were found to be incapable of self-assembling into liquid crystalline mesophases, despite the presence of 7-hydroxyl groups. Probably, intermolecular H-bonding network is too weak to induce long-range orders. On the other hand, the two amphiphilic compounds **6** and **7** with smaller alkyl groups (methyl **6** and ethyl **7**) at O2-positions displayed thermotropic liquid crystalline behavior, with the compound **6** being the most mesogenic. From differential scanning calorimetry, compound **6** revealed five endothermic phase transitions upon heating at 50, 60, 64.5, 68, and 110 °C, and four exothermic transitions upon cooling at 95, 56, 40.5 and 37.5 °C (Fig. 7.15). From polarized optical microscope, the heating phase transition observed at 110 °C was associated to the transition to the isotropic liquid phase. However, after cooling to 95 °C, compound **6** displayed fluidity and the characteristic fan-shaped smectic A pattern became increasingly bright and colorful upon further cooling while the material became highly viscous.

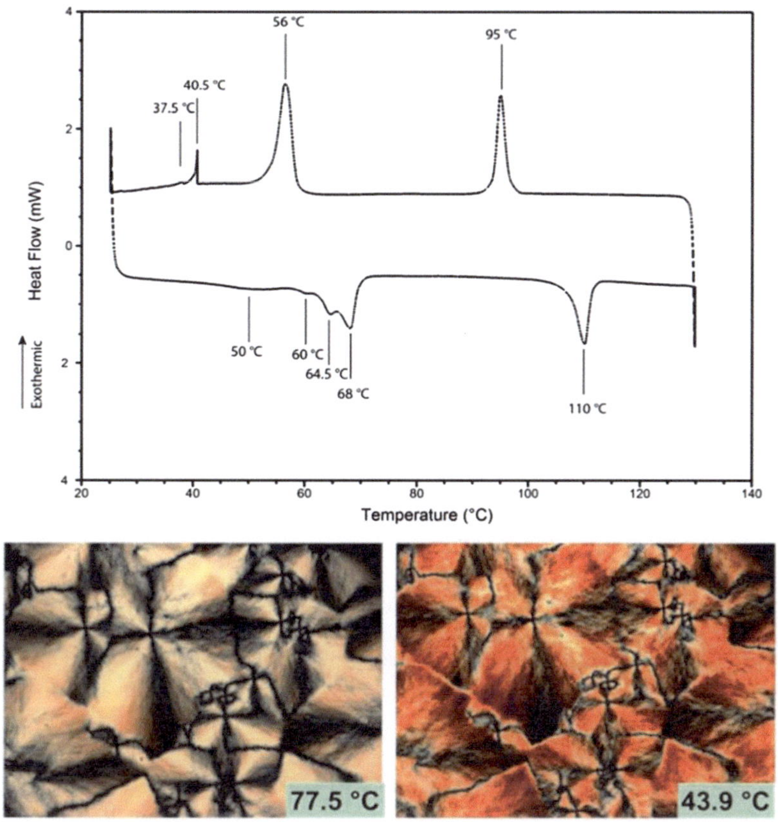

Fig. 7.15 DSC thermogram of compound **6** (Top) and its POM snapshots at 77.5 and 43.5 C (Bottom) "Reprinted with permission from (Ward et al. 2014). *DSC* differential scanning calorimetry, *POM* cross-polarized optical microscopy. Copyright 2014 Royal Chemical Society

On the other hand, compound **7** which had seven O2-ethyl groups - just one carbon longer than the O2-methyl groups in **6** and one carbon shorter than the O2-allyl group in **8**, displayed only two broad endothermic transitions at 54 and 61 °C upon heating, three broad and unresolved exothermic transitions at 43.5, 49 and 51.5 °C during the cooling cycle from the differential scanning calorimetry thermogram (Fig. 7.16). The heating transition observed at 61 °C was determined to be the clearing point of the material into isotropic liquid by cross-polarized optical microscopy. Upon cooling, the characteristic fan-shaped texture of smectic A gradually appeared, but the domains formed were generally much smaller than those of compound **6**, suggesting weakening ability for compound **7** to self-assemble.

Powder X-ray analysis was carried out for compound **6** which confirmed the presence of long range order in the smectic phase. For example, a bilayer thickness of 40.7 Å was determined; this value is significantly smaller than the calculated length (62.2 Å) of head-to-head dimer, suggesting significant interdigitation of the

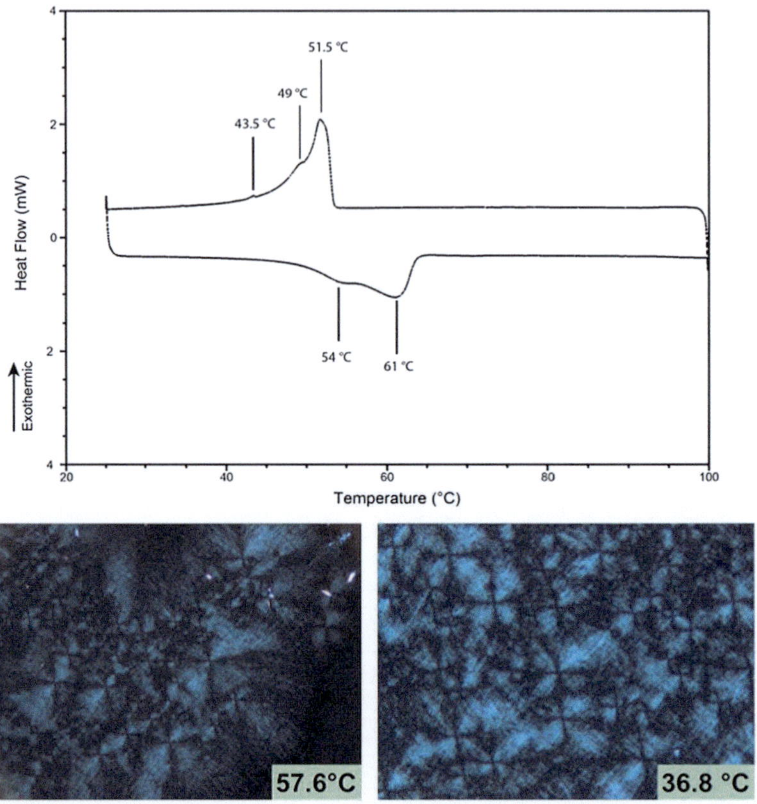

Fig. 7.16 DSC thermogram of β-CD **7** (Top) POM picture at 57.6 and 36.8 C (Bottom) "Reprinted with permission from (Ward et al. 2014). *DSC* differential scanning calorimetry, *POM* cross-polarized optical microscopy. Copyright 2014 Royal Chemical Society

hydrophobic chains in the highly ordered solid states. The presence of broad peaks in the wider angles region were also detected, which suggests the presence of short range orders within the layers at lower temperatures but no distinct phase could be identified. However, the short-range orders disappeared at higher temperatures, as those broad peaks vanished in the determined X-ray diffractograms.

The investigation by Ling and co-workers (Ward et al. 2014) provided great insights on the importance of H-bonding strength on the formation of liquid crystalline mesophases. As can be seen, the per-2-*O*-methyl-β-cyclodextrin **6** had only half number of secondary hydroxyl group compared to compound **3e**, thus intermolecular H-bond network established by compound **6** would be expected to be weaker. This substantially lowered the mesophases temperature (50–110 °C for compound **6** vs 215–280 for **3e**). The ability to form smectic mesophases quickly diminishes from compound **6** to **7** and was completely lost when the chains length attached to O2 reached 3 carbons or longer. A substantially narrower temperature range was observed for compound **7** (range: 54–61 °C) which reflects the transition.

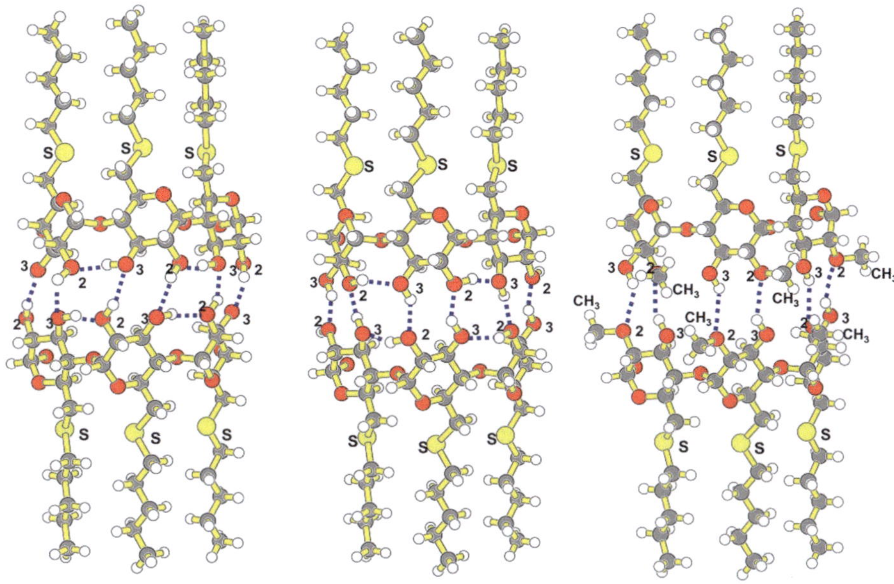

Fig. 7.17 Graphical representation of the two distinct H-bonding networks present in β-cyclodextrin liquid crystals

Previously, through *ab initio* and semiempirical quantum calculation, (Avakyan et al. 2001) proposed models of two types of inter and intramolecular H-bonding networks at the secondary faces of β-cyclodextrin dimers (Fig. 7.17). In type I model, OH-2 groups of one β-cyclodextrin act as H-bond acceptor to the neighboring OH-3 within the same molecule, while the same OH-2 groups simultaneously acts as a hydrogen bond donor to OH-3's of another cyclodextrin molecule. In another type II model, both OH-2's and OH-3's assume a reversed role within inter- and intramolecular H-bond networks. The computational analysis confirmed type I to be thermodynamically preferred as the torsion angle of OH-2 favors the formation of intermolecular H-bonding network. Based on this model, Ling and co-workers found that for both compound **6** and **7**, their per-substitutions at the O2 forbids the formation of the stronger type I network; thus, they concluded that within the mesophases formed by compounds **6** and **7**, type II network was established with OH-3 acting dynamically as H-bond donor through inter and intramolecular interactions within the bilayer. Such model can be effectively used to explain the difference on the clearing point temperatures between compounds **3d/3e** which did not display any clearing point, instead, they slowly decomposed at temperatures above 280 °C, confirming the very strong inter- and intramolecular H-bond networks present in such systems. Conversely, for compounds **6/7**: only weaker inter- and intramolecular H-bond networks could be established. This explains well their lower clearing points at 110 °C and 56 °C for β-cyclodextrin **6** and **7** respectively.

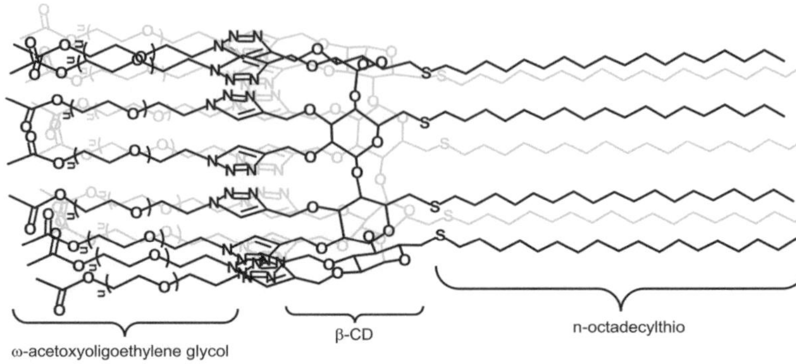

27 n=1
28 n=2
29 n=3

Fig. 7.18 Graphical representation of the stuctrure of cyclodextrin-based liquid crystalline materials **27–29** forming smectic mesophases

7.2.1.2 Dipole-Dipole Intermolecular Force Mediated Mesophase Formation Based on Oligoethylene Glycol Functionalized β-Cyclodextrin

For more than a decade, H-bond network was considered to be very important to generate cyclodextrin-based liquid crystals. This intermolecular force has been found to play a crucial role in most published designs of cyclodextrin-based thermotropic liquid crystals (Ling et al. 1993; Shaikh et al. 2007; Yang et al. 2013). Recently, Ling and co-workers (Champagne et al. 2016) reported the first β-cyclodextrin-based liquid crystals (Fig. 7.18, **27–29**) relying on using dipole-dipole interactions as the primary intermolecular force to induce self-assembly. The novelty of their design can be appreciated by the absence of any polar hydroxyl groups, but the presence of 14 *ω*-acetoxy-oligoethylene glycol residues of different lengths (2–4 repeating ethoxy groups) at the secondary face of β-cyclodextrin that are rich with polar C-O bonds. At the primary face, they kept the flexible, apolar *n*-octadecylthio functional groups. Thus, this class of compounds does show amphiphilic properties, but in solid states, due to lack of H-bond donors, the molecules are unable to interact with each other via H-bond network, thus the dipole-dipole interactions between the numerous C-O bonds of the oligoethylene glycol groups would be expected to play the most fundamental role to induce long-range orders in solid states.

Compounds **27**–**29** were synthesized from the per-6-*O*-*tert*-butyldimethylsilyl-β-cyclodextrin **20,** which was first peralkylated at all O2 and O3 positions with propargyl bromide using sodium hydride as base in anhydrous dimethylformamide. The obtained per-2,3-*O*-dipropargylated compound **30** was then subjected to a full *O*-desilylation under acidic conditions (HCl) in a methanol-dichloromethane mixture, to yield desired heptol **31**. After a per-6-mesylation with mesyl chloride in a mixture of anhydrous pyridine and dichloromethane at 0 °C, a highly versatile intermediate **32** was obtained. The alkynic

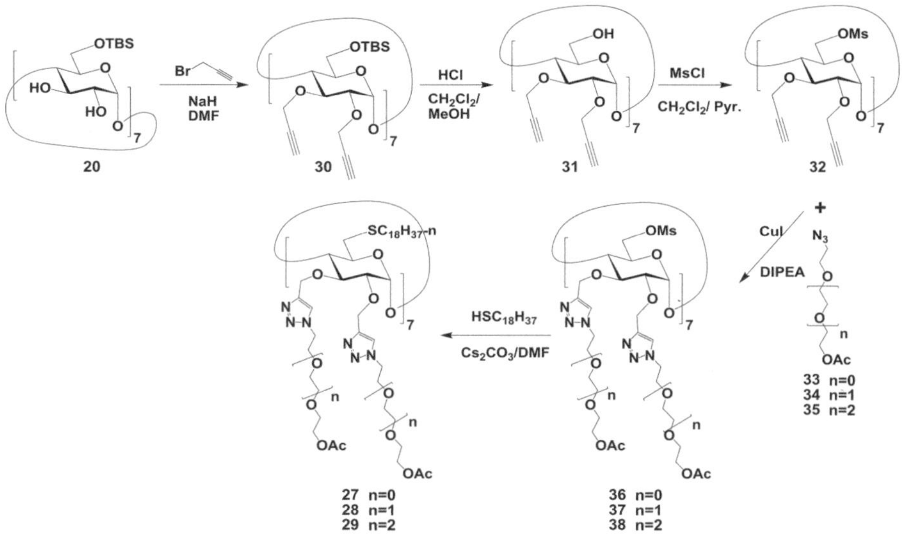

Fig. 7.19 Synthesis of CD-based liquid crystalline materials **27–29** containing polar ω-acetoxyoligoethylene glycol groups

functionalities were subjected to copper(I)-mediated 1,3-dipolar cycloaddition with 3 different ω-acetoxyoligoethylene glycol azides **33–35** of different lengths to obtain the corresponding conjugates **36–38.** The final step involved the nucleophilic substitution of all the 6-mesylates in compounds **36–38** by *n*-octadecylthiolate to afford the desired targets **27–29** (Fig. 7.19).

Thermogravimetric analysis revealed that all compounds **27–29** have excellent thermal stability (up to 283 °C, ca. 5% mass loss). The mesogenic behavior of compounds **27–29** was studied by cross-polarized optical microscopy, differential calorimetry, and powdered X-ray diffractometry. Remarkably, all three derivatives showed strong liquid crystalline properties over a wide range of temperatures. For example, under cross-polarized optical microscopy, compounds **27–29** were observed to clear into isotropic liquid phase at 234.3, 194.8, and 157.7 °C respectively (Fig. 7.20). The compound with the shortest oligoethylene glycol chains **27** cleared at the highest temperature, which was unexpected. Upon slow cooling, all three compounds gradually entered into liquid crystalline phases initially forming small bâtonnet domains that gradually grew and fused into larger bâtonnets which eventually fused together and formed fan-shaped textures. The bâtonnets are commonly observed textures at the transition from the isotropic phase to fluid smectic mesophases. Compared to the mesophases formed via stronger H-bonding network as the intermolecular forces (**3d-3d**, **6–7**), Champagne et al. reported much enhanced fluidity for mesophases formed by compounds **27–29**.

Differential scanning calorimetry thermoanalysis revealed a transition at 54.8, 45.9 and 39.3 °C for compounds **27–29**, respectively; they correspond to a relatively large enthalpy change (24.3 Jg^{-1} for **27**, 19.9 Jg^{-1}, for **28**, and 23.3 Jg^{-1} for **29**); the

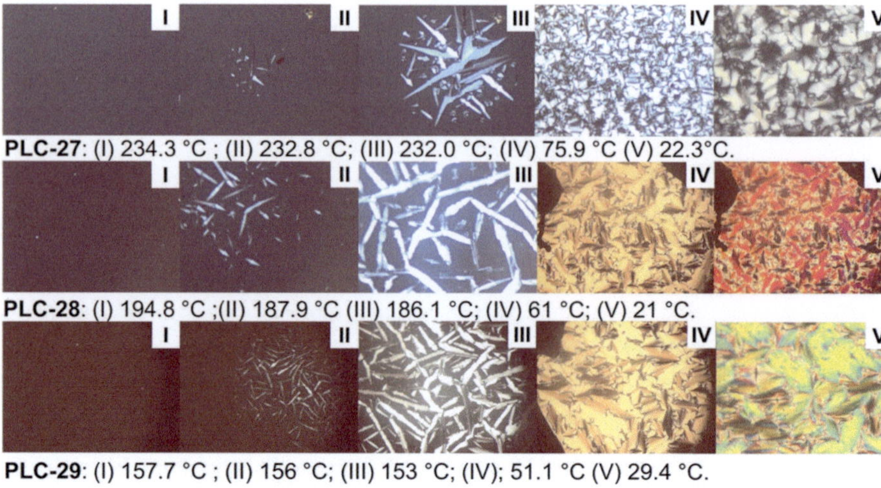

Fig. 7.20 Cross-polarized optical microscopy (POM) images images of β-cyclodextrin **27–29** recorded at various temperatures, highlighting their behavior from isotropic transition (I) to room temperature (V)

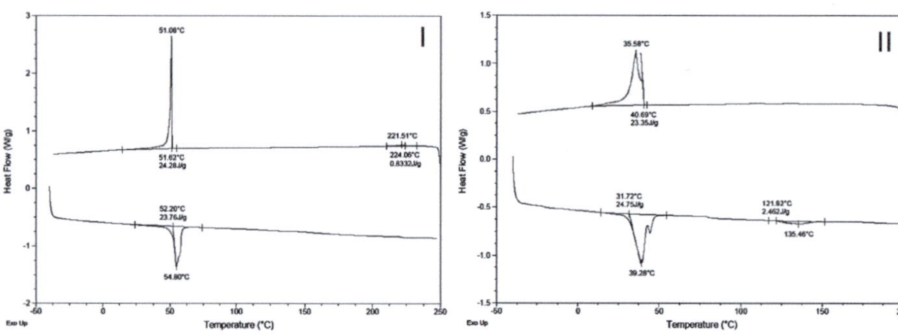

Fig. 7.21 Differential scanning calorimetry (DSC) thermograms recorded for β-cyclodextrin derivatives **27** (I) and **29** (II) as well as the enthalpies associated with the recorded transitions

authors attributed them to the liquid crystalline mesophases to crystalline solid transition for each compound. Thus, all three compounds **27–29** form mesophases over an extremely wide range (179.5 °C for **27**, 148.9 °C for **28**, and 118.4 °C **29**), which is remarkable, making this class of compounds unique. On the other hand, differential scanning calorimetry also revealed very small enthalpic variations (< 2.5 Jg^{-1}) for all three compounds **27–29** ascribed to the liquid crystalline-isotropic liquid transitions (Fig. 7.21). These enthalpic changes are much smaller when compared to the values associated with previously reported systems that relied on H-bond network (**3d-3e**, **6–7**). For instance, compound **6** has an enthalpy change of ~19–20 Jg^{-1} corresponding to its isotropic liquid-smectic liquid crystal phase transition. The higher fluidity, lower clearing temperature, and better thermostability are attributed to the absence of stronger H-bonding network. All these also contributed to the improved liquid crystalline properties of compounds **27–29**.

Solid state Molecular Packing Properties
Champagne et al. reported the powder X-ray diffractograms of **27–29** at 20 °C and 50 °C below their respective clearing point. Sharp peaks were observed in the low angle region for all three compounds, corresponding to layer spacing of 55–60 Å, confirming the formation of bilayer structures for all three compounds. This suggested that all three compounds formed smectic liquid crystalline mesophases. On the other hand, broad peaks were also observed in the higher angle region at higher temperatures, which was attributed to a lack of high intra-layer orders. However, when the X-ray diffractograms of compounds **27** and **29** were recorded at room temperature, the peaks in the higher angle region became significantly sharper, which suggested higher positional order within the layers.

Figure 7.22 shows molecular models of packing patterns for compounds **27–29**. All three compounds have a conical geometry with calculated length of the molecules varying between 43.6–51.4 Å. The X-ray diffraction data revealed a periodicity much shorter than twice the molecular length in each case, suggesting significant interdigitation in solid states for each compound. Based on the wedge-shaped geometry of each derivative, the authors believed the interdigitation occurs at the hydrophobic chain region (primary face).

7.2.2 Thermotropic Liquid Crystals Based on Cyclodextrin Derivatives Containing Conventional Mesogenic Groups

Mesogens consist of a group of organic compounds that strongly induce liquid crystallinity. These are often rigid molecules such as cholesterol, biphenyl, triphenylene that adopt the shape such as rods, disks etc. When chemically modified with flexible chains, their derivatives usually exhibit mesogenic behavior. Although cyclodextrin molecules represent a class of relatively rigid scaffolds, their mesogenic properties appear to be quite weak. For example, all compounds **4**, **5**, **8**, and **9** did not show any liquid crystalline properties, despite the incorporation of several long alphatic chains at the primary face of the molecules. However, several research groups have reported examples of cyclodextrin-based liquid crystals by incorporating commonly known mesogenic groups to a cyclodextrin scaffold, and the obtained compounds usually displayed mesophases other than smectic.

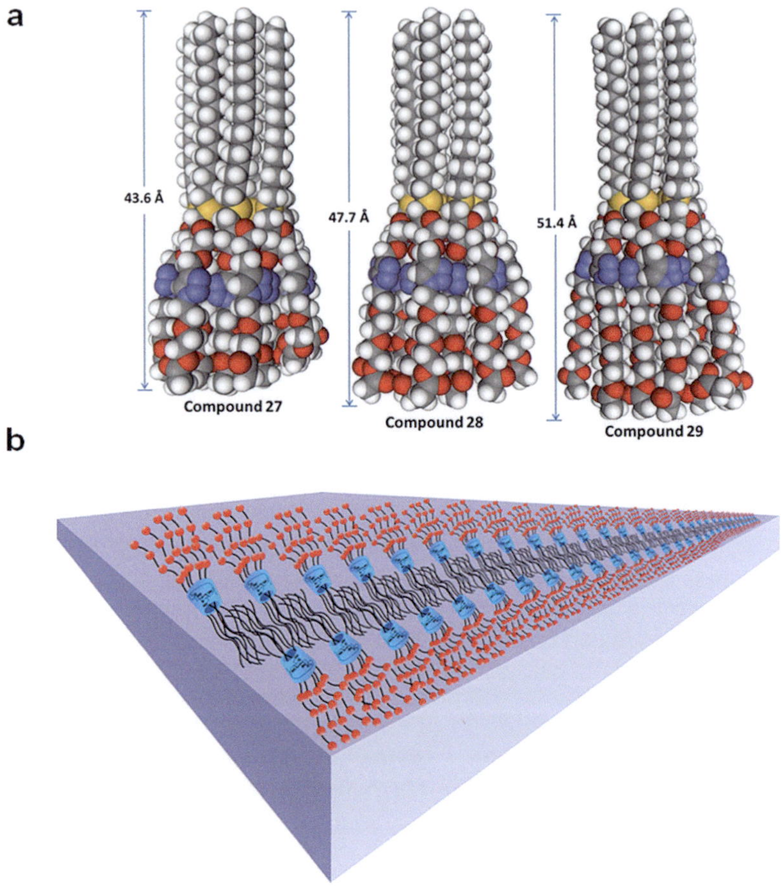

Fig. 7.22 (a) Corey-Pauling-Koltun models displaying the inherited wedge-shaped geometry of compounds **27–29,** with a length of 43.6, 47.7 and 51.4 Å respectively, in extended conformation. (b) Proposed model of smectic packing in solid states for all three compounds, with interdigitation at the hydrophobic chain regions

7.2.2.1 Amphiphilic Cyclodextrin Derivatives Containing Cholesterol Residues

Shaikh et al. (2007) reported a first example of amphiphilic β-cyclodextrin-based thermotropic liquid crystal in 2007 by covalently linking β-cyclodextrin to mono-cholesteryl succinate **39** (Fig. 7.23). The synthesis began with an esterification of cholesterol **40** with succinic anhydride in presence of pyridine; this afforded the corresponding cholesterol succinate **41** with a free carboxylic acid. The available carboxylic acid was then activated with thionyl chloride to obtain the corresponding succinyl chloride **42**, which was subsequently used to esterify β-cyclodextrin in anhydrous dimethylformamide at approximately 70 °C for 72 h. The β-cyclodextrin

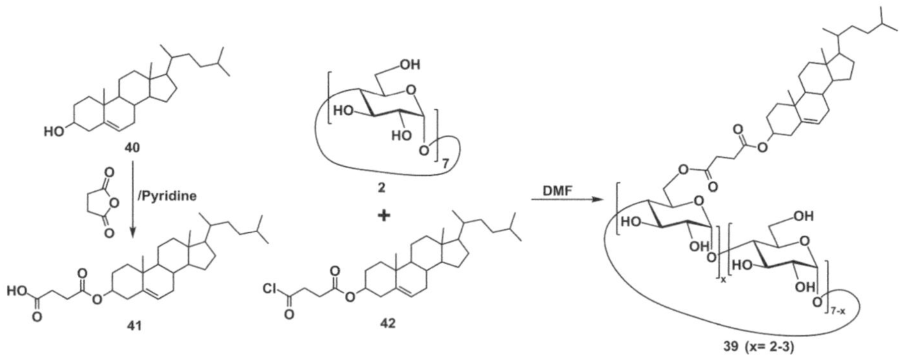

Fig. 7.23 Synthesis of cholesterol appended β-cyclodextrin liquid crystal derivatives **39**

derivative **39** was obtained, unfortunately as a mixture. The degrees of cholesteryl substitution were estimated to be ~2–3, and most substitutions probably occurred at the primary face of the β-cyclodextrin, due to the higher nucleophilicity of OH-6 groups.

The thermotropic mesomorphic behaviour of product mixture **39** was investigated by the authors using various techniques, including Fourier-transform infrared spectroscopy, NMR, differential scanning calorimetry, and cross-polarized optical microscopy. The mixture exhibited birefringence above 130 °C (first heating), and became isotropic at about 180 °C. The differential scanning calorimetry thermogram showed a transition at 120–125 °C associated to the glass-liquid transition. This was well compared to native β-cyclodextrin, which does not exhibit any liquid crystalline properties and was reported to be stable up to 299 °C without experiencing any significant mass loss (Song and Xu 2008; Song et al. 2008). The cross-polarized optical microscopy analysis revealed unclear textures which were not fully characterized. Nevertheless it was interesting to see the synthesized materials displayed birefringence, even with such a low degree of substitutions. Cholesterol is known to induce cholesteric mesophases, thus the liquid crystalline properties of compound **39** was clearly due to the incorporation of cholesterol residues to β-cyclodextrin, and this had significantly reduced the strength of both the inter- and intramolecular H-bonding network of the materials in solid states.

7.2.2.2 Amphiphilic Cyclodextrin Derivatives Containing 4-methoxybiphenyl Residues

Chen et al. (2010) reported another amphiphilic β-cyclodextrin derivative **43** (Fig. 7.24) that was modified with a hydrophobic tail derived from the propargyl 6-(4′-methoxybiphenyl-4-yloxy)hexanoate group **46** at all the primary positions. The efficient copper (I)-mediated 1,3-dipolar cycloaddition was used to complete the hydrophobic modification of β-cyclodextrin core via the per-6-azide intermediate **45** in dimethylformamide. Thus, in the final molecule **43**, Chen et al.

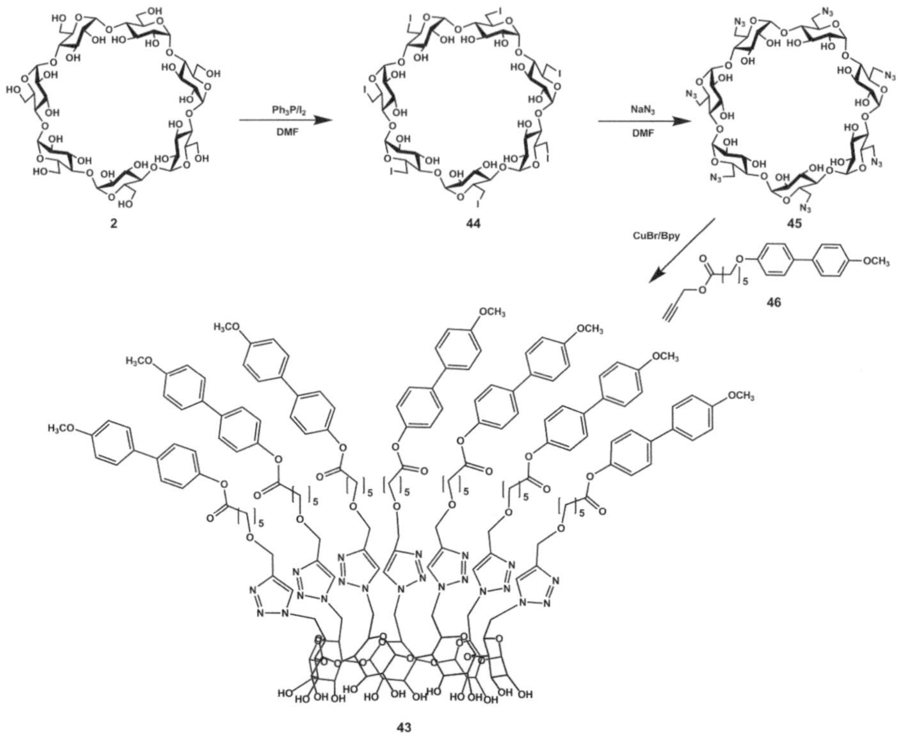

Fig. 7.24 Synthesis of cyclodextrin-biphenyl thermotropic liquid crystal **43** forming smectic mesophases

incorporated 7 copies of 4′-methoxybiphenyl group which is known to induce liquid crystallinity because of its rigidity and rode-like geometry. In solid states, the 4′-methoxybiphenyl groups can interact with each other via π-π interactions. The purity of the final compound **43** was verified by ^{1}H NMR spectroscopy, Fourier-transform infrared as well as MALDI-TOF mass spectrometry.

Thermogravimetric analysis revealed the amphiphilic compound was quite stable at high temperature, as only a loss of 5% mass was observed at 318.5 °C. Differential scanning calorimetry thermoanalysis showed several phase transitions. For example, during the second heating cycle, endothermic transitions at 130.1 °C, 188.7 °C and 217.2 °C were observed with the last one correlating to the clearing temperature to isotropic liquid. Using cross-polarized optical microscopy, the birefringent properties of compound **43** were investigated, which revealed very interesting patterns. For example, at 50 °C, fan-shaped textures were observed, which transition into conic focal patterns at 180 °C and subsequently into schlieren lines at 195 °C that persisted until the material completely cleared into isotropic phase at 230 °C. The ability to form various mesophases revealed the great self-assembly ability of compound **35**. This also differentiates it from other amphiphilic cyclodextrin derivatives that show liquid crystallinity.

Wide angle X-ray diffractometry was used to analyze the mesophases formed by compound **43**. At 30 °C, 3 peaks in the small-angle region were observed with q ratios of 1:2:3; this allowed the authors to assign the structure of formed mesophase to be lamellar. In the wide-angle region, 3 peaks were also observed with 1 very strong, 1 strong, and 1 weak. Based on the characteristic patterns, Chen et al. concluded that compound **43** formed highly ordered smectic E mesophase at low temperatures. At 160 °C, the material showed similar diffraction patterns at the low-angle region (virtually unchanged), suggesting that the layered smectic structure was maintained. However, peaks in the wide-angle region did become broader, suggesting a slow transition to smectic A mesophase (Koltzenburg et al. 1998). Upon increasing temperature further to 200 °C, wide angle X-ray diffractometry detected no peaks within the low angle region, while only a diffused halo appeared in the high-angle region, corresponding to a phase transition to nematic phase: losing the layered structures (Dong et al. 2004). The material eventually entered its isotropic liquid phase at ~230 °C. Figure 7.19 depicts the proposed self-assembled supramolecular packing model of compound **43**. As can be seen, the strongest intermolecular force that dictates the molecules to self-assemble was the H-bonding network occurring at the secondary face of β-cyclodextrin, resulting in dimerization. Such network persists from room temperature to the high clearing point into isotropic liquid state thus is responsible for the formation of lamellar structures in solid states. The lamellar organization of molecules is further promoted by the 4′-methoxybiphenyl units which interact with one another via intermolecular π-π-interactions. Chen et al. proposed that 3 or 4 of the biphenyl mesogenic units were oriented in opposite directions with respect to the primary face of the β-cyclodextrin; this adds a second level of orders that are responsible for formation of smectic E mesophases. The π-π-interactions between the 4′-methoxybiphenyl units became significantly disrupted at higher temperatures (during the smectic A and nematic phases). The authors confirmed such mode of organization through molecular modelling. For example, the calculated *d* spacing between opposing 4′-methoxybiphenyl units at the primary face of β-cyclodextrin is 38.6 Å, which matches almost perfectly the determined *d* spacing of 38.4 Å by wide angle X-ray diffractometry at room temperature (Fig. 7.25).

7.2.2.3 Cyclodextrin Derivatives Containing a Triphenylene Mesogenic Group

In 2013, another β-cyclodextrin-based thermotropic liquid crystal was reported (Yang et al. 2013). The authors conjugated a mesogenic triphenylene group to a β-cyclodextrin scaffold (Fig. 7.26). Triphenylene derivatives represent one of the most widely studied discotic mesogens due to their potential applications in many areas including light-emitting diodes, organic photovoltaic cells, organic field-effect transistors, gas sensors, and photocopying machines (Zhang et al. 1992; Kumar 2004; Kato et al. 2006; Laschat et al. 2007; Sergeyev et al. 2007; Cammidge and Gopee 2009). The synthesis began with a monotosylation of β-cyclodextrin under

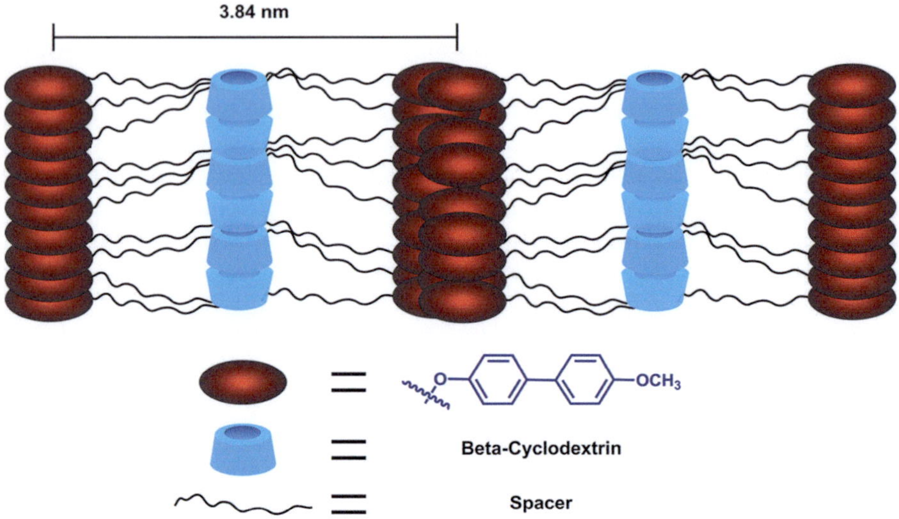

Fig. 7.25 Suggested lamellar organization of compound **43** in the smectic E mesophases

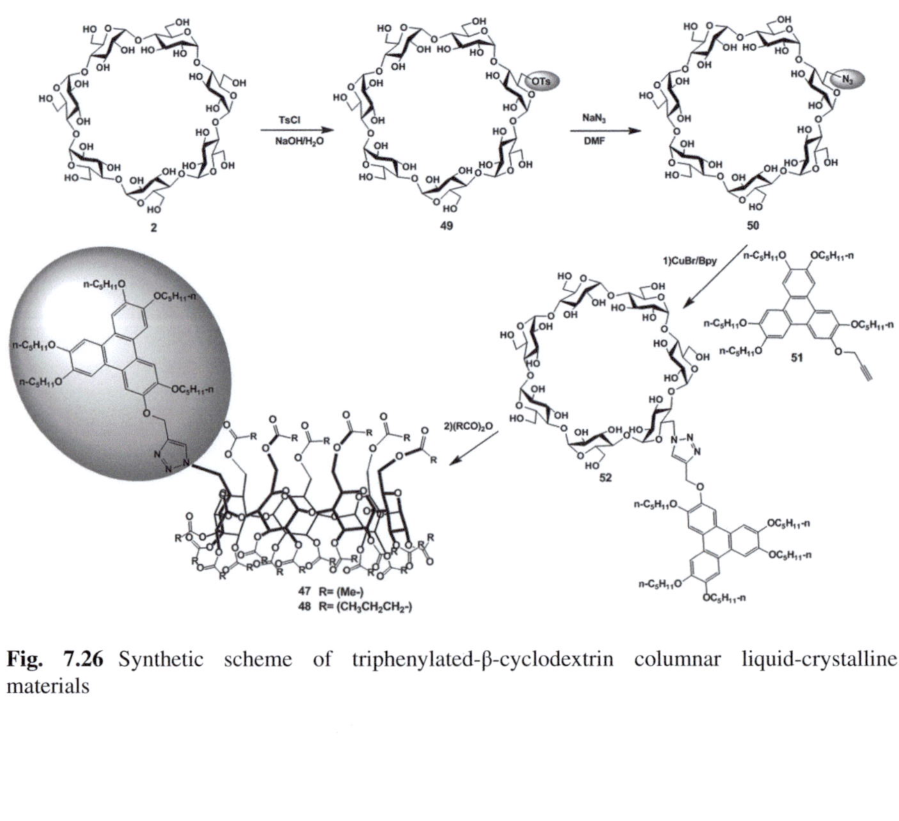

Fig. 7.26 Synthetic scheme of triphenylated-β-cyclodextrin columnar liquid-crystalline materials

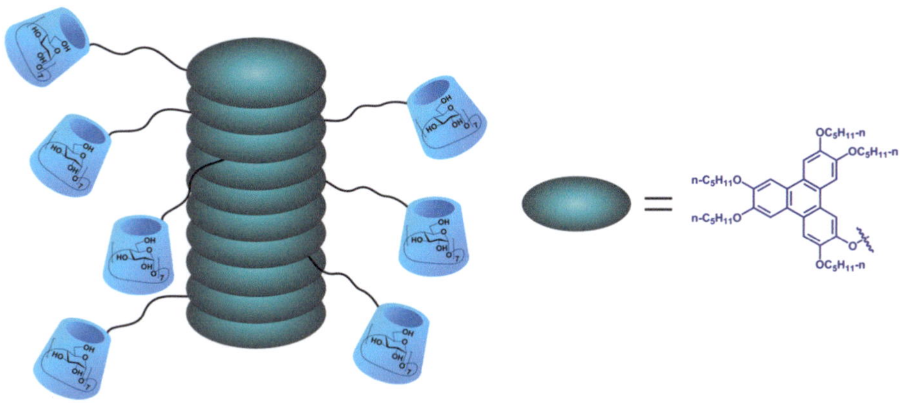

Fig. 7.27 The proposed columnar packing model of triphenylene-β-cyclodextrin conjugate **47**

basic conditions (aqueous NaOH) to obtain compound **49**, followed by a conversion to monoazide **50** in a mixture of dimethylformamide and water, and subsequently, conjugation with triphenylene derivative **38** containing five flexible *O*-n-pentyl chains and one *O*-propargyl group via the copper (I) promoted 1,3-dipolar cycloaddition. The obtained amphiphilic triphenylene-β-cyclodextrin conjugate **52** was then investigated using cross-polarized optical microscopy, but unfortunately, no mesophase was identified. In fact, compound **52** slowly decomposes without transitioning into isotropic liquid. Compound **52** was subsequently esterified, with acetic anhydride or *n*-butanoic anhydride, to afford the corresponding non-amphiphilic conjugates **47** and **48** respectively.

Interestingly, using cross-polarized optical microscopy, only the per-*O*-acetylated compound **47** was found to display mesogenic properties, with the characteristic fan-shaped textures associated to columnar mesophase observed (Prasad et al. 2003; Wan et al. 2003; Ba et al. 2003; Paraschiv et al. 2007). The more flexible compound **48** did not exhibit any birefringence upon heating, as a direct transition from solid state to isotropic phase was observed. For compound **47**, Differential scanning calorimetry only revealed two transitions upon heating at 115.2 °C and 160.4 °C, which correspond to crystalline-columnar mesophase-isotropic liquid transitions.

X-ray powder diffraction pattern of cyclodextrin derivative **47** revealed characteristic patterns of a triphenylene-based columnar mesophase. Peaks at the low angle and high angle regions were observed, and corresponded to the diameter of the columnar triphenylene groups (16.8 Å), the average distance of the molten alkyl chains (4.9–3.7 Å) and the intracolumnar spacing (3.6 Å). Yang et al. concluded that the presence of the strong H-bonding network was an obstacle for the formation of cyclodextrin-based liquid crystals. This H-bond network was eliminated through esterification, thus providing the molecules with the ability to self-assemble via π-π-interactions through the triphenylene moieties. However, the roles of β-cyclodextrin played during mesophase formation appear to be less important (Fig. 7.27).

7.2.3 Polymeric Cyclodextrin-Based Thermotropic Liquid Crystals

Recently, liquid crystalline polymers have received considerable attention due to their potential use as biomedical materials, photoelectric materials, switches, smart and stimuli responsive materials (Verploegen et al. 2009; Kim et al. 2014; Wang et al. 2014; Gündüz 2015; Mamiya et al. 2015). Liquid crystal polymers containing cyclodextrin molecules are of particular interests because of their cavities which can be used for host guest interactions with different properties. There have been two types of cyclodextrin-based liquid crystalline polymers reported in the literature so far: the first is related to cyclodextrin-based star-shaped polymers and the second is related to cyclodextrin-based rotaxane polymers.

7.2.3.1 Cyclodextrin-Core Star-Shaped Thermotropic Liquid Crystals

In 2009, He and co-workers (He et al. 2009) reported the synthesis and characterization of an α-cyclodextrin-based star-shaped polymer via atom transfer radical polymerization. Owing to circular geometry and multivalent functionality (18–24 hydroxyl groups), cyclodextrin molecules possess great potential to be used as scaffold to generate star-shaped polymers (Ohno et al. 2001; Stenzel-Rosenbaum et al. 2001; Karaky et al. 2005). Materials of such kind could find applications as switchable windows, displays, color projectors, and other electric-optical systems (Doane et al. 1986; Lin et al. 1995; Bouteiller and Barny 1996; Petti et al. 2003; Serhatli and Kacar 2006). Azobenzene is the moiety that attracted the utmost interest due to its reversible photoisomerization (Han et al. 2006). To synthesize the desired copolymer, α-cyclodextrin was first esterified with 2-bromoisobutyryl bromide, to generate a star-shaped core (incomplete substitutions were recorded, average degree of substitution: 13) containing numerous radical initiating sites; which was subsequently polymerized with a previously synthesized methacrylate derivative containing a flexible hexyl chain terminated with a 4-methoxy-4-oxy-azobenzene mesogenic group. The radical polymerizations were mediated by CuBr/Spartein using four different ratios of monomer methacrylate vs α-cyclodextrin core. All synthesized polymers showed a polydispersity index of approximately 1.5.

The thermotropic liquid crystalline properties of the synthesized materials were investigated and they all displayed both smectic-nematic phase transitions. The observed temperatures associated with smectic to nematic as well as nematic to isotropic phase transitions increased with molecular weights. For instance, the polymer obtained from combining methacrylate: α-cyclodextrin core of 60:1 was observed to display characteristic schlieren nematic textures after being heated to isotropic liquid and subsequently cooled below 125 °C; further cooling to 80 °C yielded the representative smectic textures.

The photoresponsive behavior of the material was investigated in tetrahydrofuran by UV-Vis spectroscopy. After irradiation at 360 nm, the intensity of the $\pi \rightarrow \pi^*$ transition band decreased, which contrasted with a noticeable increase of the $n \rightarrow \pi^*$ transition. The complete isomerization of the azobenzene moieties was observed after irradiating the sample for 5 min. As expected, the reverse trans-cis isomerization happened thermally and /or photochemically in which the trans isomer being thermodynamically favored by approximately 48 kJ/mol (Corruccini and Gilbert 1939).

7.2.3.2 Polyrotaxane-Based Thermotropic Liquid Crystals

Polyrotaxanes consist of a class of supramolecular polymers that have non-covalently linked cyclic organic structures such as cyclodextrins or crown ethers, threaded on a linear polymer such as polyethylene glycol chains. The cyclic molecules are mechanically locked with the threading polymer by bulky caps placed at each end of linear polymer (Gibson et al. 1994). Since their synthesis was first reported by Harada et al. in 1992, polyrotaxanes have attracted a considerable interest from the research community (Harada et al. 1992). One of their most interesting properties is the ability of cyclic structures to rotate and slide over the linear polymer chain. This feature allowed the preparation of fascinating materials such as molecular tubes, obtained by cross-linking macrocycles onto a single thread (Harada et al. 1993). Polytotoxanes have also found various applications (Okumura and Ito 2001) in designing stimuli-responsive properties such as antiscratching, self-healing (Noda et al. 2014), pressure sensitive drug releasing systems (Katsuno et al. 2013). The advantages of designing liquid crystalline materials based on polyrotaxanes are the absence of covalent linkages, resulting in materials with great fluidity, a useful parameter in thermotropic liquid crystals.

In 2007, Ito et al. (Kidowaki et al. 2007) reported a very interesting polyrotaxane-based thermotropic liquid crystal based on α-cyclodextrin. In their design, they conjugated a derivative of hexanoyl chloride which was functionalized with a mesogenic 4′-cyanobiphenyl residue at the end of the chain, to a rotaxane, pre-formed by threading α-cyclodextrin onto polyethylene glycols. 1-Amdantanamine was used as the cap for the polyrotaxane due to its unusually high binding affinity to α-cyclodextrin (Cromwell et al. 1985; Eftink et al. 1989; Gelb and Schwartz 1989; Palepu and Reinsborough 1990; Kwak and Gomez 1996; Harries et al. 2005). The average molecular weight of the rotaxane was ~35,000 (~110 copies α-cyclodextrin per thread) which represented ~28% coverage of the polyethylene glycol axis. Figure 7.26 shows a schematic representation of the synthesized polyrotaxane polymer. Using ^{1}H NMR spectroscopy, it was estimated that about 42% of the hydroxyl groups of α-cyclodextrin reacted with the hexanoyl chloride derivative, which translates into approximately 7.6 cyanobiphenlyl units per α-cyclodextrin (Fig. 7.28).

Differential scanning calorimetry revealed one endothermic transition at 136 °C upon heating, and two exothermic transitions respectively at 70 and 129 °C upon

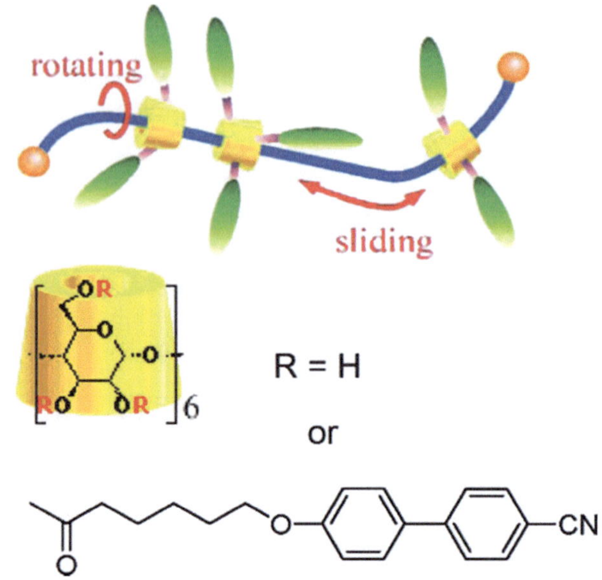

Fig. 7.28 Schematic representation of 6-(4′-cyanobiphenyl-4-yloxy)hexanoyl-functionalized polyrotaxanes based on α-cyclodextrin. "(Kidowaki et al. 2007)". Copyright 2007 American Chemical Society

Fig. 7.29 Schlieren-like texture observed for 6-(4′-cyanobiphenyl-4-yloxy)hexanoyl-functionalized polyrotaxanes. "(Kidowaki et al. 2007)". Copyright (2007) American Chemical Society

cooling. The phase transition at 70 °C was attributed to a glass transition while the phase transition at 129 °C was attributed to the biphenyl mesogenic side chains.

Using cross-polarized optical microscopy, schlieren-like texture was observed at 100 °C, indicating the presence of a nematic mesophase (Fig. 7.29). The brightness of the texture gradually increases with decreasing temperature, and were preserved

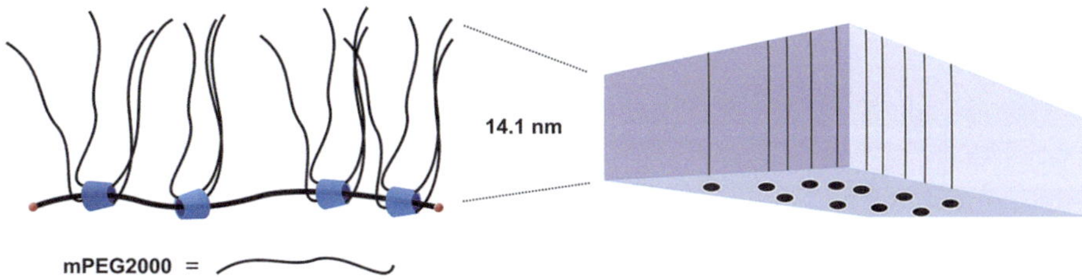

Fig. 7.30 Schematic representation of formed smectic A mesophases by α-cyclodextrin-based polyrotaxane grafted with mPEG2000 chains. The free sliding and rotation of each mPEG2000 grafted α-cyclodextrin subunit on the polyethylene glycol axis facilitates the self-organization of the mPEG2000 side chains into large organized domains

at room temperature which is indicative of a liquid crystalline glass state at ambient temperature. X-ray diffraction patterns measured at 25, 80 and 180 °C further confirmed the presence of a nematic mesophase at temperatures below 130 °C.

Following the work of Ito et al., Araki and co-workers (Araki et al. 2014) later reported another α-cyclodextrin-polyrotaxanes-based thermotropic liquid-crystal by simply grafting mPEG2000 dicarboxylic acid to similar polyrotaxne as above, using 1,1-carbonylbis-1*H*-imidazole as the coupling reagent to obtain polyrotaxane-mPEG2000 copolymer with an estimated 75 copies of mPEG2000 per chain. The authors describe the geometry of their polymer as "sliding graft copolymer" which has a 'rope-curtain like' structure (Fig. 7.28).

Cross-polarized optical microscopy confirmed the fluidity of the material at 65 °C and formation of Grandjean terrace texture, typically associated with the formation of smectic mesophases (Oh 1977; Oswald and Pieranski 2005). On the other hand, X-ray diffraction at ambient temperature revealed several peaks associated with crystalline PEG (Barnes and Ross 1936; Bortel et al. 1979). At temperatures above 65 °C, X-ray diffraction also confirmed the formation of smectic A mesophase with a layer distance of 14.1 nm, corresponding to the predicted contour length of mPEG2000 (13.18 nm). Thus, despite the absence of mesogenic groups, this material was capable of self-organize into smectic A mesophases. The authors attributed the properties of the materials to the one-directional free sliding and rotation of each mPEG2000 grafted α-cyclodextrin subunit on the polyethylene glycol axis (Fig. 7.30).

Additionally, Hu and co-workers (Hu et al. 2011) also prepared a series of α-cyclodextrin-polyrotaxane derivatives conjugated with mesogenic azobenzene moieties by varying the spacer lengths between azobenzene and α-cyclodextrin. Their experimental results suggested that azobenzene-functionalized polyrotoxanes failed to show liquid crystalline behavior with spacer length of 2 and 4 while the analogous polyrotaxanes with spacer length of 6 displayed columnar nematic mesophase. However, when the spacer length was increased to 11, additional high ordered structures were formed by the azobenzene mesogens. Figure 7.31 illustrates a model of molecular organization in solid states for the azobenzene-functionalized polyrotaxanes with a spacer length of 11 carbons (**Azo-PR-11**).

More recently, Guo and co-workers (Guo et al. 2016) reported a new kind of α-cyclodextrin-polyrotaxanes by grafting 4-phenylazobenzoyl moieties directly to

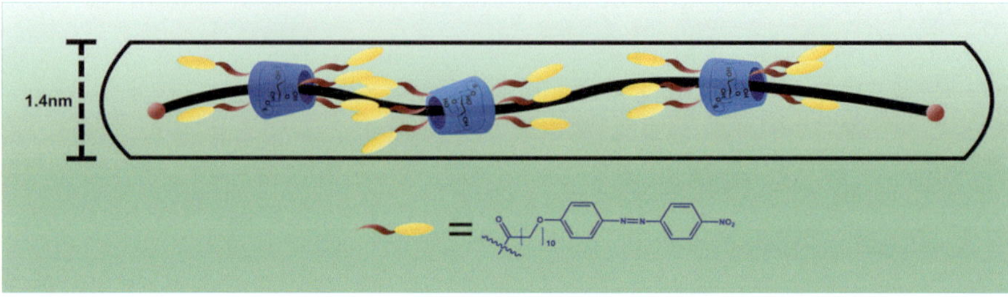

Fig. 7.31 Graphical representation of polyrotaxane Azo-PR-11, molecular arrangement allowing enhanced β-cyclodextrin mobility on the polyrotaxane glycol thread

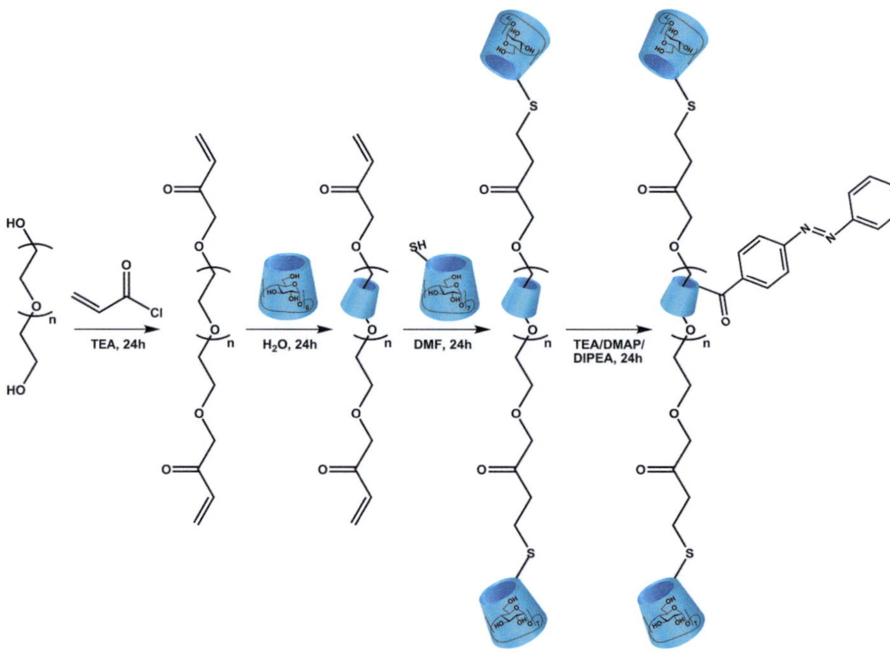

Fig. 7.32 Synthetic scheme of azobenzene-functionalized columnar nematic liquid-crystalline polymeric materials.

cyclodextrin scaffolds without spacer. The parent α-cyclodextrin-polyrotaxane was prepared by mixing vinyl-polyethylene glycol and α-cyclodextrin in aqueous solution and subsequently caped with thiolated β-cyclodextrin. The azobenzene mesogen was then added directly to the free hydroxyl groups of cyclodextrins. The introduction of azobenzene moieties reduces the strength of inter and intramolecular H-bonding networks thus, successfully destruct the channel-like crystalline structure of the parent α-cyclodextrin-polyrotaxane, to provide enhanced mobility/rotation for cyclodextrin subunits on the thread. The introduced mesogenic groups engage intermolecular interactions to induce the thermotropic liquid crystalline behavior of the material (Fig. 7.32).

Cross-polarized optical microscopy revealed schlieren-like textures when the material was heated above 180 °C. In comparison, the parent unmodified α-cyclodextrin-polyrotaxanes did not display liquid crystalline behavior at any temperature. Interestingly, this 4-phenylazobenzoyl-functionalized α-cyclodextrin-polyrotaxane also showed lyotropic liquid crystalline behaviour in dimethyl sulfoxide solution at concentration of 5–0.2 wt%. With the help of wide angle X-ray diffraction patterns, the columnar nematic mesophase of the material was confirmed.

7.3 Cyclodextrin-Based Lyotropic Liquid Crystals

As discussed earlier, amphiphilic cyclodextrin derivatives have excellent ability to self-assemble in solvents, forming different self-assembled systems including lyotropic liquid crystals. In this development, Gulik and co-workers (Gulik et al. 1998) investigated the lyotropic liquid crytstal properties of a class of amphiphilic cyclodextrins, esterified at all secondary hydroxyl groups with acyl chains of different lengths (β-cyclodextrin-C*n*; n = 6, 8, 10, 12, and 14; α-cyclodextrin-C_{14}, and γ-cyclodextrin-C_{14}). Both X-ray scattering and freeze fracture electron microscopy were used in their experiments. The structural analysis of the obtained cyclodextrins revealed that aliphatic chains are stacked similarly to lipid systems and arrange themselves in 2-D hexagonal phases. β-cyclodextrin-C6, -C8, -C10, and -C12 and γ-cyclodextrin-C14 were optically birefringent, while β-cyclodextrin-C14 and α-cyclodextrin-C14 were optically isotropic. X-ray diffractions revealed seven sharp reflections for β-cyclodextrin-Cn (n = 6, 8, 10, and 12) with dimensions of 25.9, 30.8, 33.0 and 35.3 Å, respectively. Interestingly, the birefringent materials produced were found to organize into polar columns of 2-D hexagonal lattice. Such results contrasted with γ-cyclodextrin-C14 which only displayed broad reflections associated to 2-D hexagonal lattice with dimension of 37.6 Å. On the other hand, α-cyclodextrin-C14 and β-cyclodextrin-C14 exhibited sharp reflections corresponding to body-centered cubic lattice of optically isotropic materials with dimension of 42.6 Å and 43 Å, respectively. The 2-D hexagonal phase of β-cyclodextrin-Cn (n = 6, 8, 10, and 12) and cubic lattice of α-cyclodextrin-C14 and β-cyclodextrin-C14 were further supported by freeze fracture micrographs (Fig. 7.33A–D). Compounds β-cyclodextrin-Cn (n = 6, 8, and 10) in hexagonal phases and α-cyclodextrin-C14 and β-cyclodextrin-C14 in cubic phases showed ordered and organized domains, respectively while less ordered domains were observed for β-cyclodextrin-C12. The polar columns of hexagonal phase comprised of stacked cyclodextrin units in which polar heads were pointed toward the interior of the self-assembled structure, while paraffinic units were projected toward the exterior (Fig. 7.33E). The respective cubic lattice of α-cyclodextrin-C14 and β-cyclodextrin-C14 consists of 12 molecules per unit cell. It was proposed that the cubic lattice comprised of either polar or paraffinic globules at the center and edges of the cell (Fig. 7.33F, G). The polar globules were comprised of six molecules of α-cyclodextrin-C14 or

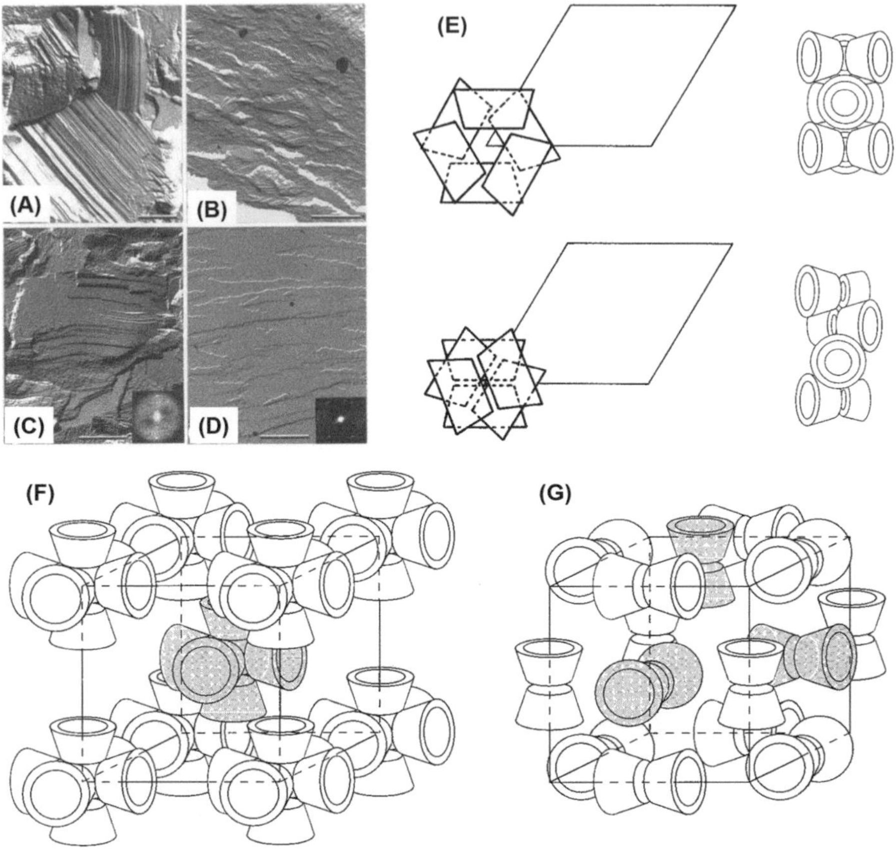

Fig. 7.33 Freeze fracture electron micrographs of (**A**) β-cyclodextrin-C6, (**B**) β-cyclodextrin-C8, (**C**) β-cyclodextrin-C10, (**D**) β-cyclodextrin-C14 (Scale bar = 200 nm). (**E**) Two possible associations of 2D hexagonal structure models with polar columns of stacked cyclodextrin units. (**F**, **G**) The possible cubic structure models of α-cyclodextrin-C14 or β-cyclodextrin-C14 containing either polar (**F**) or paraffinic globules (**G**) at the vertex (in gray color). The paraffinic chains are omitted for clarity. (**A–G**) are adapted with permission from (Gulik et al. 1998). Copyright (1998) American Chemical Society

β-cyclodextrin-C14 (Fig. 7.33F) while paraffinic globules consist of dimers of *α*-cyclodextrin-C14 or *β*-cyclodextrin-C14 connected by their primary hydroxyl groups (Fig. 7.33G). The combined analysis of X-ray scattering, freeze fracture electron microscopy, and electron density distribution confirmed the paraffinic globules arrange themselves in the cubic lattice (Fig. 7.33G). The orientation of dimers, while occupying the middle of each edge, was parallel to axis of cubic cell and orthogonal to adjacent dimer. The formation of cubic lattice for such amphiphilic cyclodextrins was found to be unusual and was rationalized by the large surface area of the aliphatic chains.

Cyclodextrins are well known to bind to various types of surfactants by inclusion of their hydrophobic tails into cyclodextrin's cavity. Surfactants are widely used in daily practical applications including cleaning, emulsifying, detergents, paints, biocides, and insecticides (Rosen and Kunjappu 2012). Upon interactions with cyclodextrins, significant changes occurred in inherent properties and structural dynamics of surfactants. For instance, sodium dodecyl sulfate (SDS), a common surfactant abundantly used in cleaning, cosmetics, and hygiene products, exhibited different self-assembly behavior in the presence of β-cyclodextrin. Sodium dodecyl sulfate forms highly stable water soluble inclusion complexes with two β-cyclodextrin molecules (sodium dodecyl sulfate@2β-cyclodextrin) in 1:1 and 1:2 stoichiometry (Dorrego et al. 2000; Jiang et al. 2011). Generally, sodium dodecyl sulfate@2β-cyclodextrin complexes disfavor self-assembly in aqueous medium due to their hydrophilic outer surface (Saenger and Müller-Fahrnow 1988). However, self-assembled aggregates such as vesicular and lamellar mesophases were observed for sodium dodecyl sulfate@2β-cyclodextrin complex in aqueous solution. To gain deeper insight into such aggregation behavior, the concentration, H-bonds, and electrostatic interactions of the cyclodextrin/sodium dodecyl sulfate complex were investigated. It was observed that by changing the mass concentration of sodium dodecyl sulfate@2β-cyclodextrin complexes, three types of aggregates were formed, including: region I (50–25 wt%), region II (25–6 wt%) and region III (6–4 wt%) (Jiang et al. 2011). In mass concentration range of 50–25 wt%, semi-transparent self-assembled lamellar structures were identified, exhibiting static birefringence observed under cross-polarized optical microscopy, a suggestive of an anisotropic phase (Fig. 7.34A). The lamellar lattice was further confirmed by freeze-fracture transmission electron microscopic and X-ray scattering profiles. Images of uniform parallel lines were obtained from the vitrified sample which showed similar pattern to typical lamellar structures (Fig. 7.34B, C). Further, small and wide angle X-ray scattering evaluations confirmed channel type crystalline bilayers associated in head to head pattern stabilized by strong H-bonding networks (Fig. 7.34F). Lamellar liquid-crystalline mesophases appeared in one direction while in-plane solid-crystalline orders appeared in other two directions. Such observations are characteristic features of transition phase between liquid crystal and solid.

Upon decreasing the mass concentration to 25–6 wt% of region II, large aggregates corresponding to microtubular structures were observed (Jiang et al. 2010a). Generally, tubular assembly is governed by non-directional hydrophobic interactions resulting in less-ordered soft aggregates. However, microtubes formed by the nonamphiphilic self-assembly of sodium dodecyl sulfate@2β-cyclodextrin with mass concentration of 25–6 wt% was driven by directional and ordered H-bonds instead of hydrophobic interactions. The formed microtubes appeared rigid, monodispersed and highly persistent in molecular dimensions owing to strong H-bond interactions. The microtubes were analyzed by confocal laser scanning microscopy using nile red as fluorescent dye that exhibits partial inclusion in the cavity of β-cyclodextrin (Hazra et al. 2004) (Fig. 7.35A, D). The fluorescent walls with non-fluorescent center of microtubes demonstrate the uniform pattern of hollow tubular structures (Fig. 7.35A, D). Similar pattern was observed in the absence of nile red by confocal laser scanning microscopy and transmission electron microscopy (Fig. 7.35B, C). The mean diameter and length of microtubes were 1.1 μm and

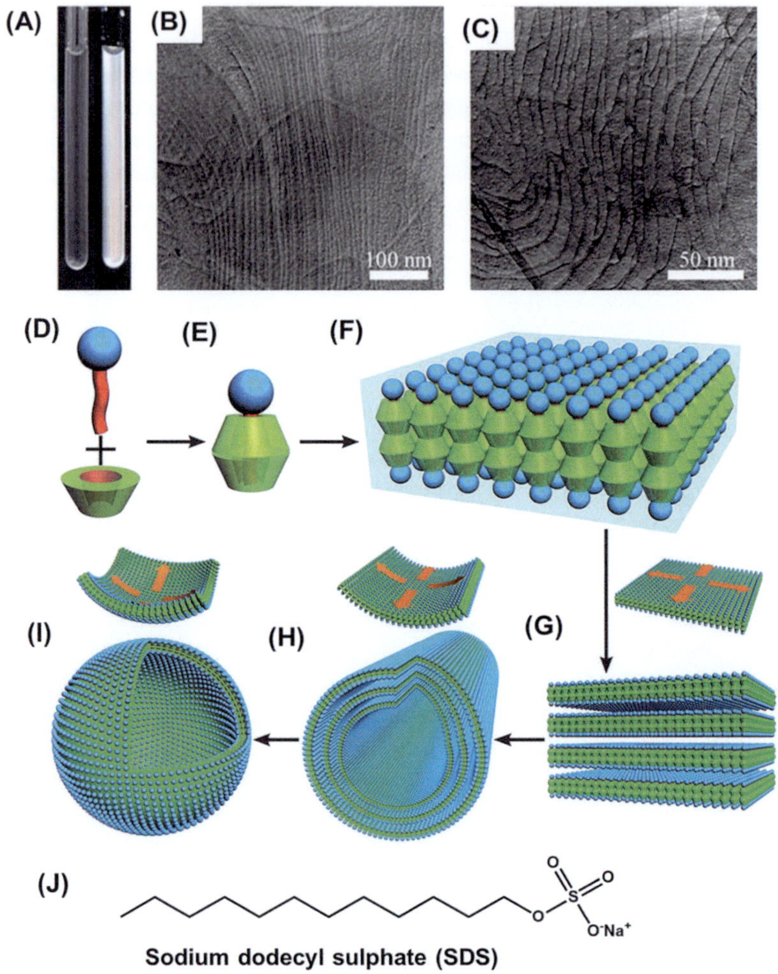

Fig. 7.34 (**A**) NMR sample photographs of 30 wt% sodium dodecyl sulfate@2β-cyclodextrin without (left) and with (right) crossed polarizer. (**B**) and (**C**) freeze-fracture transmission electron micrographs of 30 wt% SDS@2β-cylodextrin complex. (**D–I**) Graphical illustration of sodium dodecyl sulfate@2β- cyclodextrin self-assembly: (**D**) Sodium dodecyl sulfateand *β*-cyclodextrin monomers, (**E**) sodium dodecyl sulfate@2β-cyclodextrin complex, (**F**) Bilayer membrane of sodium dodecyl sulfate@2β- cyclodextrin channels structure. (**G–I**) Transition of lamellae to microtubes and vesicles. (**J**) Structure of sodium dodecyl sulphate (SDS). (**A–I**) are adapted with permission from "(Jiang et al. 2011)". Copyright 2011 Royal Society of Chemistry

40 μm, respectively with wall separation of 0.7 μM. Microtubes appeared quite inflexible as they preferred to be straight rather than bending (Fig. 7.35E). The multilamellar structure of microtubes displayed unique "annular ring" formation, when resolved by freeze-fracture transmission electron microscopy using vitrified sample of sodium dodecyl sulfate@2β-cyclodextrin. Figure 7.35F, H–K illustrates the

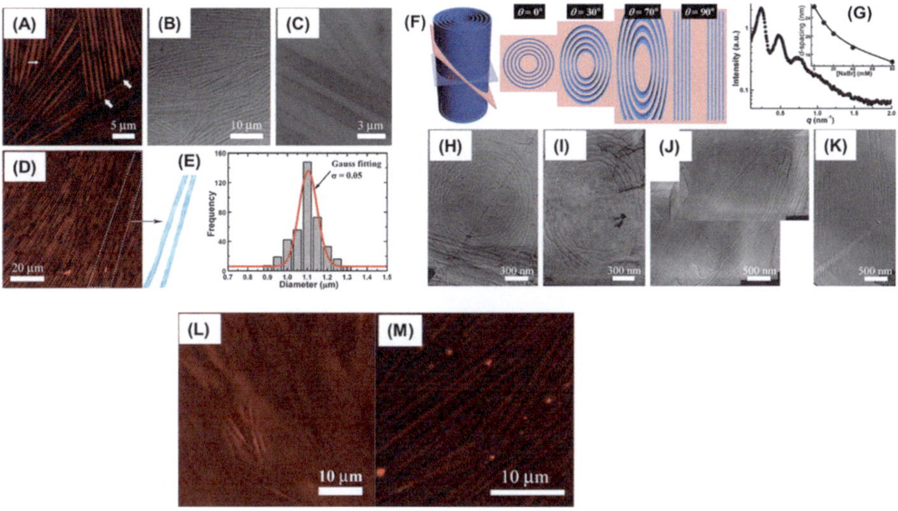

Fig. 7.35 (**A**, **D**) Confocal laser scanning microscopy images of 10 wt% sodium dodecyl sulfate@2β-cyclodextrin solution microtubes, (**B**) Differential interference contrast microscopy image, (**C**) Transmission electron microscopy image, (**E**) Diameter distribution of the microtubes, (**F–K**) Annular ring structure of the microtubes. (**F**) Cross sections view of microtube upon changing angle (θ) from 0 to 90° between radical plane and fractured surface, (**G**) Small-angle X-ray scattering profile of the microtubes, and (**H–K**) freeze-fracture transmission electron micrography images of the microtubes at different angle (θ) 0° (**H**), 30° (**I**), 70° (**J**), and 90° (**K**). (**A–K**) are adapted with permission from (Jiang et al. 2010a). Copyright 2010 Royal Society of Chemistry. (**L** and **M**) are adapted with permission from (Jiang et al. 2010b) and (Jiang et al. 2011). Copyright 2011 Royal Society of Chemistry.

appearance of annular ring changed to circles, ellipses, elongated ellipses and parallel lines upon changing the angle from 0 to 90° between radical plane and fractured surface. Further analysis by atomic-force microscopy and wide-angle X-ray scattering revealed bilayer architecture of microtubes. The nonamphiphilic self-assembly of sodium dodecyl sulfate@2β-cyclodextrin complex to microtubules was attributed to the H-bonding and electrostatic interactions between each complex. Such results were quite interesting since recently, tubular assembled architectures have gained considerable attention in different applications including sensing, synthesis, and drug delivery (Shimizu et al. 2005; Lee et al. 2009).

Structural transformations were also observed upon diluting the mass concentration of sodium dodecyl sulfate@2β-cyclodextrin (6–4%) that led to transition of microtubular to vesicular phases containing unilamellar and bilamellar hollow vesicles with concave/ convex surfaces. Absence of diffraction pattern in small-angle X-ray scattering indicated the formation of unilamellar and bilamellar vesicles instead of multilamellar vesicles. Despite concentration dependency, the recurrent feature of lamellar, microtubular, and vesicle mesophases was the presence of bilayer membrane structures. At the intermediate margins of region I/II and region II/III, concurrence of lamella-microtube for 20 wt% and microtube-vesicle for

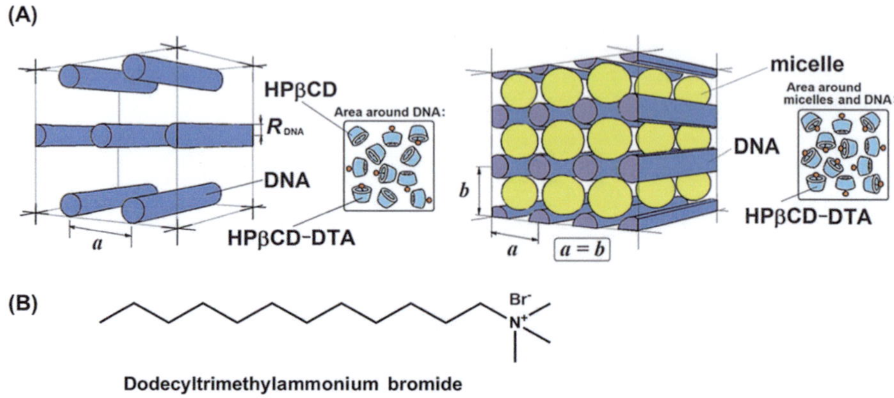

Fig. 7.36 (**A**) Pictorial representation of hexagonal (left) and tetragonal (right) phases of the microstructures of dodecyltrimethylammonium-DNA-(2-Hydroxypropyl)-β-cyclodextrin, respectively. (**B**) Structure of dodecyltrimethylammonium bromide. (**A**) is adapted with permission from (Bilalov et al. 2012). Copyright 2012 Royal Society of Chemistry

7 wt%, respectively were observed using confocal laser scanning microscopy images of entrapped nile red (Fig. 7.35L, M). Figure 7.35L illustrates non-fluorescent center corresponding to hollow microtubes with fluorescent walls of parallel lines as lamellar pattern. Similarly, spherical vesicles were found in the matrix of microtubes validating the cohabitation of lamellae, microtubes and vesicles (Fig. 7.35M). Such nonamphiphilic self-assembly driven by H-bonds and electrostatic interactions instead of hydrophobic interactions promoted the development of new designs to explore self-assembly behavior of cyclodextrin-based liquid crystalline systems.

The lyotropic liquid crystal properties of polyelectrolyte such as deoxyribonucleic acid (DNA) in the biological system depends upon its dispersion in lipids and interaction with counter ions (Dias and Lindman 2008). To this account, the aggregation of polyelectrolytes (DNA) and amphiphilic counter ions dodecyltrimethylammonium was investigated in the absence as well as in the presence of *β*-cylclodextrin and (2-hydroxypropyl)-β-cyclodextrin (Bilalov et al. 2012). In aqueous medium, DNA and dodecyltrimethylammonium essentially form self-assembled water-insoluble micelles and consequently change the liquid crystal phases of DNA. However, the presence of β-cyclodextrin or (2-hydroxypropyl)-β-cyclodextrin, as dispersion agents, redefines the liquid crystalline ordering of DNA by forming inclusion complexes with dodecyltrimethylammonium. The resulted amphiphiles are characterized by a greater degree of dispersion and increased solubility in water. With molar ratio $R < 1$ (R = [dodecyltrimethylammonium]/[cyclodextrin]) and increased dodecyltrimethylammonium-DNA concentration, 2D hexagonal liquid crystalline phases were observed (Fig. 7.36A, left). Upon further increasing the $R > 1.5–2$ and dodecyltrimethylammonium-DNA concentration, anisotropic tetragonal liquid crystal lattice possessing square packing was obtained

(Fig. 7.36A, right). Interestingly, both 2D-tetragonal and 2D-hexagonal phases displayed parallel ordering of DNA duplexes.

The rigidity of polyelectrolyte dictates its phase behavior in aqueous medium, hence lyotropic liquid crystal properties. To validate the rigidity effect, a flexible electrolyte, polyacrylate was investigated and compared with DNA, which is relatively rigid (Carlstedt et al. 2012). An isotropic phase was observed in dodecyltrimethylammonium– polyacrylate with molar ratio R = 1 which was similar to dodecyltrimethylammonium–DNA system. With 20 weight percent of (2-hydroxypropyl)-β-cyclodextrin and R ≈ 2, dodecyltrimethylammonium–polyacrylate demonstrated optically isotropic and viscous cubic phase which was further validated by singlet in ^{2}H-NMR spectra. Under similar composition of 20 wt %, hexagonal phase was also observable as optically anisotropic, and viscous with quadrupolar splitting pattern in ^{2}H-NMR spectra. The cubic phase order of dodecyltrimethylammonium–polyacrylate system, and hexagonal and tetragonal phases of dodecyltrimethylammonium–DNA system were attributed to flexible nature of polyacrylate and stiffness of DNA, respectively while keeping the *β*-cyclodextrin and (2-hydroxypropyl)-β-cyclodextrin as the dispersion agents.

7.4 Applications of Cyclodextrin Based Liquid Crystals

7.4.1 Biosensing Based Inclusion and Cholesterol Detection

Over the years, liquid crystals have been used in numerous applications owing to their optical and dielectric anisotropic properties. These include liquid crystal displays, (Hee Lee et al. 2012; Bremer et al. 2013) technical (Goossens et al. 2016) and biological areas (Brake et al. 2003a; Mushenheim et al. 2014). Exploring the liquid crystal behavior at the interface of immiscible phases gave valuable insight on the self-assembled interactions of biomolecules such as proteins (Brake and Abbott 2002) and lipids (Hartono et al. 2008), as well as synthetic molecules including polymers (Brake et al. 2003b) and surfactants (Bai and Abbott 2011).

At the interface, the specific orientational order of liquid crystals changes into different order in presence of interfacial events. These transitions have been successfully applied as imaging tool for the investigating the biomolecular interactions at the cellular level (Park and Abbott 2008; Bi et al. 2009; Lowe and Abbott 2012). Zuo et al. (2014) used this strategy to investigate the host-guest interactions between β-cyclodextrin and sodium dodecylsulfate as well as methylene blue by employing a known mesogen, 4-cyano-4′-pentylbiphenyl (5CB), as nematic liquid crystals analyzed by cross-polarized optical microscopy. At aqueous interface, 4-cyano-4′-pentylbiphenyl self-assemble into nematic liquid crystals with bright images indicating their planer orientations (Fig. 7.37A). With addition of sodium dodecylsulfate, its hydrophobic tail was observed to interact with 4-cyano-4′-pentylbiphenyl that changed their orientations from planer to homeotropic, resulting in dark images of liquid crystals (Fig. 7.37A, B). Further, with the introduction of β-cyclodextrin, the

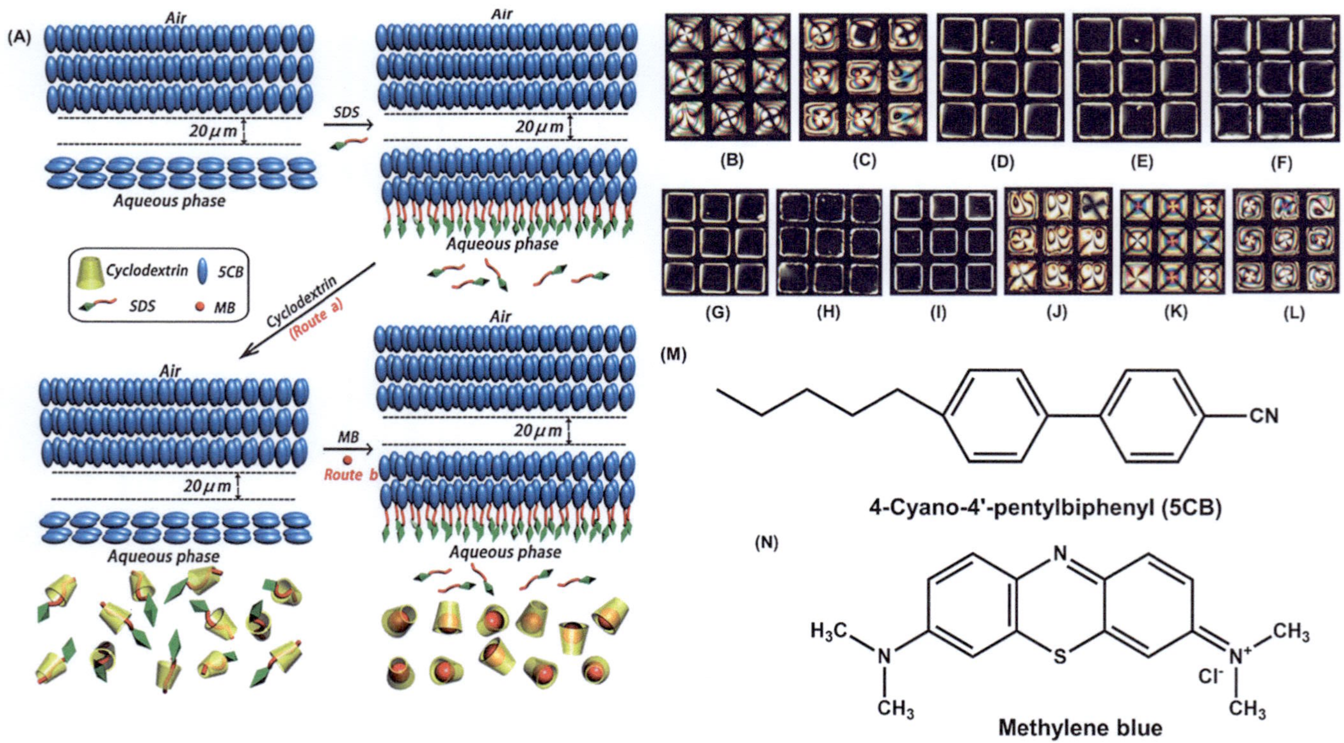

Fig. 7.37 (**A**) Graphical illustration of host–guest complexation between β-cyclodextrin and sodium dodecylsulfate (or methylene blue) using a 4-cyano-4′-pentylbiphenyl LC sensor. (**B-F**) Cross-polarized images of copper grid containing 4-cyano-4′-pentylbiphenyl upon addition of various concentrations of sodium dodecylsulfate (mM): (**B**) 0, (**C**) 0.1, (**D**) 0.25, (**E**) 1.0, (**F**) 2.5. (**G-L**) Cross-polarized images of copper grid containing 5CB-SDS after addition of various concentrations of β-cyclodextrin (mM): (**G**) 0, (**H**) 0.1, (**I**) 0.25, (**J**) 0.8, (**K**) 1.0, (**L**) 2.5, concentration of sodium dodecylsulfate used = 0.25 mM. (**M** and **N**) Structure of 4-cyano-4′pentylbiphenyl (5CB) and methylene blue (**A-L**) are adapted with permission from (Zuo et al. 2014). Copyright 2013 Royal Society of Chemistry

sodium dodecylsulfate molecules showed better inclusion binding with β-cyclodextrin cavity leaving 4-cyano-4′-pentylbiphenyl to reorients themselves from homeotropic to planer order (Fig. 7.37A, C). This resulted into the reappearance of bright images (Fig. 7.37C). No significant change in optical appearance of 4-cyano-4′-pentylbiphenyl was observed upon addition of β-cyclodextrin in the absence of sodium dodecylsulfate except minor color change attributed to weaker 4-cyano-4′-pentylbiphenyl and β-cyclodextrin interactions. To validate the host-guest inclusion interactions, they employed methylene blue that exhibit stronger binding to β-cyclodextrin than sodium dodecylsulfate. Upon exposure of methylene blue to the interfacial solution of 4-cyano-4′-pentylbiphenyl+β-cyclodextrin+sodium dodecylsulfate on the copper grid, the dark images corresponding to homeotropic order reappeared after 30 mins. This is due to the greater binding affinity of methylene blue for the β-cyclodextrin cavity, resulting in the expulsion sodium dodecylsulfate molecules which interact with 4-cyano-4′-pentylbiphenyl again to form dark images.

The LC behavior of ternary complex 4-cyano-4′-pentylbiphenyl + β-cyclodextrin+sodium dodecylsulfatein the presence of methylene blue and dopamine was further investigated by quantum mechanical calculations using Gaussian 09 program package (Liu et al. 2016). PM3 semi-empirical method was used to simulate inclusion complexes of 4-cyano-4′-pentylbiphenyl, β-cyclodextrin, sodium dodecylsulfate, methylene blue, and dopamine which were again re-augmented at DFT B3LYP/6-31G(d) level. The theoretical analysis revealed that sodium dodecylsulfate formed strong inclusion complex with β-cyclodextrin at the interface of 4-cyano-4′-pentylbiphenyl/aqueous surface which resulted in the planer reorientation of liquid crystals. The addition of methylene blue excludes sodium dodecylsulfate from the cyclodextrin cavity to allow the formation of a stronger β-cyclodextrin-methylene blue complex. The simulated energy and complexation profile corroborated 1:2 (methylene blue:2β-cyclodextrin) complex formation through the wider secondary face of cyclodextrin which was more stabilized than sodium dodecylsulfate complex. In addition, it was observed that inclusion complex of dopamine+β-cyclodextrin was less stable than sodium dodecylsulfate+β-cyclodextrin or methylene blue+β-cyclodextrin, therefore no significant change in 4-cyano-4′-pentylbipheyl liquid crystal orientation was found upon addition of dopamine. The theoretical results were in compliance with experimental results observed by Zuo et al.

The inclusion interactions were further extended to examine the specific binding of guest molecule with appropriate host molecule using 4-cyano-4′-pentylbipheyl as liquid crystal imaging tool (Liao et al. 2015). α-cyclodextrin and β-cyclodextrin inherit different cavity sizes, thereby exhibiting different binding affinity for guest molecules. To validate this, they used cetyltrimethyl ammonium bromide as surfactant which exhibits different binding abilities toward α-cyclodextrin and β-cyclodextrin. The orientational order and optical appearance of 4-cyano-4′-pentylbiphenyl in the presence of cetyltrimethyl ammonium bromide was similar to sodium dodecylsulfate pattern. In the presence of cetyltrimethyl ammonium bromide, the planer orientation and bright images of copper grid coated with 4-cyano-4′-pentylbiphenyl changed to homeotropic order and its characteristic dark images. The reverse effect, regeneration of bright images and planer order, was observed upon exposure of 4-cyano-4′-pentylbiphenyl + cetyltrimethyl ammonium bromide films to α-cyclodextrin and β-cyclodextrin solutions. Interestingly, the rate of that transition was much faster for α-cyclodextrin than β-cyclodextrin. It was observed that, at the same concentration and time interval, the cetyltrimethyl ammonium bromide exhibited stronger affinity towards α-cyclodextrin and β-cyclodextrin as visualized by cross-polarized optical microscopy images, surface tension analysis and isothermal titration curve.

For the practical utility of this approach, competitive host-guest interactions at the liquid crystal-aqueous phases have been exploited for the detection of cholesterol and cholic acid. Cholesterol is a valuable component of cell membrane (Ikonen 2008) involved substantially in the biosynthesis of vitamin D and bile acids (Ram et al. 2001a). The high cholesterol level in the blood is implicated in cardiovascular and other diseases by plaque formation in the arteries. Thus, cholesterol detection by simple and effective method is always desirable despite currently available methods including electrochemical, (Dey and Raj 2010) enzyme-based biosensors, (Devadoss and Burgess 2002) and fluorescence spectroscopy (Mondal and Jana 2012). Park et al. developed a cholesterol biosensor by utilizing the sodium dodecyl sulfate/β-cyclodextrin complex on the interface of liquid crystal and aqueous medium (Munir and Park 2015). At the interface between aqueous medium and nematic liquid crystal, the small perturbation can make diverse changes in the orientation of nematic liquid crystal. Such behavior was utilized for cholesterol detection. Sodium dodecyl sulfate forms inclusion complex with β-cyclodextrin owing to host guest interactions (*vide supra*, Fig. 7.37A). The exposure of cholesterol at the interface between aqueous solution and nematic liquid crystal excludes the sodium dodecyl sulfate from sodium dodecyl sulfate/β-cyclodextrin complex due to stronger inclusion of cholesterol into β-cyclodextrin cavity. The excluded sodium dodecyl sulfate gets adsorbed at the interface, thereby changed the orientation of 4-cyano-4′-pentylbiphenyl as demonstrated by cross-polarized optical microscopy. The inclusion concentration of sodium dodecyl sulfate (340 μM) in sodium dodecyl sulfate/β-cyclodextrin complex was carefully analyzed using pH and high-performance liquid chromatography studies. The polarized optical microscope images of 4-cyano-4′-pentylbiphenyl anchored transmission electron microscopy grid cells showed in-plane birefringence illustrating their planer orientation at the interface. Similar planer orientations and bright coloring of cross-polarized optical microscope images were observed in the absence of cholesterol. The addition of cholesterol solution of varying concentrations changed the orientation of 4′-pentylbiphenyl from planer to homeotropic with appearance of dark patches attributed to exclusion of sodium dodecyl sulfate from sodium dodecyl sulfate/β-cyclodextrin complex. The calculated detection limit of cholesterol using this method was 3 μM which is biological relevant detection concentration. The insensitive liquid crystal behavior of 4-cyano-4′-pentylbiphenyl to common interfering analytes present in blood

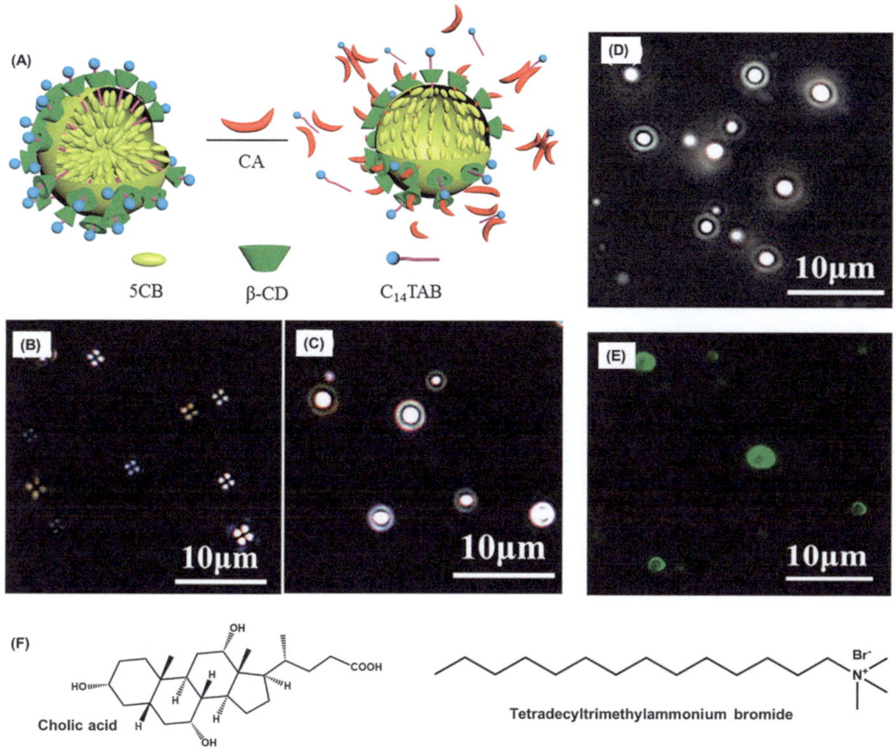

Fig. 7.38 (**A**) Pictorial representation of cholic acid-induced transition of 4-cyano-4′-pentylbiphenyl-β-cyclodextrin-C_{14}TAB(5CB-β-CD-C_{14}TAB) droplets from radial-to-bipolar configuration host–guest recognition. (**B**, **C**) Polarized optical microscope images of 4-cyano-4′-pentylbiphenyl-β-cyclodextrin-C_{14}TAB droplets before (**B**) and after (**C**) addition of 20 mM cholic acid at pH 7.4. (**D**, **E**) Polarized (**D**) and fluorescence (**E**) images of 4-cyano-4′-pentylbiphenyl-β-cyclodextrin-C_{14}TAB droplets after addition of cholyl-lysyl-fluorescein, a fluorescent cholic acid. (**F**) Structure of cholic acid and tetradecyltrimethylammonium bromide. (**A**–**E**) are adapted with permission from (Deng et al. 2015). Copyright 2015 Royal Society of Chemistry

including sodium chloride, hemoglobin, ascorbic acid, and glucose validated the practical utility of this method.

The liquid crystal-aqueous interface self-assembled system was also used to detect cholic acid, a primary component (31%) of bile acids excreted by liver (Hofmann and Hagey 2008). The abnormal production of cholic acid directly correlates with the progress of liver disease (Rani et al. 2004). The concentration analysis of cholic acid in serum and urine provides valuable diagnosis of liver disease and associated treatment (Jorquera et al. 2005; Griffiths and Sjövall 2010). Fang et al. used β-cyclodextrin-C_{14}TAB (tetradecyl trimethylammonium bromide) complex appended 4-cyano-4′-pentylbiphenyl droplets for the facile detection of cholic acid (Fig. 7.38) (Deng et al. 2015). C_{14}TAB formed 1:1 complex with β-cyclodextrin in aqueous system with hydrophilic head and hydrophobic tail oriented toward secondary and primary faces, respectively (Valente and Söderman 2014). The primary face of β-cyclodextrin-C_{14}TAB complex partially protrudes hydrophobic tail (about 13 Å) which procures this complex a supramolecular surfactant behavior, allowing interactions with 4-cyano-4′-pentylbiphenyl liquid crystals. Cross-polarized optical images and positive zeta potential (+17.6 mV) of β-cyclodextrin-C_{14}TAB-appended-4-cyano-4′-pentylbiphenyl films illustrated the formation of radial configuration in which primary and secondary faces of β-cyclodextrin were oriented toward 4-cyano-4′-pentylbiphenyl and water phases, respectively. The addition of cholic acid changed the radial configuration of β-cyclodextrin-C_{14}TAB-appended-4-cyano-4′-pentylbiphenyl films to bipolar configuration (Fig. 7.38B, C) attributed to the inclusion of cholic acid into cavity of β-cyclodextrin through its primary face and consequently exclusion of C_{14}TAB. This was validated by observing decrease in surface tensions of 20 μM CA solution (52.6 mN m^{-1} for) compared to 1.6 mM solution of β-cyclodextrin-C_{14}TAB (54.8 mN m^{-1}) suggesting 27.5% displacement of C_{14}TAB. The complexation of cholic acid with β-cyclodextrin-C_{14}TAB-appended-4-cyano-4′-pentylbiphenyl films was further verified by their bipolar fluorescent images measured by confocal microscope using cholyl-lysyl-fluorescein, a fluorescence derivative of cholic acid (Fig. 7.38D, E). The detection limit of the cholic acid was found to be 20 μM using liquid crystal imaging method. Furthermore, cholic acid was detected (detection limit = 60 μM) in synthetic urine sample (3.16 mM urea, 42.6 μM $MgSO_4$, 90.1 μM $CaCl_2$, and 1.36 mM NaCl) in the presence of interfering uric acid and urea, validating the practical utility of this method. Similar liquid crystalline behavior of 4-cyano-4′-pentylbiphenyl was observed upon interaction with (2-hydroxylpropyl)-β-cyclodextrin-C_{16}TAB complex highlighting the uniformity of detection method.

7.4.2 CD Capped Nanoparticle and Liquid Crystal Displays

From last few decades, utilization of liquid crystal displays (LCD) and their electro-optic properties have flourished significantly in the electronic and communication industries. Despite their huge utility in revolutionizing modern technology, liquid crystal displays have been associated with slow electro-optical response. Various modifications have been done by doping liquid crystals with semiconductor nanoparticles (Du and Toshima 2007), carbon nanotubes (Chen and Lee 2006), metal nanoparticles (Shiraishi et al. 2002) and fullerene (Suzuki et al. 2001) to enhance response time and contrast, and lowering the driving voltage of liquid crystal displays. Cyclodextrin-capped SiO_2 nanoparticles (Cyclodextrin = α-cyclodextrin, β-cyclodextrin, γ-cyclodextrin and poly-γ-cyclodextrin) have also been utilized for doping liquid crystals of 4-cyano-4′-pentylbiphenyl to enhance their response time. (Shiraishi et al. 2012). The dispersion of cyclodextrin-SiO_2 into the liquid crystals of 4-cyano-4′-pentylbiphenyl resulted into twisted nematic liquid crystal (TN-LCD) with comparative faster response time than pure 4-cyano-4′-pentylbiphenyl. The capped cyclodextrin-SiO_2 formed homogenous nanoparticles with average diameter of 9.4, 8.4, 10.6, and 6.4 nm for α-cyclodextrin, β-cyclodextrin, γ-cyclodextrin and poly-γ-cyclodextrin, respectively corresponding to their protecting abilities (β-cyclodextrin > α-cyclodextrin > γ-cyclodextrin) (Komiyama and Hirai 1983). Doping of cyclodextrin-SiO_2 with 4-cyano-4′-pentylbiphenyl liquid crystals resulted in formation of twisted nematic liquid crystal with similar phase transition temperature (4-cyano-4′-pentylbiphenyl = 34.2 °C, α-cyclodextrin = 34.2 °C, β-cyclodextrin = 34.1 °C, and γ-cyclodextrin = 34.1 °C). The response time (τ_{on}: rise time of 90% -10% transmittance, and τ_{off}: fall time of 10%–90% transmittance) of twisted nematic liquid crystal of cyclodextrin-SiO_2 appended 4-cyano-4′-pentylbiphenyl were obtained by measuring the transmission of liquid crystals. Faster response time was observed for twisted nematic liquid crystal of cyclodextrin-SiO_2 + 4-cyano-4′-pentylbiphenyl (76.7, 71.5, and 90.7 msec for α-cyclodextrin, β-cyclodextrin, and γ-cyclodextrin respectively) compared to 96.2 msec of 4-cyano-4′-pentylbiphenyl alone attributed to capping, protection and solubilizing properties of cyclodextrin. Comparatively, β-cyclodextrin-SiO_2 + 4-cyano-4′-pentylbiphenyl liquid crystals showed faster response time over α-cyclodextrin, and γ-cyclodextrin due to better inclusion of 4-cyano-4′-pentylbiphenyl in β-cyclodextrin cavity. To visualize the practical utility of doped system, a liquid crystal device was developed using M03 (a commonly used practical liquid crystal in display industries) as host liquid crystal and poly-γ-cyclodextrin as stabilizer owing to its higher stability for longer time and better fall time. An enhanced response time of poly-γ-cyclodextrin-SiO_2 doped M03 liquid crystal display was observed validating its industrial application.

Further experiments were performed to investigate the doping effect of different metal nanoparticles such as barium titanate ($BaTiO_3$) (Shiraishi et al. 2015) and ZrO_2/Au (Shiraishi et al. 2016) capped with cyclodextrin on liquid crystals performance. Barium titanate exhibits high dielectric constant and ferroelectric properties, explaining its numerous applications such as electro-optical displays, and thermistors. Doping of nematic liquid crystals with $BaTiO_3$ were found to have strong electromechanical polarization fluctuations (Basu 2014) and pretransitional effects on the isotropic to nematic phase transitions (Čopič et al. 2007). To increase the compatibility and response time of liquid crystals, barium titanate was capped with γ-cyclodextrin and mixed with liquid crystals of 4-cyano-4′-pentylbiphenyl. Particles of γ-cyclodextrin appended barium titanate (2.1 nm diameter) prepared using microwave/ ultra-sonication showed better response time compared to solvothermal method (2.3 nm, 2.7 nm and 3.9 nm, at 240 °C, 270 °C and 300 °C, respectively) due to their higher dispersion with liquid crystals of 4-cyano-4′-pentylbiphenyl. Faster response of γ-cyclodextrin- $BaTiO_3$ doped 4-cyano-4′-pentylbiphenyl with τ_{on} = 58.55 msec and τ_{off} = 14.49 msec was found compared to pure 4-cyano-4′-pentylbiphenyl with τ_{on} = 60.57 msec and τ_{off} = 15.45 msec. The field sequential-color (FSC) specific liquid crystal NTN01 was doped with γ-cyclodextrin-$BaTiO_3$ nanoparticles and transmission was recorded. Response time of 3.99% was enhanced for NTN01 at 25 °C in the presence of γ-cyclodextrin-$BaTiO_3$ nanoparticles (τ_{on} = 10.43 msec and τ_{off} = 4.86 msec; τ_{total} = 15.29 msec) compared to homologous system in absence of γ-cyclodextrin (τ_{on} = 9.82 msec and τ_{off} = 4.86 msec; τ_{total} = 14.68 msec). On decreasing the temperature to 0 °C, 7.44% faster response time was observed. These results illustrated that doping liquid crystals with γ-cyclodextrin capped nanoparticles gave an enhanced activity of electro-optical properties yielding better display devices.

Undoubtedly, the use of metal based nanoparticle dopant in liquid crystalline materials produces favorable physicochemical impact on their display properties. Various dopants induce perturbation and polarization of the orientation pattern of liquid crystal materials, thereby, display different electro-optical properties. Among different kinds of dopants, mono- and bimetallic nanoparticles have shown interesting applicability due to their quantum size and higher surface area (Cao et al. 2010; Corain et al. 2011). On comparing both, bimetallic nanoparticles advantageously possess large and diverse surface area for inducing higher catalytic and capping abilities (Wieckowski et al. 2003). To develop such system, poly-γ-cyclodextrin has been used as capping agent for generating ZrO_2/ Au nanoparticles using ultra-sonication and microwave methods (Shiraishi et al. 2016). Different ratios of ZrO_2/Au (1:1, 2:1, 4:1, and 9:1) were used to formulate nanoparticles of different sizes (4.6 nm, 4.0 nm, 3.1 nm, and 3.1 nm, respectively) as demonstrated by dynamic light scattering (DLS) and transmission electron microscopy images. The addition of poly-γ-cyclodextrin-ZrO_2/Au dopant to NTN01 resulted in transition of isotropic phase to twisted nematic phase. Response times of twisted nematic liquid crystals obtained by mixing poly-γ-cyclodextrin-ZrO_2/Au and NTN01 were evaluated for different ratio of ZrO_2/Au. Compared to pure NTN01, Poly-γ-cyclodextrin-ZrO_2/Au doped twisted nematic liquid crystal exhibited enhanced response time, illustrating the structural diversity effect of poly-γ-cyclodextrin and ZrO_2/Au particles on liquid crystalline behavior. Further, improved response rate was observed for ZrO_2/Au (4:1) ratio compared to lower ratios of ZrO_2/Au (2:1) and ZrO_2/Au (1:1) ratios. Thus, developing better liquid crystal displays by doping available liquid crystals with different dopants would be an advantageous strategy for reducing ill effects of modern technologies.

7.4.3 Discrete Applications

Using a combination strategy of cyclodextrin and liquid crystals, the discrimination of enantiomers discrimination has been reported. Normally, NMR spectroscopy is the technique used to discriminate enantiomers with the help of a chiral auxiliary, by taking advantage of the diastereomeric interaction between the added chiral auxiliary and each enantiomer. In order to observe appreciable chemical shifts, strong diastereomeric interactions are required. An alternative method was developed to discriminate enantiomers using cyclodextrin (β-cyclodextrin and (2-hydroxypropyl)-β-cyclodextrin) as chiral cages in non-chiral liquid crystal solvent (Péchiné et al. 2002). The different inclusion and orientation of enantiomers in chiral cages

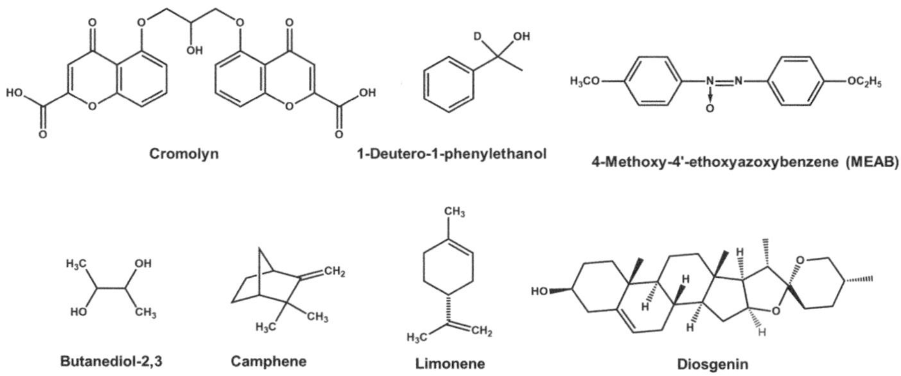

Fig. 7.39 Structures of different compounds used in various studies

produce significant and easily observable chemical shifts in the NMR spectra. Cromolyn was used as nematic liquid crystal phase in water and 1-deutero-1-phenylethanol was used as chiral compound with racemic and (S)-enriched species (Fig. 7.39). The NMR spectrum of racemic mixture showed 3 deuterium quadrupolar splittings when dissolved in cromolyn-β-cyclodextrin mesophase. Two inner doublets corresponding to encaged enantiomers with quadrupolar splittings were observe; the outer doublets correspond to outside cavity enantiomers. Thus, the different orientation of the encaged enantiomer in β-cyclodextrin cavity in liquid crystal phase allowed their differentiation.

In another application, chiral nematic liquid crystals with helically twisted mesophase were obtained as chiral sorbent, by mixing the achiral nematic liquid crystals of 4-methoxy-4′-ethoxyazoxybenzene (MEAB) with acetyl-β-cyclodextrin as chiral auxiliary (Fig. 7.39) (Onuchak et al. 2012). Higher melting enthalpy of sorbent by 2.57% compared to 4-methoxy-4′-ethoxyazoxybenzene was observed, confirming intermolecular interactions between them in twisted mesophase. The sorption properties, including separation and thermodynamic functions of these chiral sorbent, were investigated toward different organic molecules including alkanes, cycloalkanes, camphene, limonene alcohols, heterocycles, butanediol-2,3 and aromatic molecules (Fig. 7.39). The structural and enantiomeric selectivity of investigated molecules depend upon the transition temperature range of sorbent, i.e. 95–120 °C for structural and 91–100 °C for enantiomeric, respectively. The polar and aromatic sorbates showed positive sorption contributions (δ) owing to their host guest interactions with acetyl-β-cyclodextrin. However, negative sorption contributions for other sorbates were observed, due to highly ordered structure and solvation effect between liquid crystal-cyclodextrin that hindered their interactions. Further, (+) isomers of butanediol-2,3, camphene, and limonene showed better enantioselectivity compared to (−) isomers due to a better compatibility with the D-(+)-glucopyranose units of acetyl-β-cyclodextrin in temperature window of 91–100 °C.

For biological applications, cyclodextrin are among the most suitable scaffolds owing to their biocompatibility and low cytotoxicity profile. Diosgenin (Fig. 7.39), (Final Report of the Amended Safety Assessment of Dioscorea Villosa (Wild Yam) Root Extract, 2004) a steroidal saponin, is a multifunctional molecule with numerous applications including plasma glucose (Pari et al. 2012) and cholesterol inhibitor (Gong et al. 2010), source of steroid hormones, dietary supplement and melanogenesis inhibitor (Lee et al. 2007). The bioavailability of diosgenin for medical applications is very limited due to its poor aqueous solubility; however, upon complexation with β-cyclodextrin, its solubility improves significantly. Okawara et al. (2014) used a combination of diosgenin, liquid crystalline amphipathic lipids: glyceryl monooleate and phytantriol and β-cyclodextrin, to investigate their enhanced solubility profile. Glyceryl monooleate and phytantriol exhibited self-assembled and dispersed liquid crystalline properties. The small-angle X-ray scattering analysis of glyceryl monooleate-diosgenin and phytantriol-diosgenin mixtures confirmed the formation of hexagonal and cubic mesophases of 100 and 200 nm respectively. The solubility and plasma concentration of diosgenin significantly increased upon mixing with glyceryl monooleate or phytantriol lipids illustrating enhanced permeation across intestinal mucosa. Comparatively, these effects are better for glyceryl monooleate than phytantriol, which were further enhanced upon addition of β-cyclodextrin. Pharmacokinetic analysis of oral administration of glyceryl monooleate-diosgenin- β-cyclodextrin and phytantriol-diosgenin-β-cyclodextrin to Wistar rats revealed higher diosgenin distribution across skin compared to diosgenin suspension alone with minimal undesired toxicity. Thus, these results validated the benefit of using β-cyclodextrin to enhance the drug solubility and bioavailability, resulting in better pharmacokinetics.

7.5 Conclusion

Since the first report of cyclodextrin-based thermotropic liquid crystals, over the years, several families of cyclodextrin-based liquid crystal materials have been reported. However, comparing to other types of materials, the research in developing cyclodextrin-based liquid crystal materials has been relatively slow. Overall, the design of self-assembled cyclodextrin materials can take advantages of fundamental intermolecular forces such as H-bonding, dipole-dipole interactions, as well as π-π interactions. Interestingly, the use of weaker interaction forces appears to generate materials with improved fluidity and richer phase transitions. Although native cyclodextrin molecules are incapable of forming thermotropic liquid crystals, the introduction of mesogenic groups can produce various mesophases such as nematic, smectic, and columnar. Photoresponsive systems were also obtained by conjugating azobenzene moieties to cyclodextrin scaffolds. Several polymeric liquid crystalline materials were demonstrated which possess very interesting properties. Even after all these investigations, several unanswered questions remain. More investigations are needed to tune the materials in order to be commercially viable.

Acknowledgements The financial support from Alberta Innovates – Technology Futures, the Natural Sciences and Engineering Research Council of Canada, and the University of Calgary are greatly acknowledged.

References

Araki J, Sato H, Takagi Y, Ohta K (2014) "Molecular rope curtain" type of liquid crystals based on a sliding graft copolymer having mobile peg side chains. Mol Cryst Liq Cryst 592:99–105. https://doi.org/10.1080/07315171.2013.840758

Ashton PR, Ellwood P, Staton I, Stoddart JF (1991) Per-3,6-anhydro-alpha-cyclodextrin and per-3,6-anhydro-beta-cyclodextrin. J Org Chem 56:7274–7280. https://doi.org/10.1021/jo00026a017

Ashton PR, Boyd SE, Gattuso G, Hartwell EY, Koeniger R, Spencer N et al (1995) A novel approach to the synthesis of some chemically-modified cyclodextrins. J Org Chem 60:3898–3903. https://doi.org/10.1021/jo00117a049

Avakyan VG, Nazarov VB, Alfimov MV, Bagatur AA, Voronezheva NI et al (2001) The role of intra-and intermolecular hydrogen bonds in the formation of β-cyclodextrin head-to-head and head-to-tail dimers. The results of ab initio and semiempirical quantum-chemical calculations. Russ Chem Bull 50:206–216. https://doi.org/10.1023/A:1009557729668

Ba C-Y, Shen Z-R, Gu H-W, Guo G-Q, Xie P, Zhang R-B et al (2003) A triphenylene-containing side chain liquid crystalline ladder-like polysiloxane and its highly ordered superstructure. Liq Cryst 30:391–397. https://doi.org/10.1080/0267829031000089924

Bai Y, Abbott NL (2011) Recent advances in colloidal and interfacial phenomena involving liquid crystals. Langmuir 27:5719–5738. https://doi.org/10.1021/la103301d

Barnes WH, Ross S (1936) The diffraction of X-rays by the higher polyethylene glycols and by polymerized ethylene oxides1. J Am Chem Soc 58:1129–1131. https://doi.org/10.1021/ja01298a018

Basu R (2014) Soft memory in a ferroelectric nanoparticle-doped liquid crystal. Phys Rev E 89:022508. https://doi.org/10.1103/PhysRevE.89.022508

Bi X, Hartono D, Yang K-L (2009) Real-time liquid crystal pH sensor for monitoring enzymatic activities of penicillinase. Adv Funct Mater 19:3760–3765. https://doi.org/10.1002/adfm.200900823

Bilalov A, Carlstedt J, Krivtsova E, Lindman B, Olsson U (2012) DNA with amphiphilic counterions: tuning colloidal DNA with cyclodextrin. Soft Matter 8:4988. https://doi.org/10.1039/c2sm25058f

Bonini M, Rossi S, Karlsson G, Almgren M, Lo Nostro P, Baglioni P (2006) Self-assembly of β-cyclodextrin in water. Part 1: Cryo-TEM and dynamic and static light scattering. Langmuir 22:1478–1484. https://doi.org/10.1021/la052878f

Bortel E, Hodorowicz S, Lamot R (1979) Relation between crystallinity degree and stability in solid state of high molecular weight poly(ethylene oxide)s. Macromol Chem Phys 180:2491–2498. https://doi.org/10.1002/macp.1979.021801023

Bouteiller L, Barny PL (1996) Polymer-dispersed liquid crystals: preparation, operation and application. Liq Cryst 21:157–174. https://doi.org/10.1080/02678299608032820

Brake JM, Abbott NL (2002) An experimental system for imaging the reversible adsorption of amphiphiles at aqueous–liquid crystal interfaces. Langmuir 18:6101–6109. https://doi.org/10.1021/la011746t

Brake JM, Daschner MK, Luk Y-Y, Abbott NL (2003a) Biomolecular interactions at phospholipid-decorated surfaces of liquid crystals. Science 302:2094–2097. https://doi.org/10.1126/science.1091749

Brake JM, Mezera AD, Abbott NL (2003b) Active control of the anchoring of 4'-pentyl-4-cyanobiphenyl (5CB) at an aqueous–liquid crystal interface by using a redox-active ferrocenyl surfactant. Langmuir 19:8629–8637. https://doi.org/10.1021/la034469u

Bremer M, Kirsch P, Klasen-Memmer M, Tarumi K (2013) The TV in your pocket: development of liquid-crystal materials for the new millennium. Angew Chem Int Ed 52:8880–8896. https://doi.org/10.1002/anie.201300903

Cammidge AN, Gopee H (2009) Synthesis and liquid crystal properties of mixed alkynyl-alkoxy-triphenylenes. Liq Cryst 36:809–816. https://doi.org/10.1080/02678290903063000

Cao A, Lu R, Veser G (2010) Stabilizing metal nanoparticles for heterogeneous catalysis. Phys Chem Chem Phys 12:13499–13510. https://doi.org/10.1039/C0CP00729C

Carlstedt J, Bilalov A, Olsson U (2012) Aqueous phase behavior of polyelectrolytes with amphiphilic counterions modulated by cyclodextrin: the role of polyion flexibility. Phys Chem Chem Phys 14:9574. https://doi.org/10.1039/c2cp41353a

Challa R, Ahuja A, Ali J, Khar RK (2005) Cyclodextrins in drug delivery: an updated review. AAPS PharmSciTech 6:E329–E357. https://doi.org/10.1208/pt060243

Champagne P-L, Ester D, Ward S, Williams V E, Ling C-C (2016) A family of amphiphilic cyclodextrin liquid crystals governed by dipole–dipole interactions. ChemPlusChem n/a-n/a doi:https://doi.org/10.1002/cplu.201600556

Chen H-Y, Lee W (2006) Suppression of field screening in nematic liquid crystals by carbon nanotubes. Appl Phys Lett 88:222105. https://doi.org/10.1063/1.2208373

Chen L, Hu T-H, Xie H-L, Zhang H-L (2010) A mixed cyclodextrin-biphenyl thermotropic liquid crystal: synthesis, liquid-crystalline properties, and supramolecular organization. J Polym Sci A Polym Chem 48:2838–2845. https://doi.org/10.1002/pola.24060

Collings PJ, Hird M (1997) Introduction to liquid crystals: chemistry and physics. CRC Press

Čopič M, Mertelj A, Buchnev O, Reznikov Y (2007) Coupled director and polarization fluctuations in suspensions of ferroelectric nanoparticles in nematic liquid crystals. Phys Rev E 76:011702. https://doi.org/10.1103/PhysRevE.76.011702

Corain B, Schmid G, Toshima N (2011) Metal nanoclusters in catalysis and materials science: the issue of size control. Elsevier

Corruccini RJ, Gilbert EC (1939) The heat of combustion of cis-and trans-azobenzene. J Am Chem Soc 61:2925–2927

Cramer F, Mackensen G, Sensse K (1969) On ring compounds. XX. ORD-spectra and conformation of the glucose ring in cyclodextrins. Chem Ber 102:494–508. https://doi.org/10.1002/cber.19691020217

Cromwell WC, Bystrom K, Eftink MR (1985) Cyclodextrin-adamantanecarboxylate inclusion complexes: studies of the variation in cavity size. J Phys Chem 89:326–332. https://doi.org/10.1021/j100248a029

Davis ME, Brewster ME (2004) Cyclodextrin-based pharmaceutics: past, present and future. Nat Rev Drug Discov 3:1023–1035

Deng J, Lu X, Constant C, Dogariu A, Fang J (2015) Design of β-CD–surfactant complex-coated liquid crystal droplets for the detection of cholic acid via competitive host–guest recognition. Chem Commun 51:8912–8915. https://doi.org/10.1039/C5CC01561H

Devadoss A, Burgess JD (2002) Detection of cholesterol through electron transfer to cholesterol oxidase in electrode-supported lipid bilayer membranes. Langmuir 18:9617–9621. https://doi.org/10.1021/la0258594

Dey RS, Raj CR (2010) Development of an amperometric cholesterol biosensor based on graphene–pt nanoparticle hybrid material. J Phys Chem C 114:21427–21433. https://doi.org/10.1021/jp105895a

Dias R, Lindman B (2008) DNA interactions with polymers and surfactants. Wiley, Hoboken

Doane JW, Vaz NA, Wu B-G, Žumer S (1986) Field controlled light scattering from nematic microdroplets. Appl Phys Lett 48:269–271. https://doi.org/10.1063/1.96577

Dong Y, Lam JW, Peng H, Cheuk KK, Kwok HS, Tang BZ (2004) Syntheses and mesomorphic and luminescent properties of disubstituted polyacetylenes bearing biphenyl pendants. Macromolecules 37:6408–6417. https://doi.org/10.1021/ma049094d

Dorrego AB, García-Río L, Hervés P, Leis JR, Mejuto JC, Pérez-Juste J (2000) Micellization versus cyclodextrin–surfactant complexation. Angew Chem Int Ed 39:2945–2948. doi:10.1002/1521-3773(20000818)39:16<2945::AID-ANIE2945>3.0.CO;2-6

Du Y, Toshima N (2007) Low driving voltage of twisted nematic liquid crystal displays doped with CdS nanoparticles. Bull Chem Soc Jpn 80:2446–2450. https://doi.org/10.1246/bcsj.80.2446

Eftink M, Andy M, Bystrom K, Perlmutter H, Kristol D (1989) Cyclodextrin inclusion complexes - studies of the variation in the size. J Am Chem Soc 111:6765–6772. https://doi.org/10.1021/ja00199a041

Fügedi P (1989) Synthesis of heptakis (6-O-tert-butyldimethylsilyl) cyclomaltoheptaose and octakis (6-O-tert-butyldimethylsilyl) cyclomalto-octaose. Carbohydr Res 192:366–369. https://doi.org/10.1016/0008-6215(89)85197-3

Fujita K, Nagamura S, Imoto T (1984) Convenient preparation and effective separation of the C-2 and C-3 tosylates of α-cyclodextrin. Tetrahedron Lett 25:5673–5676. https://doi.org/10.1016/S0040-4039(01)91409-1

Gadelle A, Defaye J (1991) Selective halogenation at primary positions of cyclomaltooligosaccharides and a synthesis of per-3,6-anhydro cyclomaltooligosaccharides. Angew Chem Int Ed EngL 30:78–80. https://doi.org/10.1002/anie.199100781

Gelb RI, Schwartz LM (1989) Complexation of adamantane-ammonium substrates by beta-cyclodextrin and its O-methylated derivatives. J Incl Phenom Mol Recognit Chem 7:537–543. https://doi.org/10.1007/BF01080464

Gibson HW, Bheda MC, Engen PT (1994) Rotaxanes, catenanes, polyrotaxanes, polycatenanes and related materials. Prog Polym Sci 19:843–945. https://doi.org/10.1016/0079-6700(94)90034-5

Gong G, Qin Y, Huang W, Zhou S, Wu X, Yang X et al (2010) Protective effects of diosgenin in the hyperlipidemic rat model and in human vascular endothelial cells against hydrogen peroxide-induced apoptosis. Chem Biol Interact 184:366–375. https://doi.org/10.1016/j.cbi.2010.02.005

Goossens K, Lava K, Bielawski CW, Binnemans K (2016) Ionic liquid crystals: versatile materials. Chem Rev 116:4643–4807. https://doi.org/10.1021/cr400334b

Gray GW (1962) Molecular structure and the properties of liquid crystals. Academic Press Inc, New York

Griffiths WJ, Sjövall J (2010) Bile acids: analysis in biological fluids and tissues. J Lipid Res 51:23–41. https://doi.org/10.1194/jlr.R001941-JLR200

Gulik A, Delacroix H, Wouessidjewe D, Skiba M (1998) Structural properties of several amphiphile cyclodextrins and some related nanospheres. An X-ray scattering and freeze-fracture electron microscopy study. Langmuir 14:1050–1057. https://doi.org/10.1021/la971019r

Gündüz B (2015) Sensing and surface morphological properties of a poly [(9, 9-dioctylfluorenyl-2, 7-diyl)-co-bithiophene] liquid-crystalline polymer for optoelectronic applications. J Appl Polym Sci 132. Available at: http://onlinelibrary.wiley.com/doi/10.1002/app.41659/pdf. Accessed 15 Feb 2017

Guo A, Yan Z, Ye L, Zhang A, Feng Z (2016) The synthesis and characterization of spacer-free liquid crystal polyrotaxane by virtue of the mobility of threaded α-cyclodextrins. Macromol Chem Phys 217:646–653. https://doi.org/10.1002/macp.201500415

Han X, Chai L, Shanks R, Pavel D (2006) Monte Carlo simulations of properties of side-chain liquid-crystal polymers. Polym Int 55:1323–1329. https://doi.org/10.1002/pi.2087

Harada A, Li J, Kamachi M (1992) The molecular necklace: a rotaxane containing many threaded alpha-cyclodextrins. Nature 356:325. https://doi.org/10.1038/356325a0

Harada A, Li J, Kamachi M (1993) Synthesis of a tubular polymer from threaded cyclodextrins. Nature 364:516–518. https://doi.org/10.1038/364516a0

Harries D, Rau DC, Parsegian VA (2005) Solutes probe hydration in specific association of cyclodextrin and adamantane. J Am Chem Soc 127:2184–2190. https://doi.org/10.1021/ja045541t

Hartono D, Bi X, Yang K-L, Yung L-YL (2008) An air-supported liquid crystal system for real-time and label-free characterization of phospholipases and their inhibitors. Adv Funct Mater 18:2938–2945. https://doi.org/10.1002/adfm.200800424

Hazra P, Chakrabarty D, Chakraborty A, Sarkar N (2004) Intramolecular charge transfer and solvation dynamics of Nile Red in the nanocavity of cyclodextrins. Chem Phys Lett 388:150–157. https://doi.org/10.1016/j.cplett.2004.02.078

He T, Hu T, Zhang X, Zhong G, Zhang H (2009) Synthesis and characterization of a novel liquid crystalline star-shaped polymer based on α-CD core via ATRP. J App Polymer Sci 112:2120–2126. https://doi.org/10.1002/app.29657

Hee Lee S, Sarathi Bhattacharyya S, Seok Jin H, Jeong K-U (2012) Devices and materials for high-performance mobile liquid crystal displays. J Mater Chem 22:11893–11903. https://doi.org/10.1039/C2JM30635B

Hiltrop K (1994) Lyotropic liquid crystals. In: Stegemeyer PDH, Behret GSDH, DB-G für PC e.V (Steinkopff) (ed) Liquid crystals topics in physical chemistry, pp 143–171. https://doi.org/10.1007/978-3-662-08393-2_4

Hirayama F, Uekama K (1999) Cyclodextrin-based controlled drug release system. Adv Drug Deliv Rev 36:125–141. https://doi.org/10.1016/S0169-409X(98)00058-1

Hoare TR, Kohane DS (2008) Hydrogels in drug delivery: progress and challenges. Polymer 49:1993–2007. https://doi.org/10.1016/j.polymer.2008.01.027

Hoeben FJM, Jonkheijm P, Meijer EW, Schenning APHJ (2005) About supramolecular assemblies of π-conjugated systems. Chem Rev 105:1491–1546. https://doi.org/10.1021/cr030070z

Hofmann AF, Hagey LR (2008) Bile acids: chemistry, pathochemistry, biology, pathobiology, and therapeutics. Cell Mol Life Sci 65:2461–2483. https://doi.org/10.1007/s00018-008-7568-6

Phases of Liquid Crystals (n.d.) Available at: http://plc.cwru.edu/tutorial/enhanced/files/lc/phase/phase.htm. Accessed 13 Mar 2016

Hu T, Xie H, Chen L, Chen S, Zhang H (2011) Intriguing liquid crystalline behavior of liquid crystalline polyrotaxane containing azobenzene mesogens. Polym Bull 67:937–950

Ikonen E (2008) Cellular cholesterol trafficking and compartmentalization. Nat Rev Mol Cell Biol 9:125–138. https://doi.org/10.1038/nrm2336

Jiang L, Peng Y, Yan Y, Deng M, Wang Y, Huang J (2010a) "Annular Ring" microtubes formed by SDS@2β-CD complexes in aqueous solution. Soft Matter 6:1731. https://doi.org/10.1039/b920608f

Jiang L, Yan Y, Huang J, Yu C, Jin C, Deng M et al (2010b) Selectivity and stoichiometry boosting of β-cyclodextrin in cationic/anionic surfactant systems: when host−guest equilibrium meets biased aggregation equilibrium. J Phys Chem B 114:2165–2174. https://doi.org/10.1021/jp911092y

Jiang L, Peng Y, Yan Y, Huang J (2011) Aqueous self-assembly of SDS@2β-CD complexes: lamellae and vesicles. Soft Matter 7:1726–1731. https://doi.org/10.1039/C0SM00917B

Jorquera F, Monte MJ, Guerra J, Sanchez-Campos S, Merayo JA, Olcóz JL et al (2005) Usefulness of combined measurement of serum bile acids and ferritin as additional prognostic markers to predict failure to reach sustained response to antiviral treatment in chronic hepatitis C. J Gastroenterol Hepatol 20:547–554. https://doi.org/10.1111/j.1440-1746.2005.03725.x

Karaky K, Reynaud S, Billon L, François J, Chreim Y (2005) Organosoluble star polymers from a cyclodextrin core. J Polym Sci A Polym Chem 43:5186–5194. https://doi.org/10.1002/pola.21012

Kato T, Mizoshita N, Kishimoto K (2006) Functional liquid-crystalline assemblies: self-organized soft materials. Angew Chem Int Ed 45:38–68

Katsuno C, Konda A, Urayama K, Takigawa T, Kidowaki M, Ito K (2013) Pressure-responsive polymer membranes of slide-ring gels with movable cross-links. Adv Mater 25:4636–4640. https://doi.org/10.1002/adma.201301252

Khan AR, Forgo P, Stine KJ, D'Souza VT (1998) Methods for selective modifications of cyclodextrins. Chem Rev 98:1977–1996. https://doi.org/10.1021/cr970012b

Kidowaki M, Nakajima T, Araki J, Inomata A, Ishibashi H, Ito K (2007) Novel liquid crystalline polyrotaxane with movable mesogenic side chains. Macromolecules 40:6859–6862. https://doi.org/10.1021/ma070785u

Kim H, Kim YH, Lee S, Walba DM, Clark NA, Lee SB et al (2014) Orientation control over bent-core smectic liquid crystal phases. Liq Cryst 41:328–341. https://doi.org/10.1080/02678292.2013.817618

Koltzenburg S, Wolff D, Springer J, Nuyken O (1998) Novel study on the liquid crystalline behavior of poly (methacrylate) s with biphenyl side groups. J Polym Sci A Polym Chem 36:2669–2679. doi:10.1002/(SICI)1099-0518(19981115)36:15<2669::AID-POLA1>3.0.CO;2-4

Komiyama M, Hirai H (1983) Colloidal rhodium dispersions protected by cyclodextrins. Bull Chem Soc Jpn 56:2833–2834. https://doi.org/10.1246/bcsj.56.2833

Kumar S (2004) Recent developments in the chemistry of triphenylene-based discotic liquid crystals. Liq Cryst 31:1037–1059. https://doi.org/10.1080/02678290410001724746

Kumar S (2010) Chemistry of discotic liquid crystals: from monomers to polymers. CRC Press, Boca Raton

Kwak E-S, Gomez FA (1996) Determination of the binding of β-cyclodextrin derivatives to adamantane carboxylic acids using capillary electrophoresis. Chromatographia 43:659–662. https://doi.org/10.1007/BF02292984

Laschat S, Baro A, Steinke N, Giesselmann F, Hägele C, Scalia G et al (2007) Discotic liquid crystals: from tailor-made synthesis to plastic electronics. Angew Chem Int Ed 46:4832–4887. https://doi.org/10.1002/anie.200604203

Lee J, Jung K, Kim YS, Park D (2007) Diosgenin inhibits melanogenesis through the activation of phosphatidylinositol-3-kinase pathway (PI3K) signaling. Life Sci 81:249–254. https://doi.org/10.1016/j.lfs.2007.05.009

Lee E, Kim J-K, Lee M (2009) Reversible scrolling of two-dimensional sheets from the self-assembly of laterally grafted amphiphilic rods. Angew Chem Int Ed 48:3657–3660. https://doi.org/10.1002/anie.200900079

Liao Z, Du S, Qin Z, Wang J, Zuo F, Luo J (2015) Use of liquid crystals for imaging different inclusion abilities of α-cyclodextrin and β-cyclodextrin toward cetyltrimethyl ammonium bromide. Chem Phys Lett 637:189–194. https://doi.org/10.1016/j.cplett.2015.08.021

Lin H, Ding H, Kelly JR (1995) The mechanism of switching a PDLC film. Mol Cryst Liq Cryst 262:99–109. https://doi.org/10.1080/10587259508033516

Ling C-C, Darcy R, Risse W (1993) Cyclodextrin liquid crystals: synthesis and self-organisation of amphiphilic thio-β-cyclodextrins. J Chem Soc Chem Commun 438–440. doi:https://doi.org/10.1039/C39930000438.

Linlin D, Wei L, Jun C, Ban L, Shouxin L (2015) Formation, tuning and application of chiral nematic liquid crystal phase based on nanocrystalline cellulose. Prog Chem 27:861–869. https://doi.org/10.7536/PC141239

Liu Q-Y, Zuo F, Chong Y-Y, Zhao Z-G, Kwon Y, Chen J-X et al (2016) Molecular simulation of liquid crystal sensor based on competitive inclusion effect. J Incl Phenom Macrocycl Chem. https://doi.org/10.1007/s10847-016-0678-0

Lowe AM, Abbott NL (2012) Liquid crystalline materials for biological applications. Chem Mater 24:746–758. https://doi.org/10.1021/cm20263 2 m

Mamiya J, Kuriyama A, Yokota N, Yamada M, Ikeda T (2015) Photomobile polymer materials: photoresponsive behavior of cross-linked liquid-crystalline polymers with mesomorphic diarylethenes. Chem Eur J 21:3174–3177. https://doi.org/10.1002/chem.201406299

McArdle CB (1990) Side chain liquid crystal polymers. Springer Science & Business Media, New York

Mondal A, Jana NR (2012) Fluorescent detection of cholesterol using β-cyclodextrin functionalized graphene. Chem Commun 48:7316. https://doi.org/10.1039/c2cc33410k

Munir S, Park S-Y (2015) The development of a cholesterol biosensor using a liquid crystal/aqueous interface in a SDS-included β-cyclodextrin aqueous solution. Anal Chim Acta 893:101–107. https://doi.org/10.1016/j.aca.2015.08.051

Mushenheim PC, Trivedi RR, Weibel DB, Abbott NL (2014) Using liquid crystals to reveal how mechanical anisotropy changes interfacial behaviors of motile bacteria. Biophys J 107:255–265. https://doi.org/10.1016/j.bpj.2014.04.047

Nagarajan R (2002) Molecular packing parameter and surfactant self-assembly: the neglected role of the surfactant tail. Langmuir 18:31–38. https://doi.org/10.1021/la010831y

Noda Y, Hayashi Y, Ito K (2014) From topological gels to slide-ring materials. J Appl Polym Sci 131:n/a–n/a. https://doi.org/10.1002/app.40509

Oh CS (1977) Induced smectic mesomorphism by incompatible nematogens. Mol Cryst Liq Cryst 42:1–14. https://doi.org/10.1080/15421407708084491

Ohno K, Wong B, Haddleton DM (2001) Synthesis of well-defined cyclodextrin-core star polymers. J Polym Sci A Polym Chem 39:2206–2214. https://doi.org/10.1002/pola.1197.

Okawara M, Hashimoto F, Todo H, Sugibayashi K, Tokudome Y (2014) Effect of liquid crystals with cyclodextrin on the bioavailability of a poorly water-soluble compound, diosgenin, after its oral administration to rats. Int J Pharm 472:257–261. https://doi.org/10.1016/j.ijpharm.2014.06.032.

Okumura Y, Ito K (2001) The polyrotaxane gel: a topological gel by figure-of-eight cross-links. Adv Mater 13:485–487

Onuchak LA, Burmatnova TS, Spiryaeva EA, Kuraeva YG, Belousova ZP (2012) Sorption and selective properties of binary liquid-crystalline sorbent based on 4-Methoxy-4′-ethoxyazoxybenzene and a cetylated β-Cyclodextrin. Russ J Phys Chem A 86:1308–1317. https://doi.org/10.1134/S0036024412080080.

Oswald P, Pieranski P (2005) Smectic and columnar liquid crystals: concepts and physical properties illustrated by experiments. CRC Press. Available at: https://books.google.ca/books?hl=en&lr=&id=f4Q1m9cLEaEC&oi=fnd&pg=PP1&dq=Grandjean+terrace+mesophase+characteristics+of+smectic+liquid+crystals&ots=eRfiz3-CaM&sig=gd31p7B_9AcOzg3BokWm_7Vr-h0. Accessed 20 Feb 2017

Palepu R, Reinsborough VC (1990) β-cyclodextrin inclusion of adamantane derivatives in solution. Aust J Chem 43:2119–2123. https://doi.org/10.1071/ch9902119.

Paraschiv I, Tomkinson A, Giesbers M, Sudhölter EJ, Zuilhof H, Marcelis AT (2007) Amide, urea and thiourea-containing triphenylene derivatives: influence of H-bonding on mesomorphic properties. Liq Cryst 34:1029–1038. https://doi.org/10.1080/02678290701609640

Pari L, Monisha P, Mohamed Jalaludeen A (2012) Beneficial role of diosgenin on oxidative stress in aorta of streptozotocin induced diabetic rats. Eur J Pharmacol 691:143–150. https://doi.org/10.1016/j.ejphar.2012.06.038.

Park J-S, Abbott NL (2008) Ordering transitions in thermotropic liquid crystals induced by the interfacial assembly and enzymatic processing of oligopeptide amphiphiles. Adv Mater 20:1185–1190. https://doi.org/10.1002/adma.200702012.

Péchiné J-M, Meddour A, Courtieu J (2002) Monitoring the differential ordering of enantiomers included into cyclodextrins through deuterium NMR in lyotropic liquid crystals. Chem Commun:1734–1735. https://doi.org/10.1039/B205256C.

Petti L, Mormile P, Blau WJ (2003) Fast electro-optical switching and high contrast ratio in epoxy-based polymer dispersed liquid crystals. Opt Lasers Eng 39:369–377. https://doi.org/10.1016/S0143-8166(01)00119-1.

Prasad SK, Rao DS, Chandrasekhar S, Kumar S (2003) X-Ray studies on the columnar structures of discotic liquid crystals. Mol Cryst Liq Cryst 396:121–139. https://doi.org/10.1080/15421400390213258.

Rajewski RA, Stella VJ (1996) Pharmaceutical applications of cyclodextrins. 2. in vivo drug delivery. J Pharm Sci 85:1142–1169. https://doi.org/10.1021/js960075u.

Ram MK, Bertoncello P, Ding H, Paddeu S, Nicolini C (2001a) Cholesterol biosensors prepared by layer-by-layer technique. Biosens Bioelectron 16:849–856

Rani K, Garg P, Pundir CS (2004) Measurement of bile acid in serum and bile with arylamine-glass-bound 3α-hydroxysteroid dehydrogenase and diaphorase. Anal Biochem 332:32–37. https://doi.org/10.1016/j.ab.2004.02.039.

Rasheed A (2008) Cyclodextrins as drug carrier molecule: a review. Sci Pharm 76:567–598. https://doi.org/10.3797/scipharm.0808-05

Rosen MJ, Kunjappu DJT (2012) Surfactants and interfacial phenomena. Wiley, Hoboken

Saenger W, Müller-Fahrnow A (1988) Cyclodextrins increase surface tension and critical micelle concentrations of detergent solutions. Angew Chem Int Ed Eng 27:393–394. https://doi.org/10.1002/anie.198803931.

Saenger W, Noltemeyer M, Manor PC, Hingerty B, Klar B (1976) "Induced-fit"-type complex formation of the model enzyme α-cyclodextrin. Bioorg Chem 5:187–195. https://doi.org/10.1016/0045-2068(76)90007-9

Sergeyev S, Pisula W, Geerts YH (2007) Discotic liquid crystals: a new generation of organic semiconductors. Chem Soc Rev 36:1902–1929. https://doi.org/10.1039/B417320C

Serhatli IE, Kacar T (2006) Synthesis of liquid crystalline–amorphous block copolymers by combination of CFRP and ATRP mechanisms. J Appl Polym Sci 99:3187–3194. https://doi.org/10.1002/app.22036.

Shaikh VAE, Lonikar SV, Dhobale DA, Pawar GM (2007) Cholesterol-linked β-cyclodextrin—a thermotropic liquid-crystalline derivative. Bull Chem Soc Jpn 80:1975–1980. https://doi.org/10.1246/bcsj.80.1975

Shimizu T, Masuda M, Minamikawa H (2005) Supramolecular nanotube architectures based on amphiphilic molecules. Chem Rev 105:1401–1444. https://doi.org/10.1021/cr030072j.

Shiraishi Y, Toshima N, Maeda K, Yoshikawa H, Xu J, Kobayashi S (2002) Frequency modulation response of a liquid-crystal electro-optic device doped with nanoparticles. Appl Phys Lett 81:2845–2847. https://doi.org/10.1063/1.1511282

Shiraishi Y, Sugihara K, Okamura N, Sawai H, Kobayashi S, Toshima N (2012) Fast electro-optic switching of twisted nematic lcd doped with cyclodextrin capped silica nanoparticles. Macromol Symp 317–318:28–33. https://doi.org/10.1002/masy.201100066

Shiraishi Y, Tsujihata R, Sawai H, Kobayashi S, Toshima N (2015) Effect of particle size on electro-optic properties of liquid crystal devices doped with γ-cyclodextrin stabilized barium titanate nanoparticles. Mol Cryst Liq Cryst 611:100–108. https://doi.org/10.1080/15421406.2015.1028000.

Shiraishi Y, Watanabe C, Sawai H, Asano H, Toshima N (2016) Electro-optic function of liquid crystal displays doped with Poly(γ-cyclodextrin)-Protected ZrO_2 /Au nanoparticles. Macromol Symp 364:56–61. https://doi.org/10.1002/masy.201500150.

Song LX, Xu P (2008) A comparative study on the thermal decomposition behaviors between β-cyclodextrin and its inclusion complexes of organic amines. J Phys Chem A 112:11341–11348. https://doi.org/10.1021/jp806026q

Song LX, Teng CF, Xu P, Wang HM, Zhang ZQ, Liu QQ (2008) Thermal decomposition behaviors of β-cyclodextrin, its inclusion complexes of alkyl amines, and complexed β-cyclodextrin at different heating rates. J Incl Phenom Macrocycl Chem 60:223–233. https://doi.org/10.1007/s10847-007-9369-1

Stenzel-Rosenbaum MH, Davis TP, Chen V, Fane AG (2001) Synthesis of poly(styrene) star polymers grown from sucrose, glucose, and cyclodextrin cores via living radical polymerization mediated by a half-metallocene iron carbonyl complex. Macromolecules 34:5433–5438. https://doi.org/10.1021/ma0021803.

Suzuki M, Furue H, Kobayashi S (2001) Polarizerless nanomaterial doped guest-host LCD exhibiting high luminance and good legibility. Mol Cryst Liq Cryst Sci Technol Sect 368:191–196. https://doi.org/10.1080/10587250108029946.

Szejtli J, Huber O (2012) Proceedings of the fourth international symposium on cyclodextrins: Munich, West Germany, April 20–22, 1988. Springer Science & Business Media Available at: https://books.google.ca/books?hl=en&lr=&id=BCTpCAAAQBAJ&oi=fnd&pg=PA1&dq=Szejtli,+J.+ Cyclodextrin+Technology,+Kluwer+Academic+Publisher:+Dordrecht,+The+Netherlands,+1988.&ots = _snXLJcmEP&sig = BJZuOXtRoZ--QZPB2N3xKH30q2U. Accessed 13 Mar 2016

Takeo K, Ueraura K, Mitoh H (1988) Derivatives of α-cyclodextrin and the synthesis of 6-O-α-D-glucopyranosyl-α-cyclodextrin. J Carbohydr Chem 7:293–308. https://doi.org/10.1080/07328308808058926.

Tanaka MYI, Matsumoto M, Nakamura T, Yabe A, Nakanishi H, Kawabata Y (1987) Host–guest complexes of an amphiphilic β-cyclodextrin and azobenzene derivatives in Langmuir–Blodgett films. Kawabara H Takahashi Tamura W Tagaki H Nakahara K Fukuda. Chem Lett 1307. doi:https://doi.org/10.1246/cl.1987.1307.

Taneva S, Ariga K, Okahata Y, Tagaki W (1989) Association between amphiphilic cyclodextrins and cholesterol in mixed insoluble monolayers at the air-water interface. Langmuir 5:111–113. https://doi.org/10.1021/la00085a020.

Tolédano P, Neto AMF (1998) Phase transitions in complex fluids. World Scientific, Singapore

Uekama K (2004) Design and evaluation of cyclodextrin-based drug formulation. Chem Pharm Bull (Tokyo) 52:900–915. https://doi.org/10.1248/cpb.52.900

Ueno A, Breslow R (1982) Selective sulfonation of a secondary hydroxyl group of β-cyclodextrin. Tetrahedron Lett 23:3451–3454. https://doi.org/10.1016/S0040-4039(00)87639-X.

Valente AJM, Söderman O (2014) The formation of host–guest complexes between surfactants and cyclodextrins. Adv Colloid Interf Sci 205:156–176. https://doi.org/10.1016/j.cis.2013.08.001.

Verploegen E, Soulages J, Kozberg M, Zhang T, McKinley G, Hammond P (2009) Reversible switching of the shear modulus of photoresponsive liquid-crystalline polymers. Angew Chem Int Ed 48:3494–3498. https://doi.org/10.1002/anie.200900583.

Wan W, Monobe H, Tanaka Y, Shimizu Y (2003) Mesomorphic properties of copolyesters of 3, 6-linked triphenylene-based units and polymethylene spacers. Liq Cryst 30:571–578. https://doi.org/10.1080/0267829031000099608.

Wang S-J, Zhao R-Y, Yang S, Yu Z-Q, Chen E-Q (2014) A complex of poly (4-vinylpyridine) and tolane based hemi-phasmid benzoic acid: towards luminescent supramolecular side-chain liquid crystalline polymers. Chem Commun 50:8378–8381. https://doi.org/10.1039/C4CC03113J

Ward S, Calderon O, Zhang P, Sobchuk M, Keller SN, Williams VE et al (2014) Investigation into the role of the hydrogen bonding network in cyclodextrin-based self-assembling mesophases. J Mater Chem C 2:4928–4936. https://doi.org/10.1039/C4TC00448E

Wieckowski A, Savinova ER, Vayenas CG (2003) Catalysis and electrocatalysis at nanoparticle surfaces. CRC Press, New York

Yagai S, Kitamura A (2008) Recent advances in photoresponsive supramolecular self-assemblies. Chem Soc Rev 37:1520. https://doi.org/10.1039/b703092b.

Yallapu MM, Jaggi M, Chauhan SC (2010) Poly(β-cyclodextrin)/curcumin self-assembly: a novel approach to improve curcumin delivery and its therapeutic efficacy in prostate cancer cells. Macromol Biosci 10:1141–1151. https://doi.org/10.1002/mabi.201000084.

Yang F, Zhang Y, Guo H, Lin J (2013) Novel supramolecular liquid crystal: synthesis of cyclodextrin–triphenylene column liquid crystal based on click chemistry. Tetrahedron Lett 54:4953–4956. https://doi.org/10.1016/j.tetlet.2013.07.018.

Zhang P, Parrot-lopez H, Tchoreloff P, Baszkin A, Ling C, de Rango C et al (1992) Self-organizing systems based on amphiphilic cyclodextrin diesters. J Phys Org Chem 5:518–528. https://doi.org/10.1002/poc.610050814.

Zuo F, Liao Z, Zhao C, Qin Z, Li X, Zhang C et al (2014) An air-supported liquid crystal system for real-time reporting of host–guest inclusion events. Chem Commun 50:1857. https://doi.org/10.1039/c3cc48885c

MIX
Papier aus verantwortungsvollen Quellen
Paper from responsible sources
FSC® C105338

If you have any concerns about our products,
you can contact us on
ProductSafety@springernature.com

In case Publisher is established outside the EU,
the EU authorized representative is:
Springer Nature Customer Service Center GmbH
Europaplatz 3, 69115 Heidelberg, Germany

Printed by Libri Plureos GmbH
in Hamburg, Germany